AF478000

METHODS IN MOLECULAR BIOLOGY

Series Editor
John M. Walker
School of Life and Medical Sciences
University of Hertfordshire
Hatfield, Hertfordshire, AL10 9AB, UK

For further volumes:
http://www.springer.com/series/7651

Cellulases

Methods and Protocols

Edited by

Mette Lübeck

Section of Sustainable Biotechnology, Department of Chemistry and Bioscience, Aalborg University, Copenhagen, Denmark

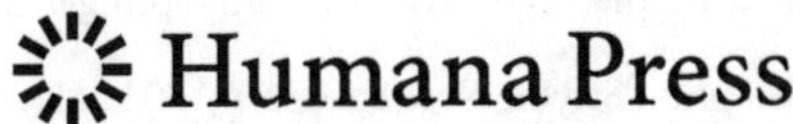 Humana Press

Editor
Mette Lübeck
Section of Sustainable Biotechnology
Department of Chemistry and Bioscience
Aalborg University
Copenhagen, Denmark

ISSN 1064-3745   ISSN 1940-6029 (electronic)
Methods in Molecular Biology
ISBN 978-1-4939-7876-2   ISBN 978-1-4939-7877-9 (eBook)
https://doi.org/10.1007/978-1-4939-7877-9

Library of Congress Control Number: 2018941816

Printed on acid-free paper

This Humana Press imprint is published by the registered company Springer Science+Business Media, LLC part of Springer Nature.
The registered company address is: 233 Spring Street, New York, NY 10013, U.S.A.

Preface

Cellulases, especially of microbial origin, have applications in various industries including pulp and paper, textile, laundry, food and feed industry, brewing, and agriculture. Cellulases are becoming increasingly important as key enzymes for biomass deconstruction within biorefineries, for the production of fuels and chemicals. Cellulases are thereby of utmost importance in the transition of the society into a more sustainable, biobased economy. Research in discovery, improvement, and applications of novel cellulases is growing rapidly, not only at the university level but also in industry. The aim of this book is to become the definite lab protocol source in this important area. The target audience is biochemists, molecular biologists, chemical engineers, industrial researchers, and others with an interest in the area.

The book contains the following topics within the area of cellulases together with detailed protocols that enable the researchers within the field to familiarize themselves with the methods and use the protocols in their research. Most importantly, each protocol has notes with tricks and tips to the methods.

The introductory chapter 1 deals with the role of cellulases for lignocellulose deconstruction for generation of a sugar platform for biofuel and biomaterial production in biorefineries. The chapter explains the different steps within biorefineries and the differences between first, second, third, and fourth generation of biofuel production. The chapter explains the composition of lignocellulosic biomass and describes the processes, focusing primarily on the role of enzymes for deconstructing lignocellulosic biomass. The chapter also includes the classification of microbial cellulases and highlights how important these enzymes are due to emerging applications in various industries. By facilitating the utilization of lignocellulosic waste, cellulases are promising and significant contributors to the circular bioeconomy and help decreasing environmental pollution.

Fungi are the organisms that have attracted most attention in terms of production of efficient cellulolytic enzymes at the industrial level. Fungi such as *Trichoderma reesei* and *Aspergillus* sp. are industrial workhorses and are the basic production organisms in enzyme companies such as Novozymes A/S and Genencor Inc. The commercial enzyme cocktails such as Celluclast, Novozyme 188, CellecCTec (generation 1–3), and the Accellerase enzyme lines are all based on fungal cellulolytic enzymes. Several studies have focused on increasing the production efficiency of cellulolytic enzymes by selection of microorganisms capable of secreting a high and diversified amount of enzymes as well as by optimizing the composition of the cellulolytic cocktail. Chapter 2 is a case study outlining an example with screening for novel fungal β-glucosidases from a variety of fungal strains, identification of a gene encoding a key β-glucosidase in the fungus with highest activity, followed by cloning and expressing the gene in a heterologous host, and finally evaluating the activity of the expressed enzyme.

An increasing number of studies show that fungi as well as many other microorganisms from a great variety of ecological niches are highly interesting as sources for novel cellulases, e.g., cellulosome-producing bacteria. Chapters 3–9 focus on protocols for the discovery of novel enzymes in bacteria or fungi from various ecological niches using classical or molecular methods. The methods presented involve different techniques for the isolation and screening of microbes including aerobic and anaerobic bacteria as well as fungi capable of degrading cellulose. Interesting habitats for discovery of novel cellulases in focus are compost, decomposed wood, rumen from ruminants, and the guts of wood-destroying insects such as termites. Among the techniques are the isolation of strictly anaerobic bacteria, and a protocol for one of the most useful techniques, the roll-tube technique, is included in Chapter 5 outlining the technique for isolating cellulolytic rumen bacteria. For isolating rumen bacteria, the environmental and nutritional conditions similar to the rumen environment should be simulated in vitro.

Molecular methods include genomics, transcriptomics, and proteomics (Chapters 6–9). These methods can be applied to single microbes as well as multiple organisms using the so-called meta-versions (metagenomics, metatranscriptomics, and metaproteomics), when they are applied to multiple organisms simultaneously. The latter has the advantage that they not only facilitate to screen multiple organisms for novel enzymes but also enable to screen uncultivable microbes for novel enzymes. One of the protocols explains a combined genomic procedure using bioinformatics and a biochemical method for identifying novel cellulosomal cellulases from a human gut bacterium (Chapter 6). These tools can be expanded into metagenomics for the discovery of novel cellulosome-producing bacteria and the description of their cellulosomal genes. Cellulosomes are large enzymatic complexes that possess scaffold proteins harboring anchors to which the catalytic cellulases and other enzymes are strongly attached.

Interesting novel cellulases can be identified within complex symbiotic communities such as the termite hindgut. The complexity of such systems warrants applying novel technologies and sequencing strategies that facilitate sequencing entire consortiums or microecosystems. Among the protocols, there is a metatranscriptomics strategy allowing detecting and measuring expressed genes from all three domains of life in a single sample simultaneously (Chapter 7). Bioinformatics then facilitates the identification of cellulases and other genes encoding carbohydrate active enzymes from such complex communities. Such strategies have applications for identifying cellulases from different communities of prokaryotes and eukaryotes.

Lignocellulosic biomasses are complex and an efficient degradation requires the synergistic action of different types of enzymes. From the genome projects, it has become clear that most cellulolytic microbes contain a large repertoire of genes encoding plant cell wall degrading enzymes in their genomes. When microbes degrade lignocellulosic biomass in order to obtain nutrients, they release a wide spectrum of these enzymes but from the genome projects it is neither possible to know the expression level nor possible to predict the power of the degradation capacity of the microbe. Characterizing the compositions of the secreted enzymes from microbes cultivated on complex lignocellulose substrates using proteomic strategies may aid in the understanding of the enzymatic degradation of the material. In Chapters 8 and 9, proteomic strategies are presented for the analysis of enzyme mixtures produced by lignocellulolytic fungi. Following identification of interesting cellulases, they can be cloned and expressed in heterologous hosts. In many cases, fungal hosts are selected for enzyme production and a protocol for this is given in Chapter 10.

There are only a few reports of cellulosomes of thermophilic origin, but as pretreatment processes in biorefineries are carried out at elevated temperatures, it is of interest to develop thermophilic "designer cellulosomes." This could allow to control the specific placement and copy-number of the enzymes within the complex and to engineer efficient cellulosomes. Advanced cloning tools for construction of such cellulosomes are given in Chapter 11. Designer cellulosomes can be comprised of mixtures of enzymes from mesophilic and thermophilic organisms and the overall thermal profile of such a complex is a result of the individual parts. A protocol for the determination of temperature stability for individual designer cellulosome components and their complexes is given in Chapter 12.

Even though it may seem trivial, there is a great challenge in identifying and optimizing enzyme assays that are suitable for characterizing the activity of cellulases. This is due to the complex and heterogenic nature of the substrate whether it be isolated cellulose material such as avicel or carboxymethyl cellulose or lignocellulose, and the cellulases therefore do not follow the traditional Michaelis-Menten kinetics when they cleave the glycoside bond between the glucose moieties. Several cellulase activity assay methods are applied and can be divided into qualitative and quantitative methods (Chapters 13–15). Often the qualitative methods are the chosen methods when it comes to isolation and screening of cellulolytic microorganisms. The qualitative methods can be used for screening already isolated and pure cultures of microorganisms as well as they can also be used for the selective isolation of them, including the use of enrichment cultivation. Several protocols for different qualitative methods are presented in the chapters. After promising strains are obtained, the quantitative methods can be employed in order to select the best cellulase producers and to optimize the enzyme production. The accurate quantification of the cellulases is challenging due to the complexity of cellulose, which along with the action of the cellulases (being composed of several different enzymes with different activities) will give changes in the enzyme-substrate ratio and product inhibition during the degradation.

The book also contains an extensive chapter on assaying the very interesting "cellulase-like" enzymes, the redox enzyme lytic polysaccharide monooxygenases (LPMOs), which have been found to boost the lignocellulosic deconstruction in a very efficient way (Chapter 16). LPMOs are copper-dependent enzymes that perform oxidative cleavage of glycoside bonds in cellulose. In contrast to cellulases, LPMOs can act directly on the crystalline cellulose and make it more accessible to the action of the cellulases by introducing strand breaks. The efficiency by which these enzymes work has had a huge effect on the commercial cellulosic cocktails now available on the market, as the inclusion of LPMOs significantly lowers the total dose of enzymes needed for lignocellulosic degradation, which has substantially lowered the price for the enzymes. Reduction of costs of lignocellulosic ethanol facilitates commercialization; however, since the current oil prices are very low, it remains to be a challenge for the industry to compete with fossil resources. The analytical tools to assay the LPMO are not trivial and the chapter includes several of the variety of analytical tools that have been developed. The action of LPMOs requires extracellular electron donors, which recently were shown to originate from excited photosynthetic pigments, which is the theme of Chapter 17.

Purification of cellulases and characterization of the total cellulase activity from microbial cultures are the theme of Chapter 18. Once superior cellulolytic organisms are discovered, it is important to address the optimal bioprocess engineering strategies to improve the production by optimizing process variables, bioreactor type, and cultivation methods. In an industrial setting this is most commonly carried out using submerged fermentation. As the cellulolytic enzymes for biomass degradation within a biorefinery or a bioethanol plant are among the most costly steps, there are attempts to pursue local on-site enzyme production using cheap substrates such as the feedstock for the biorefinery or side streams for enzyme induction and production. Using the feedstock biomass for production of enzymes could be feasible in order to produce an enzymatic complex tailored to the biodegradation of the feedstock. Such an on-site production can be carried out using submerged fermentation, solid-state fermentation, or even a combination of both which is the theme of one of the protocols within this book (Chapter 19).

After their production in production hosts, cellulase mixtures to be used for degradation of lignocellulosic biomass, requires not only assaying the enzyme activity on avicel or other cellulosic material such as the most common filter paper technique, but needs testing on real biomass samples. This is due to the complexity of the matrix, which differs from plant type to plant type and from an applied pretreatment method. The efficacy of cellulase mixtures to degrade lignocellulosic materials is influenced by many factors, and in Chapter 20, two protocols with different solids concentrations for testing the efficacy of cellulases on pretreated biomass samples are described. The protocols can also be used to evaluate different feedstocks.

Crystallization of cellulases has had a great impact in understanding the general nature of the enzyme, by analyzing the fold of the enzyme, conserved residues, catalytic tunnel/pocket as well as substrate and product binding sites. For novel cellulases including those in databases, homology modeling is a very powerful tool in the absence of such atomic structures for characterizing the enzymes and understanding their characteristics. This information can be used in developing robust enzymes especially for industrial applications. A protocol for how to do homology modeling is given in Chapter 21.

Finally, it is my hope that this compiled knowledge and protocols will be a great help to achieve more progress both with regard to the basic understanding of these important enzymes and for the development of efficient cellulases for various industrial applications.

Copenhagen, Denmark   *Mette Lübeck*

Contents

Contributors

FINN L. AACHMANN • *Department of Biotechnology and Food Science, NOBIPOL, NTNU Norwegian University of Science and Technology, Trondheim, Norway*

ALBERTO C. BADINO • *Graduate Program of Chemical Engineering, Federal University of São Carlos, São Carlos, SP, Brazil*

SCOTT E. BAKER • *Joint BioEnergy Institute, Emeryville, CA, USA; Pacific Northwest National Laboratory, Richland, WA, USA*

EDWARD A. BAYER • *Department of Biomolecular Sciences, The Weizmann Institute of Science, Rehovot, Israel*

YONIT BEN-DAVID • *Department of Biomolecular Sciences, The Weizmann Institute of Science, Rehovot, Israel*

LIZI BENSOUSSAN • *Department of Biomolecular Sciences, The Weizmann Institute of Science, Rehovot, Israel*

HEATHER BREWER • *Environmental Molecular Sciences Laboratory, Mass Spectrometry Facility, Richland, WA, USA*

SAORI AMAIKE CAMPEN • *Joint BioEnergy Institute, Emeryville, CA, USA; J. Craig Venter Institute, San Diego, CA, USA*

DAVID CANNELLA • *Department of Geoscience and Natural Resources, University of Copenhagen, Frederiksberg C, Denmark; Interfaculty School of Bioengineers, Universite Libre de Bruxelles ULB, Bruxelles, Belgium*

BAREKET DASSA • *Department of Immunology, The Weizmann Institute of Science, Rehovot, Israel*

VINCENT G. H. EIJSINK • *Biotechnology and Food Science, Faculty of Chemistry, Norwegian University of Life Sciences, Ås, Norway*

CRISTIANE S. FARINAS • *Embrapa Instrumentation, São Carlos, SP, Brazil; Graduate Program of Chemical Engineering, Federal University of São Carlos, São Carlos, SP, Brazil*

SHERAN FERNANDO • *Department of Biology, Lakehead University, Thunder Bay, ON, Canada*

CAMILA FLORENCIO • *Embrapa Instrumentation, São Carlos, SP, Brazil*

ANASTASIA P. GALANOPOULOU • *Microbiology Group, Department of Biology, National and Kapodistrian University of Athens, Athens, Greece*

JOHN M. GLADDEN • *Joint BioEnergy Institute, Emeryville, CA, USA; Sandia National Laboratory, Livermore, CA, USA*

DIMITRIS G. HATZINIKOLAOU • *Microbiology Group, Department of Biology, National and Kapodistrian University of Athens, Athens, Greece*

HENNING JØRGENSEN • *Department of Plant and Environmental Sciences, Faculty of Science, University of Copenhagen, Frederiksberg C, Denmark*

AMARANTA KAHN • *Department of Biomolecular Sciences, The Weizmann Institute of Science, Rehovot, Israel*

AYYAPPA KUMAR SISTA KAMESHWAR • *Department of Biology, Lakehead University, Thunder Bay, ON, Canada*

STJEPAN KREŠIMIR KRAČUN • *GlycoSpot IVS, Frederiksberg C, Denmark*

GUODONG LIU • *State Key Laboratory of Microbial Technology, School of Life Science, Shandong University, Jinan, Shandong, China*

JENNIFER S. M. LOOSE • *Biotechnology and Food Science, Faculty of Chemistry, Norwegian University of Life Sciences, Ås, Norway*

METTE LÜBECK • *Section of Sustainable Biotechnology, Department of Chemistry and Bioscience, Aalborg University, Copenhagen, Denmark*

PETER STEPHENSEN LÜBECK • *Section of Sustainable Biotechnology, Department of Chemistry and Bioscience, Aalborg University, Copenhagen, Denmark*

JON K. MAGNUSON • *Joint BioEnergy Institute, Emeryville, CA, USA; Pacific Northwest National Laboratory, Richland, WA, USA*

MAKOTO MITSUMORI • *Institute of Livestock and Grassland Science, National Agriculture and Food Research Organization (NARO), Tsukuba, Ibaraki, Japan*

SARAH MORAÏS • *Department of Biomolecular Sciences, The Weizmann Institute of Science, Rehovot, Israel*

ANGELA NORBECK • *Environmental Molecular Sciences Laboratory, Mass Spectrometry Facility, Richland, WA, USA*

BRITTANY F. PETERSON • *Center for Insect Science, Tucson, AZ, USA*

WENSHENG QIN • *Department of Biology, Lakehead University, Thunder Bay, ON, Canada*

JINGYAO QU • *State Key Laboratory of Microbial Technology, School of Life Science, Shandong University, Jinan, Shandong, China; National Glycoengineering Research Center, Shandong University, Jinan, Shandong, China*

YINBO QU • *State Key Laboratory of Microbial Technology, School of Life Science, Shandong University, Jinan, Shandong, China; National Glycoengineering Research Center, Shandong University, Jinan, Shandong, China*

M. SHAFIQUR RAHMAN • *Department of Biology, Lakehead University, Thunder Bay, ON, Canada; Department of Microbiology, University of Chittagong, Chittagong, Bangladesh*

MORGANN C. REILLY • *Joint BioEnergy Institute, Emeryville, CA, USA; University of California, San Francisco, CA, USA*

BRIAN ROSS • *Department of Biology, Lakehead University, Thunder Bay, ON, Canada; Northern Ontario School of Medicine, Thunder Bay, ON, Canada*

MICHAEL E. SCHARF • *Department of Entomology, West Lafayette, IN, USA*

JULIA SCHÜCKEL • *GlycoSpot IVS, Frederiksberg C, Denmark*

AMITA SHARMA • *Shaheed Udham Singh College of Research and Technology, Mohali, India*

HEM KANTA SHARMA • *Department of Biology, Lakehead University, Thunder Bay, ON, Canada*

BLAKE A. SIMMONS • *Joint BioEnergy Institute, Emeryville, CA, USA; Lawrence Berkeley National Laboratory, Berkeley, CA, USA*

RAMAN SONI • *Department of Biotechnology, D.A.V. College, Chandigarh, India*

SANJEEV KUMAR SONI • *Department of Microbiology, Panjab University, Chandigarh, India*

MORTEN SØRLIE • *Biotechnology and Food Science, Faculty of Chemistry, Norwegian University of Life Sciences, Ås, Norway*

WIMAL UBHAYASEKERA • *Department of Cell and Molecular Biology, Uppsala University, Biomedical Centre, Uppsala, Sweden*

GUSTAV VAAJE-KOLSTAD • *Biotechnology and Food Science, Faculty of Chemistry, Norwegian University of Life Sciences, Ås, Norway*

BJØRGE WESTERENG • *Biotechnology and Food Science, Faculty of Chemistry, Norwegian University of Life Sciences, Ås, Norway*

Jiangning Wu • *Department of Chemical Engineering, Ryerson University, Toronto, ON, Canada*

Chunbao (Charles) Xu • *Department of Chemical and Biochemical Engineering, Western University, London, ON, Canada*

Jing Zhu • *State Key Laboratory of Microbial Technology, School of Life Science, Shandong University, Jinan, Shandong, China*

Marta Zoglowek • *Section of Sustainable Biotechnology, Department of Chemistry and Bioscience, Aalborg University, Copenhagen, Denmark*

Part I

Introduction Chapter and a Case Example

Chapter 1

Cellulases: Role in Lignocellulosic Biomass Utilization

Sanjeev Kumar Soni, Amita Sharma, and Raman Soni

Abstract

Rapid depletion of fossil fuels worldwide presents a dire situation demanding a potential replacement to surmount the current energy crisis. Lignocellulose presents a logical candidate to be exploited at industrial scale owing to its vast availability, inexpensive and renewable nature. Microbial degradation of lignocellulosic biomass is a lucrative, sustainable, and promising approach to obtain valuable commercial commodities at gigantic scale. The enzymatic hydrolysis involving cellulases is fundamental to all the technologies needed to transform lignocellulosic biomass to valuable industry relevant products. Cellulases have enormous potential to utilize cellulosic biomass, thus reducing environmental stress in addition to production of commodity chemicals resolving the current challenge to meet the energy needs globally. The substitution of petroleum-based fuels with bio-based fuels is the subject of thorough research establishing biofuel production as the future technology to achieve a sustainable, eco-friendly society with a zero waste approach.

Key words Biofuels, Lignocellulose, Cellulases, Classification, Biomass utilization

1 Introduction

With the advent of energy and environmental crisis globally, scientific world is forced to employ novel strategies for finding alternative fuels by utilizing renewable resources, using clean technologies. Regarding this aspect, exploitation of microbial cellulases to achieve efficient utilization of lignocellulosic biomass is a promising strategy [1, 2]. The cost-effective exploitation of the lignocellulosic material as a substrate to produce biofuels and other value added compounds would exemplify a shift in the classical carbon consumption processes in industries. It was reported that 105 billion liters biofuel was produced in 2010 which contributed about 2.7% of the fuels utilized in transportation. It is anticipated that by year 2050, more than a quarter of world's need for transportation fuels could be fulfilled by biofuels [3, 4]. In India, it is estimated that around 500 million metric tons biomass (agricultural and forest biomass) is available annually [5].

Mette Lübeck (ed.), *Cellulases: Methods and Protocols*, Methods in Molecular Biology, vol. 1796, https://doi.org/10.1007/978-1-4939-7877-9_1, © Springer Science+Business Media, LLC, part of Springer Nature 2018

Microbial cellulases have established themselves as promising candidates to be utilized commercially in a myriad of industrial processes, viz., food, animal feed, brewery, laundry, textile, wine, pulp and paper, agriculture, pharmaceutical, and biorefinery [2, 6, 8]. Significant advances in the research have shed new light on different structural and functional aspects of these enzymes. Among numerous applications of cellulases, the saccharification of lignocellulose has become the crux of the future technology as this is a promising approach to solve the current energy and environmental problems in an eco-friendly way. Several industries are producing cellulases globally and the expected cellulase market in future will be US $ 400 million per year [9, 10]. Cellulases share about 8% of global enzyme demand in the industries [11, 12]. Despite enormous research underway, high cost of enzymes is the main obstacle hampering success, thus highlighting the demand to acquire economically beneficial route. In current scenario, lignocellulose bioprocessing and genetic engineering are considered as foreseeable approaches to tackle the problem of rising costs of fossil fuels. Future trends are directed toward the exploitation of lignocellulosic biomass in addition of green biotechnology to attain sustainable growth.

2 Lignocellulosic Biomass

2.1 Lignocellulosic Biomass Composition

Biomass is defined as the biological material, which is derived from living organisms like animal- and plant-derived material. It is composed of different carbon-based molecules containing C, H, O, N; minor amounts of heavy metals; and Group I and II elements. Lignocellulosic biomass includes woods, agricultural residues, and industrial wastes. The processing of lignocellulosic biomass is gaining immense attention, owing to their enormous biotechnological potential as they represent an inexpensive and renewable natural source for economic production of value-added compounds [2, 8, 13]. Various agricultural and industrial processes produce a vast quantity of lignocellulosic residues, which is underutilized and poses a threat to environment. The efficient utilization of these wastes in industrial processes will not only solve the problem of environmental pollution but also provide a cost-effective substrate to meet the ever-increasing demand of industries. Lignocellulosic biomass is recalcitrant and heterogeneous material, which is difficult to depolymerize. Microorganisms possess a repertoire of enzymes, which act sequentially to achieve complete saccharification of lignocellulosic biomass [14]. Structurally, lignocellulosic biomass comprises three constituents: cellulose, hemicellulose, and lignin.

1. Lignin is the main structural component of lignocellulose, which is composed of three different types of phenolic monomers: p-coumaryl alcohol, coniferyl alcohol, and sinapyl alcohol. It provides strength to the lignocellulosic biomass and hinders the action of hydrolytic enzymes by acting as a barrier. It is considered as the most recalcitrant constituent of the lignocellulosic material owing to its structural complexity [2, 9, 13].

2. Hemicellulose is second most abundant polymer of the world comprising short linear and branched polymers. The polymers are composed of different pentoses (D-arabinose, D-xylose), hexoses (D-galactose, D-glucose, D-mannose), and sugar acids. Hemicelluloses in hardwood (e.g., dicot angiosperms) mainly composed of xylans, whereas glucomannans are major constituents in softwood (e.g., gymnosperms). Hemicellulose is hydrolyzed quickly due to its amorphous and branched nature, in comparison to cellulose.

3. Cellulose is a well-known representative of naturally existing renewable polymer. It has been suggested that in case of terrestrial ecosystems, cellulose is synthesized by half of the fixed carbon per annum. It is produced mainly by plants; but some microorganisms, algal species, and animals can also synthesize it [15, 16]. Cellulose consists of two distinct crystalline forms Ia and Ib. The form Ia is represented primarily in bacterial and algal cellulose, whereas Ib is the dominant form present in higher plants [17]. Cellulosic network has a complex architecture composed of microfibrils arranged in bundles. Each microfibril is composed of 36–1200 cellulose chains, held together by hydrogen bonds and Van der Waal forces ultimately resulting in a highly ordered crystalline structure. Each cellulose chain is a linear chain of D-glucose monomers (100–20,000) linked together by β-1, 4 glycosidic bonds [18]. Cellobiose, a disaccharide is the repeating unit of cellulose. The 3D structure of cellulose consists of sheets of glucopyranose rings lying in a plane with sheets stacked on one another to form highly organized form. The highly ordered crystalline structures of cellulose are interspersed with disorganized amorphous regions. Hydrogen bonding interactions in the amorphous regions are suboptimal, thus accessible to water molecules and hydrolyzing enzymes [9, 17].

2.2 Sources of Lignocellulosic Biomass

Lignocellulosic biomass is basically derived from animals, plants, trees, grasses, crops, vegetables, agroforest residues, and aquatic plants (Fig. 1). Plants fix carbon dioxide by the process of photosynthesis and synthesize compounds like primary and secondary metabolites. Primary metabolites include carbohydrates (sugars, cellulose, hemicellulose, starch, etc.) and lignin. These are the

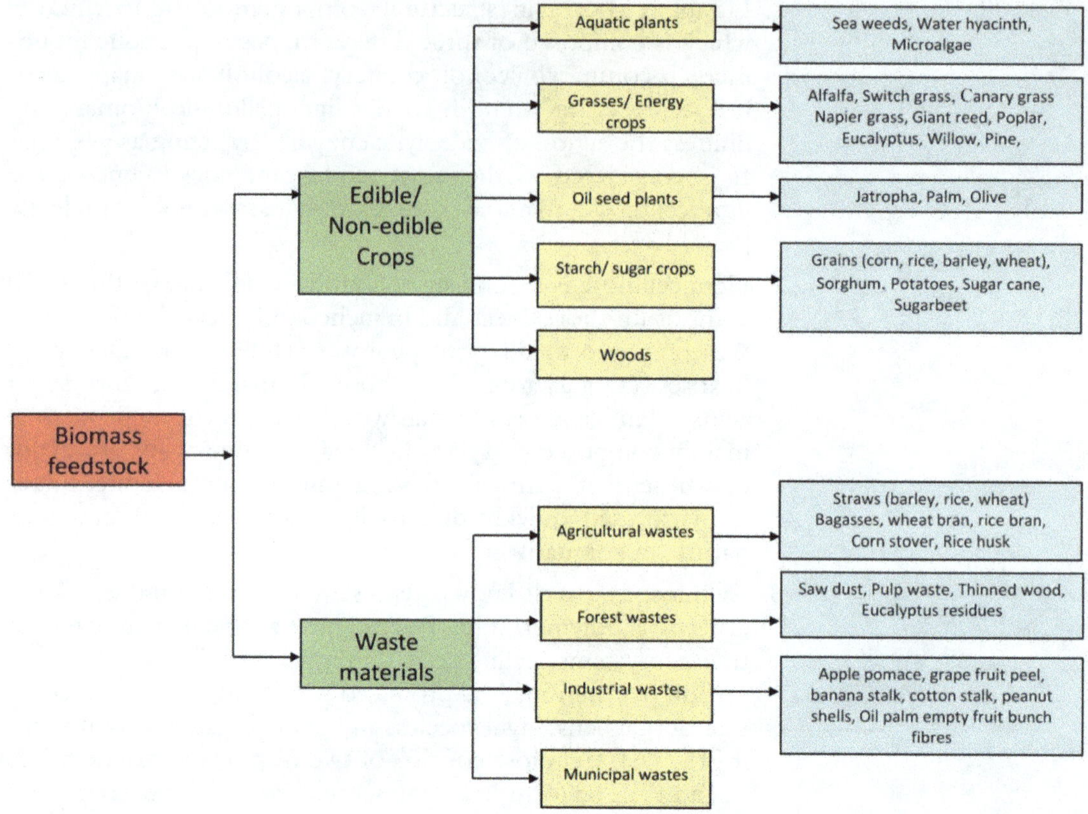

Fig. 1 Lignocellulosic biomass as potential feedstock for production of commodity chemicals

major components of biomass, and they can be utilized to produce a myriad of industry-relevant compounds.

Secondary metabolites include acids, alkaloids, flavonoids, gums, resins, rubber, steroids, tannins, terpenes, terpenoids, triglycerides, and waxes. They are produced in lesser amounts in plants in comparison to primary metabolites. Secondary metabolites have role in plant defense and nowadays they are gaining immense attention due their novel applications in animal feeds, pharmaceuticals, nutraceuticals, food flavors, etc. The biomass feedstocks for biofuels production can be classified into following types: (1) - starch-containing feedstock (e.g., corn, wheat, barley), (2) sugar-containing feedstock (e.g., sugar cane, sweet sorghum, sugar beet), and (3) lignocellulosic material (e.g., grasses, straw, wood). Biomass-derived sugars could be used directly to produce biofuels by fermentation using suitable enzyme based processes [2, 13, 19].

2.3 Lignocellulose Processing

The transformation of lignocellulose into biofuels and commercially useful products involves following steps (Fig. 2): Pretreatment, enzymatic hydrolysis, and fermentation, which are described further below.

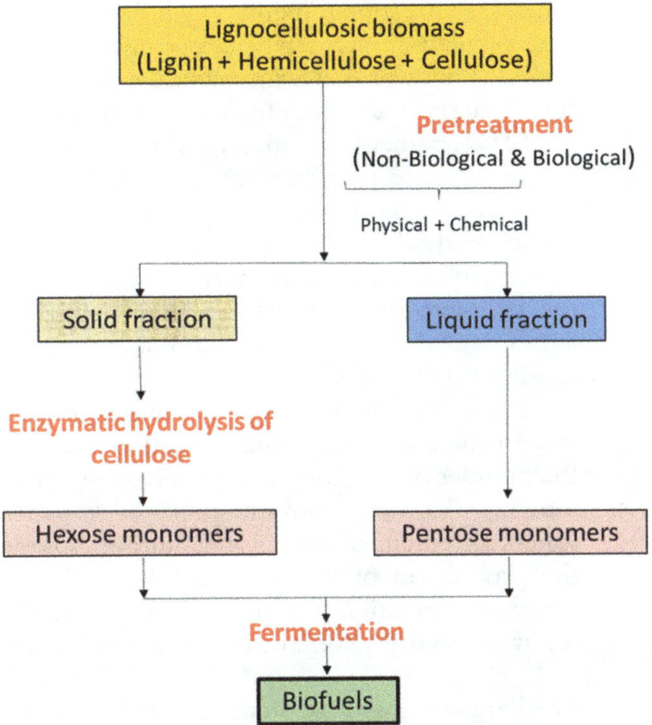

Fig. 2 Major steps involved in the biofuel production from lignocellulosic biomass

2.3.1 Pretreatment

The crystalline nature of cellulose and complexity of lignin presents a main deterrent in the path of saccharification process of lignocellulosic biomass. Numerous methods have been developed so far to enhance the yield of the sugars by pretreatment processes [20, 21]. These processes boost up the efficacy of saccharification process as saccharification of nonpretreated biomass is very slow and yields lesser products. Pretreatment results in separation of the three basic components of lignocellulose, i.e., lignin, cellulose and hemicellulose effectively, thus increasing the availability of each component to the enzymes. In the pretreatment process, most of the hemicellulose and small cellulose amount is hydrolyzed to form monomers. Various pretreatment strategies include biological, chemical, and physical processes.

The pretreatment process is classified into two broad categories: Nonbiological and Biological. Nonbiological pretreatment do not utilize microbes whereas the biological pretreatment involves the use of microorganisms. Nonbiological pretreatment is further classified into three categories: chemical, physical, and physicochemical process. The chemical methods involve treatment with dilute acids, mild alkali, ionic liquids, deep eutectic solvents, natural deep eutectic solvents, ozonolysis (ozone treatment), and organosolv (treatment with aqueous organic solvents). The physical pretreatment method includes mechanical extrusion, milling, pyrolysis, microwave, ultrasound, and pulsed electric field methods. Out of these, mechanical extrusion is the most conventional method. In the physicochemical methods: wet oxidation, SPORL treatment, steam explosion, liquid hot water treatment, ammonia-based pretreatment, CO_2 explosion, and oxidative pretreatment methods are usually utilized. The biological pretreatment method is an economic and ecofriendly process in comparison to nonbiological treatment. This method involves the use of microorganisms mainly fungi, which favors the selective lignin degradation of lignocellulosic biomass and improve its saccharification [20–23].

Integrated pretreatment processes (combination of two or more processes) are gaining attention these days as they reduce the number of operational steps and production of unwanted inhibitors. Various advantages of pretreatment include (a) enhancement of sugar yields during enzyme hydrolysis due to the production of highly digestible solids, (b) minimization of inhibitor formation for subsequent steps of fermentation, and (c) lignin recovery for conversion into value added products.

2.3.2 Enzymatic Hydrolysis

The lignocellulose post pretreatment becomes susceptible to enzymatic attack as cellulosic microfibrils are well exposed. The pretreatment process of biomass is followed by filtration to separate solids and liquids. The solid mass contains cellulose and lignin whereas the liquid fraction contains the products obtained after hydrolysis of hemicellulose. The solid mass containing cellulose is treated with lignocellulolytic enzymes to accomplish the hydrolysis process [21].

Cellulases

Cellulolytic enzymes are produced by a myriad of microbes like fungi, bacteria, and actinomycetes. Numerous microorganisms (both aerobic and anaerobic) possess the ability to degrade cellulose (Table 1) [6, 8, 24]. Aerobic fungi and bacteria are becoming the choice of interest related to cellulose hydrolysis owing to their attributes like rapid growth and better enzyme titers.

Cellulases are modular enzymes as they are composed of different distinct entities called Domains or Folds or Modules, which can fold independently. A typical cellulase enzyme comprises a C-terminal carbohydrate binding domain (CBD) linked by a short linker region to an N-terminal catalytic domain (CD). The CBM (Carbohydrate Binding Module) is considered the crucial accessory domain and they are grouped into 71 families. The important function of CBM is to facilitate cellulose hydrolysis by bringing the CD in close proximity with the cellulose network. It was proposed that role of CBM in cellulose hydrolysis is noncatalytic; i.e., it causes "sloughing off" of cellulose fibers from

Table 1
Microorganisms with cellulose-degrading ability

Fungi	*Aspergillus aculeatus, A. candidus, A. flavus, A. heteromorphus; Bulgaria* sp.; *Candida tropicalis; Chaetomium* sp.;*Cladosporium* sp.; *Coriolus* sp.; *Fusarium* sp.; *Geotrichum* sp.; *Helotium* sp.; *Myrothecium* sp.; *Paecilomyces* sp.; *Penicillium waksmanii; Phanerochaete* sp.; *Poria* sp.; *Schizophyllum* sp.; *Serpula* sp.; *Trichoderma aureoviride, T. reesei; Tricothecium roseum*
Aerobic Bacteria	*Acetobacter oboediens, A. pasteurianus; Acidothermus cellulolyticus; Acinetobacter amitratus, A. junii; Aeromonas sp.; Anoxybacillus sp.; Bacillus amyloliquefaciens, B. cereus, B. circulans, B. flexus, B. licheniformis, B. pumilus, B. subtilis, B. thuringiensis; Bacteriodes sp.; Botrytis sp.; Cellulomonas bioazotea, C. cellulans, C. fimi, C. flavigena, C. uda; Cellvibrio gilvus; Citrobacter freundii; Cytophaga hutchinsonii; Enterobacter sp.; Erwinia sp.; Escherichia coli; Eubacterium cellulosolvens; Geobacillus pallidus, G. stearothermophilus, G. thermodenitrificans; Gluconacetobacter entani, G. europaeus, G. hansenii, G. intermedius, G. xylinus; Halomonas caseinilytica, H. muralis; Kallotenue papyrolyticum; Klebsiella pneumonia; Microbispora bispora; Ochrobactrum cytisi, O. haematophilum; Paenibacillus curdlanolyticus; Paracoccus sulfuroxidans; Proteus vulgaris; Pseudomonas aeruginosa, P. cellulose, P. coleopterorum, P. fluorescens, P. nitroreducens; Rhodothermus marinus, Salinivibrio sp.; Serratia liquefaciens; Thermobispora bispora; Thermomonospora curvata, T. fusca*
Anaerobic Bacteria	*Acetivibrio cellulolyticus; Alcaligenes faecalis; Anaerobacterium chartisolvens; Butyrivibrio fibrisolvens; Bacteroides luti, B. Succinogenes; Clostridium acetobutylicum, C. Cellulolyticum, C. Papyrosolvens, C. Thermocellum; Fibrobacter succinogenes; Ornatilinea apprima; Ruminococcus albus*
Actinomycetes	*Streptomyces drozdowiczii, S. abietis, S. Lividans*

lignocellulosic biomass like cotton fibers. The CBDs and CDs of cellulase enzyme are mostly separated by flexible chain like structures called linker sequences. Linkers usually have disordered arrangement and they are 5–100 residues long. They are mostly composed of proline, hydroxyl amino acid residues (serine and threonine), alanine, and glycine. Mostly linkers have 4–7 residue-long repeated motifs [8, 25].

Classification of Cellulases

Cellulolytic enzymes produced from microbes are either membrane bound or secreted free in the medium. Cellulases have been classified on the basis of their mechanism of action [8, 26, 27].

(a) *Exoglucanase*

Cellobiohydrolase (EC 3.2.1.91): It prefers crystalline compounds like cellooligosaccharides and avicel, and catalyzes nonreducing or reducing ends of polysaccharide chain. It is further categorized into two groups based on the main product obtained due to its reaction. If it liberates glucose, it is termed Glucanohydrolase and cellobiohydrolase if the end product is cellobiose.

Cellodextrinase (EC 3.2.1.74): It does not prefer soluble forms like CMC or amorphous cellulose and mainly liberates cellobiose by hydrolyzing cellooligosaccharides.

(b) *Endoglucanase (EC 3.2.1.4)*: In contrast to cellodextrinase, this enzyme acts on amorphous cellulose or soluble forms of cellulose like CMC. This enzyme generates oligosaccharides with different lengths and exposes new chain ends by cutting at the inner sites of the amorphous region of the polysaccharide chain randomly.

(c) *β-Glucosidase (EC 3.2.1.21)*: It is inactive against both amorphous and crystalline cellulose. It prefers the nonreducing ends and produces glucose by hydrolysis of cellodextrins and cellobiose.

(d) *Cellobiose Phosphorylase (EC 2.4.1.20)*: This enzyme was first reported in the cells of *Ruminococcus flavefaciens* and also purified from a thermophile *Clostridium stercorarium*. It is also termed as orthophosphate α-D-glucosyl transferase and hydrolyses the reversible reaction of phosphorolytic breakdown of cellobiose.

(e) *Cellodextrin Phosphorylase (EC 2.4.1.49)*: This enzyme was found in the cells of *Clostridium thermocellum* and *C. stercorarium*. It is also termed as 1, 4-β-D-oligoglucan orthophosphate α-D-glucosyl transferase and hydrolyzes the reversible reaction of phosphorylytic breakdown of cellodextrins. It is not involved in breakdown of cellobiose.

(f) *Cellobiose Epimerase (EC 5.1.3.11)*: It was first reported in *Ruminococcus albus*. This enzyme is involved in formation of 4-O-β-D glucosylmannose from cellobiose.

Mechanism of Enzymatic Hydrolysis

Lignocellulose hydrolysis by cellulases represents a reaction approach, wherein enzyme present in an aqueous medium interacts with cellulose fibers (insoluble nature). Cellulose chains must be accessible to cellulases to achieve efficient hydrolysis of cellulose. Cellulose chains are arranged in the form of microfibrils, which are tightly packed with hemicellulose and lignin. They must be disrupted or loosened to make the surface area more accessible for reaction with cellulases. The possible process of enzymatic action on cellulose was suggested by Mandels and Reese and they postulated the C1-Cx model [28]. They proposed that loosening or disruption of the cellulose matrix is caused by an unknown component of enzymatic system termed swelling factor (C1) that makes the substrate accessible for the hydrolytic enzymes (Cx) during the initial stage of saccharification. Fiber swelling is observed, which is followed by fragmentation of cellulose into short fibrils. There is no release of any detectable amount of reducing sugars during saccharification and this step is known as amorphogenesis [29] (Fig. 3).

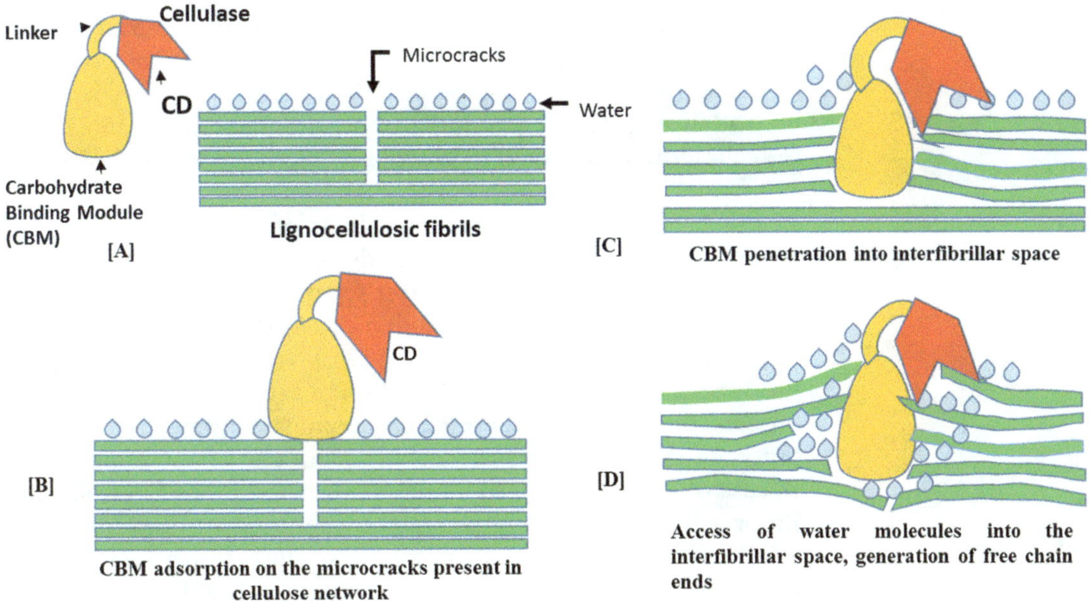

Fig. 3 Cellobiohydrolase I (CBHI) action on cellulosic fibers (amorphogenesis)

As soon as the cellulose chains become available to the cellulolytic enzymes (Cx), their activity results in hydrolysis of the cellulosic network thus liberating soluble sugars. Cellulases get adsorbed on the defects in the cellulose network (microcracks present in the compact cellulosic structure) and induce mechanical action called dispersion by penetrating into the interfibrillar space. This ultimately intensifies the pressure on the walls resulting in swelling of cellulose network. This is followed by further and further penetration between the microfibrils by water molecules, finally leading to dissociation of microfibrils due to breakage of the hydrogen bonds between the cellulose chains. All the cellulolytic enzymes act upon β-1, 4-glycosidic bonds of the cellulosic network by either endocleavage or exocleavage. Among cellulases, exoglucanases preferentially hydrolyze crystalline regions, whereas endoglucanases hydrolyze soluble and amorphous portions of cellulosic network. The hydrolysis of cellulose is achieved by cellulase system by a phenomenon known as synergism. Thus complete hydrolysis of cellulosic material is attained by activity of three enzymes including exoglucanases, endoglucanases and β-glucosidases in a synergistic manner.

Enzymatic saccharification can be explained with the help of a kinetic model taking into consideration multiple factors [30]. The sequence of the major steps of the cellulose degradation can be summed up as: (a) binding of cellulase to the substrate via cellulose binding domains (CBD); (b) recognition of the susceptible bonds by cellulases for hydrolysis (exoglucanases and β-glucosidase recognize nonreducing ends, whereas endoglucanases recognize internal bonds of the amorphous region); (c) initiation of substrate hydrolysis by the formation of the enzyme-substrate complex; (d) forward movement of the enzyme on the cellulose chain and hydrolysis of β-(1, 4) glycosidic bond simultaneously; and finally (e) cellobiose hydrolyzed to glucose by β-glucosidase [23, 30].

2.3.3 Fermentation

In this process the monomeric products obtained post hydrolysis are converted to valuable compounds. The hydrolysis and fermentation processes can be carried out separately (separate hydrolysis and fermentation or SHF or simultaneously (simultaneous saccharification and fermentation or SSF) [21]. In SHF, the hydrolyzed products are subjected to fermentation in a separate process. In SSF process, the hydrolysis of pretreated biomass is followed by fermentation of the monomers (hydrolysis products) in the same vessel. In case of a more advanced integrated method called Consolidated Bioprocessing (CBP), all the steps are reduced to one single step using one or a combination of enzymes. All these types of fermentation processes result in the production of industry relevant compounds including biofuels [31, 32].

Biofuels

Biofuels include energy products like bioethanol, biohydrogen, biomethanol, and biodiesel derived from natural biomass materials by biological processes. Although the use of lignocellulosic biomass to produce biofuels is a well-established concept but still the development of cost-effective, eco-friendly and efficient conversion of biomass is a considerable challenge. Biofuels are categorized into four classes on the basis of biomass utilized in the production of biofuels (Table 2).

Table 2
Types of biofuels

Feedstock	Products	Biofuel types
Food crops, vegetable oils or animal fats, starch, sugar	Biobutanol, bioethanol, biogas, biodiesel, vegetable oil	First generation
Nonfood crops, wood, corn, corn Stover, pine wood, newspaper, solid waste, office paper, wheat straw, energy crops like giant reed, switch grass, poplar, Miscanthus, napier grass, canary grass, alfalfa	Bioethanol, biobutanol, biohydrogen, bio-oil, wood diesel, bio-Fischer-Tropsch diesel	Second generation
Specially engineered algae like *Chlamydomonas, Chlorella, Dunaliella,* and *Scenedesmus*	Microalgal oil, biodiesel	Third generation
Metabolically engineered algae, members of Asclepiadaceae, Apocyanaceae, Euphorbiaceae, and Urticaceae	Biogasoline	Fourth generation

(a) *First generation biofuels*: These biofuels are biomass-derived products, which utilize feedstocks like sugar/starch-based crops (sugar beet, sugarcane) and oil-based crops (rapeseed oil). These crops also provide food and therefore known as first generation bio-energy crops. However, first generation biofuels have been used limited due to following factors: (a) food security issue; (b) battle for land and water utilized for production of food; (c) processing costs are high suggesting the need for government sponsorships to fight with petroleum-based fuels [33].

(b) *Second generation biofuels*: These biofuels utilize lignocellulosic biomass materials like sugarcane bagasse, forest residues, cereal straw, fruit wastes, kitchen waste, brewer's spent grain, municipal solid wastes and purposely grown energy crops like giant reed, Miscanthus, switch grass, canary grass, napier grass, and alfalfa [34–42]. These energy crops provide more energy in comparison to first generation biofuels. Another advantage of these crops is that they can be grown on a poorer quality land with limited water and nutrient content [39]. However, there are some technical barriers, which need to overcome to produce second generation biofuels with cost-efficacy. Moreover, the energy crops still compete with production of fiber and food crops with regard to land use [33].

(c) *Third generation biofuels*: Despite numerous limitations associated with first- and second-generation biofuels, they account for near about 99% of the present biofuels produced globally. Numerous novel technologies are on the brink of commercialization as first- and second-generation biofuels but are not suitable for long-term displacement of petroleum products. Third generation biofuels are those biomass-derived fuels that are synthesized from specially engineered energy crops like algae [43]. Third generation biofuels have many advantages over the previous ones including (a) wide availability of low-cost, renewable and high energy algal source; (b) no competition to food and fiber crops as algae can be flourished utilizing water and land that is unfit for producing food crops; (c) wide range of valuable products can be prepared like petrol, diesel, jet fuel etc.

(d) *Fourth generation biofuels*: Both third- and fourth-generation biofuels use same concept of algal-to-biofuels technology but methodology is different. The algae source used for production of fourth generation biofuels is metabolically engineered and it is designed such that it can capture and store large amounts of carbon [44]. Other lignocellulosic material suitable for the synthesis of fourth generation biofuels include high biomass crops, drought tolerant energy crops, acid tolerant crops and trees with very high carbon storage ability etc. The plants with ability to produce terpenoids, a type of secondary metabolites are preferred to produce gasoline-like fuels. Recently, engineered *Eucalyptus* trees are gaining attention for the production of fourth generation biofuels as they have low-lignin content thus allowing easier conversion to ethanol [45]. Other petro-crops like members of Asclepiadaceae, Apocyanaceae, Euphorbiaceae, and Urticaceae have been reported to have considerable potential to be utilized commercially. In comparison to other petro crops, *Euphorbia* sp. (*E. lathyris, E. caducifolia*) have established themselves as promising candidates for biofuel production due to the presence of higher carbon storage capacity and other extractable compounds [46].

3 Application of Cellulases in Lignocellulosic Biomass Utilization

Lignocellulosic biomass processing represents a sustainable platform to meet up the growing inevitable demand for alternate bio-based fuels worldwide. A variety of valuable bio-based fuels and other commodity chemicals can be produced at low cost by eco-friendly bioprocessing of lignocellulosic material [4, 11]. Biorefinery is one promising approach gaining attention worldwide to produce biofuels utilizing biomass with great efficacy [2]. In addition to biofuels, lignocellulosic biomass is exploited commercially to produce highly valuable chemicals with potential role as nutraceuticals and pharmaceuticals. Cellulases play crucial role in hydrolysis of lignocellulosic biomass and wide range of lignocellulolytic enzymes are available commercially to achieve biological transformation of waste biomass into valuable chemicals. Major suppliers of the commercial cellulases in India and worldwide are given in the Table 3.

3.1 Extraction of Value-Added Chemicals

3.1.1 Bioactive Compounds

These are the natural compounds extracted from plants that either confer a health advantage or are toxic in nature when consumed. Extraction of these compounds from plants has intensified dramatically in response to the accelerating need for unique bioactive compounds with novel commercial applications. Enzymes based extraction was suggested to boost up the yields of bioactive compounds from plants. Enhanced yields of a nonnutritive sweetener, stevioside were obtained by eco-friendly enzyme-assisted extraction from *Stevia rebaudiana* [47]. Enzyme-assisted extraction of bioactives offers an environment friendly, cost-effective, rapid, and reproducible process with enhanced yields. Cellulases are gaining immense attention as they are beneficial in extraction of a myriad of bioactive compounds alone or in combination with other enzymes.

Table 3
Major national and international suppliers of cellulases

Global supplier	Enzyme
Alko-EDC (New York, USA)	Econase CE
Amano enzyme (troy, Virginia, USA)	Cellulase TAP 106
Amano enzyme (troy, Virginia, USA)	Cellulase AP30K
Dyadic (Jupiter, Florida, USA)	Viscostar 150 L
Genencor-Danisco (Rochester, New York, USA)	Accelerase 1500 GC 440, GC 880, GC 220, Spezyme CP
Genencor Intl. (S. San Francisco, California, USA)	Multifect GC, Multifect CL, Spezyme#1, Spezyme#2, Spezyme#3
Logen (Ottawa, Canada)	Ultralow microbial (ULM)
50 L Lyven (Colombelles, France)	Cellulyve
Novozymes (Copenhagen, Denmark)	Bio-feed beta L Cellubrix (celluclast) Cellulase TAP 10 Energex L, Ultraflo L, Viscozyme L Novozymes 188
Quest Intl. (Sarasota, Florida, USA)	Biocellulase A
Quest Intl. (Sarasota, Florida, USA)	Biocellulase TRI
Rhodia-Danisco (Vinay, France)	Cellulase 2000L
Rohm-AB enzymes (Rajamaki, Finland)	Rohament CL
Solvay enzymes (Elkhart, Indiana, USA)	Cellulase TRL
Indian supplier	*Enzyme*
Advanced biochemicals ltd. (Pune, India)	Cellulase
Advanced enzyme technologies ltd. (thane, India)	Cellulase SacchariSEB EG SacchariSEB C 6L SacchariSEB BG
Aumgene biosciences (Surat, India)	Cellulase
Biocon India ltd. (Bangalore, India)	Cellulase
MAPS India ltd. (Ahmedabad, Gujarat, India)	Palkosoft super 720
Noor enzymes Pvt. ltd. (Kolkata, West Bengal, India)	BioConvert—ACEL
Rossari biotech ltd. (Mumbai, India)	Rosszyme super Rosszyme plus Rosscomp
TCI chemicals Pvt. ltd. (India)	Cellulase
Zytex Pvt. ltd. (Mumbai, India)	Zytex (FPU)

They are commercially attractive candidates involved in extraction of various bioactive compounds from plants like anthocyanins, carotenoids, flavonoids, glycosides, phenolics, and vanillin, which can be used as nutraceuticals, functional foods, and food supplements [48, 49].

The pigments carotenoids are gaining immense attention as natural food colorants owing to their unique biological characteristics like natural origin, nontoxic nature, and a wide range of color availability. Pigment extraction by use of enzyme combinations like cellulases and hemicellulases is a new emerging approach. It is catching the eye of scientific community as they result in extraction of pigments still attached to proteins, which is not possible in solvent aided extraction process. The pigment extraction in the conjugated form is beneficial as it helps to maintain the stability of the color by preventing the oxidation of pigments [50]. A carotenoid lycopene extracted from tomatoes possess high nutraceutical value. Lycopene in its conjugated form confers antioxidant properties but solvent based extraction results in lesser yields thus adding to the cost of the commercially available product. An enzyme preparation consisting of cellulases, hemicellulases, and pectinases boosts up (8–18 folds) the recovery of lycopene significantly from the outer layer of tomato [51]. Thus, enzyme-based extraction is becoming the method of choice to extract chemicals with high nutraceutical value. Lignocellulosic biomass like apple pomace can be utilized at industrial scale for carotenoids extraction as it provides a cost-effective and highly efficient way, which results in better yields of the product [51, 52]. Lignocellulosic feedstock like crude palm oil (CPO) and palm oil mill effluent waste (POME) was hydrolyzed by cellulases in combination with amylases revealed better extraction yields of β-carotene [53]. Tomato seeds and tomato peels also represent a cost-effective feedstock for lycopene extraction as well as potential feedstock for bioethanol production [54].

3.1.2 Olive Oil Extraction

Extraction of olive oil involves crushing followed by grinding of olives. After this, the minced olive paste is moved across the decanters. For the recovery of oil, the last step is the centrifugation at high-speed [8]. Cellulases alone or in combination with pectinases; enhanced the extraction process and resulted in better yields with improved quality of olive oil [55]. Enzyme-aided extraction of olive oil has been found to have numerous health claims and this has attracted the world market [55, 56]. The addition of enzyme formulations during olive oil extraction resulted in enhanced amount of phenolic compounds with improved antioxidant activity as well as vitamin E content [57].

3.2 Biorefinery

Fossil fuel resources are exhausting at a rapid rate, which will result in development of a terrible state worldwide. The application of cellulases in conversion of widely available lignocellulosic biomass to biofuels is the most attractive commercial strategy, which can provide a cleaner and greener solution to solve this problem. The exploitation of biomass in an economically profitable manner is of great importance to achieve sustainable development and advancement in green technology [2, 31]. Production of biofuels and other valuable products from waste cellulosic biomass by enzyme-assisted hydrolysis has gained paramount importance to solve the problem of energy crisis [2, 8, 56, 57]. Bioethanol is establishing itself as promising alternative candidate to nonrenewable fossil fuels due to its unique properties as an octane enhancer, chemical feedstock, and petrol additive. Various crop residues like corn-cob, grape fruit peel, sunflower stalks, sunflower hulls, rice straw, wheat straw, and water hyacinth have been utilized for manufacture of bioethanol [7, 11, 56]. In USA, bioethanol constitutes about 99% of the biofuels and major feedstock used is corn [58].

Biorefineries are the manufacturing facilities involved in conversion of bio-based materials (such as agricultural residues, industrial wastes, algae, municipal wastes etc.) into a myriad of valuable products, viz., animal feed, food, chemicals and biofuels with the help of lignocellulolytic enzymes [2, 58, 59] (Fig. 4). Biofuels produced from waste biomass are beneficial to the society because (a) they are sustainable and renewable, (b) reduce environmental

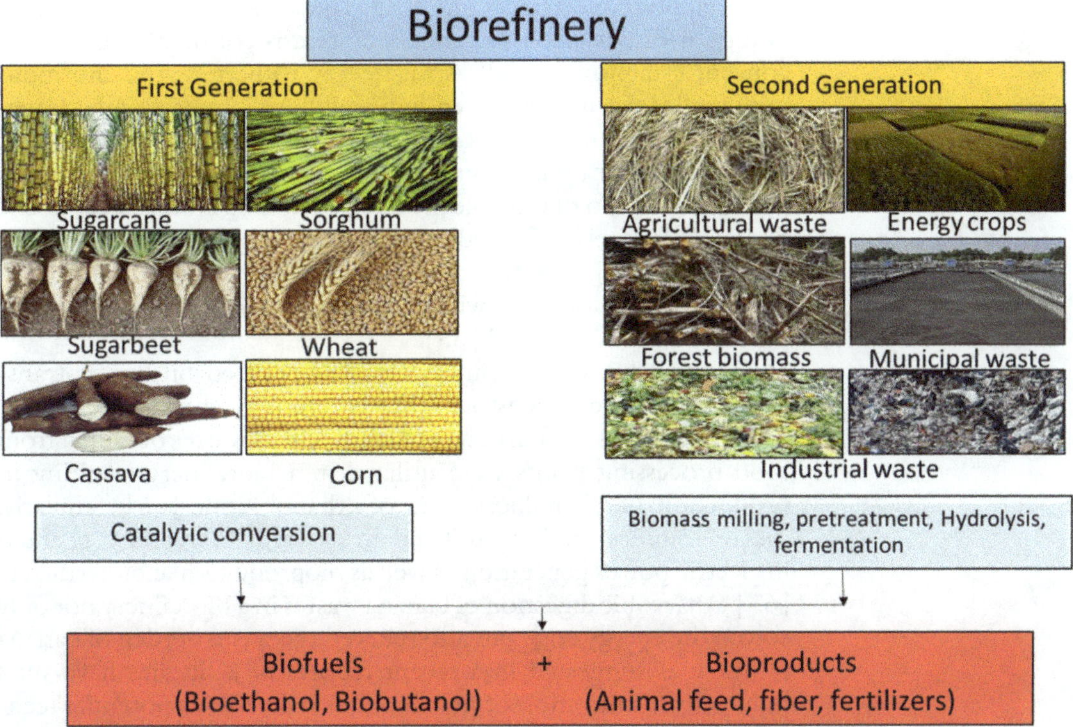

Fig. 4 Different feedstocks utilized in the biorefineries for producing industry relevant compounds

pollution, (c) help in CO_2 fixation indirectly, and (d) provide energy security to the countries. The production of biofuels utilizes waste biomass like municipal solid wastes, kitchen wastes, and agro-industrial wastes (brewer's spent grain, sugarcane bagasse, grape fruit peel, waste paper, citrus fruit waste, newspaper waste, corn stover, soybean hull, and paper sludge).

The *first generation biorefinery* utilizes barley, cassava, corn, sugar beet, soybean, sugarcane, sweet sorghum, rye, and wheat as feedstocks. All these feedstocks are processed by biological transformation using cellulases and other enzymes to obtain different food products (oils, syrups, protein = rich products), animal feed (gluten meal, processed cakes, soluble fractions, grains), chemicals (lactic acid, propionic acid, 1,3-propane diol), biofuels (butanol, ethanol) [59, 60]. Other biomass-derived products synthesized in the first generation biorefinery are cardboard, detergents, adhesives, dyes, paper, sorbents, paint pigments, polymers, etc. [59–62].

Second generation biorefinery gained much attention as it utilizes nonfood crops or nonedible feedstocks for biofuels production unlike first generation biorefinery which are based on food crops. Nonedible feedstocks include forest residues, agricultural residues, municipal solid waste, industrial wastes, and energy crops. The advanced biorefineries exploiting these feedstocks are upgraded versions of conventional biorefineries (CBR) and include green biorefinery (GBR), whole crop biorefinery (WCBR), and lignocellulosic biorefinery (LCBR). The association of green technology with advanced biorefinery system is considered as a novel concept establishing itself as future key to harness the enormous potential of yet underutilized lignocellulosic biomass with cleaner approach. Green biorefinery (GBR) represents multifunctional, complex system utilizing green biomass like grasses and beans to produce a myriad of valuable products including animal feed, fibers, fuel, and biofuels [2]. Dedicated energy crops (both herbaceous and woody), e.g., switchgrass, could provide a meaningful feedstock to be exploited in whole crop biorefinery (WCBR) for the production of bioethanol [63].

Lignocellulosic biorefinery (LCBR) is based on the biotransformation of these feedstocks into commodity chemicals and biofuels [58, 59, 64, 65]. In a recent study, citrus wastes obtained from food processing plants were utilized in a biorefinery resulting in production of 27 million liters of ethanol alone [66]. Similarly, cassava biomass residue utilized in a waste biorefinery resulted in electric power generation as well as bioproducts like biofertilizers [67]. Anaerobic digestion of banana wastes in a biorefinery not only solved the energy crisis but also relieved the waste burden of banana industry as suggested in a recent report [68]. Studies have shed light on the utilization of new biomass materials as potential feedstock in waste biorefineries for sustainable production of bioenergy. Seed residues of *Pongamia pinnata* have been successfully utilized as inedible, low cost resource for saccharification by cellulolytic enzymes to produce bioethanol [69]. Advanced biorefineries like LCBR and WCBR are very bright prospects for the future technologies from biotechnology and industrial standpoint; gaining considerable attention worldwide [2].

New nonedible energy crops have been designed with the progress in genetic engineering techniques, which can be exploited in the biorefineries. These designer plants possess unique characteristics thus significantly enhancing the cellulose to fuel conversion with great efficacy [70]. The genetic modifications in the components and architecture of the cell wall of plants represent a novel strategy to reduce recalcitrance of the biomass to lignocellulolytic enzymes. With the advancement in various fields of science, it is simple to regulate the lignin synthesis in plants in a predictable manner. Innovative strategies to improve bioenergy output from biomass include alteration of the lignin framework to boost up polysaccharide availability for hydrolytic enzymes, enhancing number of polysaccharides with improved hexose content and reduction of inhibitors derived from polymer processing [71]. Further refinement and progress in these technologies will provide tailored biomass materials to be exploited as potential feedstocks in waste based biorefineries.

3.2.1 Bio-Products	High value nontoxic bio-products like biocosmetics, bionutrients, biofertilizers biopharmaceuticals, and biomaterials are also synthesized in biorefineries in addition to biofuels [72]. It was reported that an antioxidant Robinin with potential application as anticancerous activity was extracted from Kudzu biomass and then residual mass was utilized for bioenergy production in biorefinery [73]. Biopharmaceuticals can be harnessed from raw biomass like bark of trees in biorefineries with enzyme-based hydrolysis. Phytochemicals like diarylheptanoids obtained from dry bark of alder tree possess medicinal properties like anti-inflammatory effect and preventing diseases associated with aging [74]. Similarly, poplar tree bark yields pharmaceuticals and after extraction, residual parts can be processed in biorefineries for the production of biofuels [75]. Biomaterials obtained in biorefineries include plastics, packaging materials, textiles, fibers, lubricants, surfactants, and functional materials. Bio-polyethylene, biopolyesters, polylactic acid, starch, polyhydroxyalkanoate, bio-PVC, etc. are dominating the market as they are bioplastics with properties to reduce greenhouse gas emission [76]. Biofertilizers obtained after processing of biomass in biorefineries include digestate obtained after anaerobic digestion, ashes left after biomass combustion [72].

A great diversity of waste biomass is available to be exploited in biorefineries platform, but the need of the hour is to pay attention to the limitations hampering the success of this technology. Numerous challenges including improvisation of pretreatment processes, optimization of processing steps, lowering production costs, and efficient recovery of the valuable products need to be addressed to scale up this technology.

4   Future Perspectives

Microbial cellulases are catching the eye of the scientific as well as commercial world due to emerging applications in biofuel, detergent, food, animal feed, textile, pulp and paper, pharmaceutical, and various other industries. Furthermore, they are considered as promising applicants wisely exploiting the lignocellulosic wastes resulting in decrease in environmental pollution. The significant advancement in the disciplines of biotechnology, novel approaches will turn up dealing with the present limitations in the area of industrial microbiology. The aspects, which need consideration, include strategies dealing with the pretreatment of lignocellulosic materials; increasing accessibility to enzymes; economic production of enzymes; metabolic engineering of microorganisms; and protein engineering to make the enzymes unique in terms of improved characteristics like specific activity, process tolerance and stability. Combined production of biofuels and bio products could tackle market niches as they can be produced by economical efficient and eco-friendly biorefinery based approach. Further innovation in the scientific processes concerned with the biotransformation of waste to biofuels is crucial to come up with a sustainable, greener, and cleaner way out to address the current global energy crisis.

References

1. Maitan-Alfenas GP, Visser EM, Guimarães VM (2015) Enzymatic hydrolysis of ligno-cellulosic biomass, converting food waste in valuable products. Curr Opin Food Sci 1:44–49

2. Arevalo-Gallegos A, Ahmad Z, Asgher M et al (2017) Lignocellulose: a sustainable material to produce value-added products with a zero waste approach- a review. Int J Biol Macromol 99:30–318

3. Shrank S, Farahmand F (2011) Biofuels regain momentum. http://vitalsigns.worldwatch.org/vs-trend/biofuels-regain-momentum. Accessed 29 Aug 2011

4. Kircher M (2015) Sustainability of biofuels and renewable chemicals production from biomass. Curr Opin Chem Biol 29:2631

5. MNRE (2016) Biomass and cogeneration programme, ministry of new and renewable energy. Ministry of Government of India, New Delhi. http://mnre.gov.in/related-links/grid-connected/biomass-powercogen/

6. Saini JK, Saini R, Tewari L (2015) Lignocellulosic agriculture wastes as biomass feedstocks for second-generation bioethanol production: concepts and recent developments. Biotech 5:337–353

7. Menendez E, Garcia-Fraile P, Rivas R (2015) Biotechnological applications of bacterial cellulases. Bioengineering 2:163–182

8. Sharma A, Tewari R, Rana SS et al (2016) Cellulases: classification, methods of determination and industrial applications. Appl Biochem Biotechnol 179:1346–1380

9. Gupta A, Verma JP (2015) Sustainable bio-ethanol production from agro-residues: a review. Renew Sustainable Energy Rev 41:550–567

10. Han J, Luterbacher JS, Alonso DM et al (2015) A lignocellulosic ethanol strategy via nonenzymatic sugar production: process synthesis and analysis. Bioresour Technol 182:258–266

11. Elba PSB, Maria AF (2007) Bioethanol production via enzymatic hydrolysis of cellulosic biomass. In: The role of agricultural biotechnologies for production of bioenergy in developing countries an FAO seminar held in Rome. http://www.fao.org/biotech/seminaroct 2007.htm. Accessed 12 Oct 2007

12. Ilic DD, Dotzauer E, Trygg L, Broman G (2014) Introduction of large-scale biofuel production in a district heating system - an opportunity for reduction of global greenhouse gas emissions. J Clean Prod 64:552–561

13. Yao S, Yang Y, Song H et al (2015) Quantitative industrial analysis of lignocellulosic composition in typical agro-residues and extraction of inner hemicelluloses with ionic liquid. J Sci Ind Res 74:58–63

14. Eveline QPT, Buckeridge MS (2015) Do plant cell walls have a code? Plant Sci 241:286–294

15. Zeng X, Small DP, Wan W (2011) Statistical optimization of culture conditions for bacterial cellulose production by *Acetobacter xylinum* BPR 2001 from maple syrup. Carbohydr Polym 85:506–513

16. Mohite BV, Kamalja KK, Patil V (2012) Statistical optimization of culture conditions for enhanced bacterial cellulose production by *Gluconoacetobacter hansenii* NCIM 2529. Cellulose 19:1655–1666

17. Lima M, Gomez L, Steele-King C, Simister R et al (2014) Evaluating the composition and processing potential of novel sources of Brazilian biomass for sustainable biorenewables production. Biotechnol Biofuels 7:10

18. Zhang YHP, Lynd LR (2004) Toward an aggregated understanding of enzymatic hydrolysis of cellulose: noncomplexed cellulase systems. Biotechnol Bioeng 88:797–824

19. Balat M, Balata H, Oz C (2008) Progress in bioethanol processing. Prog Energy Combust Sci 34:551–573

20. Kumar AK, Sharma S (2017) Recent updates on different methods of pretreatment of lignocellulosic feedstocks: a review. Biores Bioprocess 4:7

21. Bhatia L, Johri S, Ahmad R (2012) An economic and ecological perspective of ethanol production from renewable agro waste: a review. AMB Express 2:65

22. Dashtban M, Schraft H, Qin W (2009) Fungal bioconversion of lignocellulosic residues, opportunities & perspectives. Int J Biol Sci 5:578–595

23. Kumar P, Barrett DM, Delwiche MJ, Stroeve P (2009) Methods for pretreatment of lignocellulosic biomass for efficient hydrolysis and biofuel production. Ind Eng Chem Res 48:3713–3729

24. Mattam AJ, Kuila A, Suralikerimath N et al (2016) Cellulolytic enzyme expression and simultaneous conversion of lignocellulosic sugars into ethanol and xylitol by a new *Candida tropicalis* strain. Biotechnol Biofuels 9:157

25. Obeng EM, Adam SNN, Budiman C et al (2017) Lignocellulases: a review of emerging and developing enzymes, systems, and practices. Biores Bioproces 4:16

26. Saranraj P, Stella D, Reetha D (2012) Microbial cellulases and its applications: a review. Int J Biochem Biotech Sci 1:1–12

27. Sadhu S, Maiti TK (2013) Cellulase production by bacteria: a review. Br Microbiol Res J 3:235–258

28. Mandels M, Reese ET (1964) Fungal cellulases and the microbial decomposition of cellulosic fabric. Dev Ind Mycol Washington: Am Inst Biol Sci 5:5–20

29. Esteghalian AR, Srivastava V, Gilkes N et al (2000) An overview of factors influencing the enzymatic hydrolysis of lignocellulosic feedstocks. In: Himmel ME, Baker JO, Saddler JN (eds) Glycosyl hydrolases for biomass conversion, ACS symposium series, vol 769. American Chemical Society, Washington, DC, pp 100–111

30. Arantes V, Saddler JN (2010) Access to cellulose limits the efficiency of enzymatic hydrolysis: the role of amorphogenesis. Biotechnol Biofuels 3:1–11

31. Gaurav N, Sivasankari S, Kiran GS et al (2017) Utilization of bioresources for sustainable biofuels: a review. Renew Sustain Energy Rev 73:205–214

32. Demirbas MF (2009) Biorefineries for biofuel upgrading: a critical review. Appl Energy 86:151–161

33. Sims REH, Mabee W, Saddler JN, Taylor M (2010) An overview of second generation biofuel technologies. Bioresour Technol 101:1570–1580

34. Soni SK, Batra N, Bansal N, Soni R (2010) Bioconversion of sugarcane bagasse into second generation bioethanol after enzymatic hydrolysis with in-house produced cellulases from *Aspergillus* species S4B2F. Bioresources 5:741–757

35. Bansal N, Tewari R, Gupta JK et al (2011) A novel strain of *Aspergillus niger* producing a cocktail of hydrolytic depolymerising enzymes for the production of second generation biofuels. Bioresources 6:552–569

36. Bansal N, Soni R, Janveja C, Soni SK (2012) Production of xylanase-cellulase complex by *Bacillus subtilis* NS7 for the biodegradation of agro-waste residues. Lignocellulose 1:196–209

37. Bansal N, Tewari R, Soni R, Soni SK (2012) Production of cellulases from *Aspergillus niger* NS-2 in solid state fermentation on agricultural and kitchen waste residues. Waste Manag 32:1341–1346

38. Bansal N, Janveja C, Tewari R et al (2014) Highly thermostable and pH-stable cellulases from *Aspergillus niger* NS-2: properties and application for cellulose hydrolysis. Appl Biochem Biotechnol 172:141–156

39. Chandel AK, Singh V (2011) Weedy lignocellulosic feedstock and microbial metabolic engineering: advancing the generation of 'biofuel'. Appl Microbiol Biotechnol 89:1289–1303

40. Janveja C, Rana SS, Soni SK (2013) Environmentally acceptable management of kitchen waste residues by using them as substrates for the production of a cocktail of fungal carbohydrates. Int J Chem Environ Sys 4:20–29

41. Janveja C, Rana SS, Soni SK (2013) Kitchen waste residues as potential renewable biomass resources for the production of multiple fungal carbohydrases and second generation bioethanol. J Technol Innov Renew Energy 2:186–200

42. Rana SS, Janveja C, Soni SK (2013) Brewer's spent grain as a valuable substrate for low cost production of fungal cellulases by statistical modeling in solid state fermentation and generation of cellulosic ethanol. Int J Food Ferment Technol 3:41–55

43. Mamo G, Faryar R, Nordberg KE (2013) Microbial glycoside hydrolases for biomass utilization in biofuels application. In: Gupta VK, Tuhoy MG (eds) Biofuel technologies: recent developments, 1st edn. Springer-Verlag, Berlin, Heidelberg

44. Lu J, Sheahan C, Fu P (2011) Metabolic engineering of algae for fourth generation biofuels production. Energy Environ Sci 4:2451–2466

45. Joshi CP, Nookaraju A (2012) New avenues of bioenergy production from plants: green alternatives to petroleum. J Pet Environ Biotechnol 3:7

46. Sake K, Bugudea R, Reddy V, Khan SV (2013) Production of bioethanol from spent residues of latex yielding plants *Euphorbia antiquorum* L. and *Euphorbia caducifolia* Haines. Int J Recent Sci Res 4:4

47. Puri M, Sharma D, Barrow CJ (2011) Enzyme-assisted extraction of bioactives from plants. Trends Biotechnol 30:37–44

48. Khandare V, Walia S, Singh M, Kaur C (2011) Black carrot (*Daucus carota* ssp. *sativus*) juice: processing effects on antioxidant composition and color. Food Bioprod Process 89:482–486

49. Gil-Chavez GJ, Villa JA, Ayala-Zavala JF et al (2013) Technologies for extraction and production of bioactive compounds to be used as nutraceuticals and food ingredients: an overview. Compr Rev Food Sci Food Safety 12:5–23

50. Kuhad RC, Gupta R, Singh A (2011) Microbial cellulases and their industrial applications. Enzyme Res 2011:280696. https://doi.org/10.4061/2011/280696

51. Lavecchia R, Zuorro A (2010) Enhanced lycopene recovery from tomato processing waste by enzymatic degradation of plant tissue components. IREBIC 1:2

52. Zuorro A, Fidaleo M, Lavecchia R (2011) Enzyme-assisted extraction of lycopene from tomato processing waste. Enzyme Microb Technol 49:567–573

53. Hudiyono S, Septian A (2012) Optimization carotenoids isolation of the waste crude palm oil using alpha-amylase, beta-amylase, and cellulase. IOSR J Appl Chem 2:7–12

54. Kehili M, Schmidt LM, Reynolds W et al (2016) Biorefinery cascade processing for creating added value on tomato industrial by-products from Tunisia. Biotechnol Biofuels 9:261

55. Sharma R, Sharma PC, Rana JC, Joshi VK (2013) Improving the olive oil yield and quality through enzyme-assisted mechanical extraction, antioxidants and packaging. J Food Process Preserv 39:157–166

56. Mortabit D, Zyani M, Koraichi SI (2014) Improvement of olive oil quality of Moroccan picholine by *Bacillus licheniformis* enzyme's preparation. IJPAS 20:44–52

57. Singh R, Kumar M, Mittal A, Mehta PK (2016) Microbial cellulases in industrial applications. Ann Appl Bio Sci 3:23–29

58. Balan V (2014) Current challenges in commercially producing biofuels from lignocellulosic biomass. ISRN Biotechnol 2014:463074. https://doi.org/10.1155/2014/463074

59. Valdivia M, Galan JL, Laffarga J, Ramos J (2016) Biofuels 2020: biorefineries based on lignocellulosic materials. Microb Biotechnol 9:585–594

60. Yang S-T, Yu M (2013) Integrated biorefinery for sustainable production of fuels, chemicals, and polymers. In: Yang ST, El-Enshasy H, Thongchul N (eds) Bioprocessing technologies in biorefinery for sustainable production of

fuels, chemicals and polymers. John Wiley & Sons, New York

61. Kamm B, Kamm M, Gruber PR, Kromus S (2006) Biorefinery systems—an overview. In: Kamm B, Gruber PR, Kamm M (eds) Biorefineries-industrial processes and products: status quo and future directions. Wiley-VCH, New York

62. Cherubini F, Ulgiati S (2010) Crop residues as raw materials for biorefinery systems - a LCA case study. Appl Energy 87:47–57

63. Zhang Z, Hu S, Chen D, Zhu B (2016) An analysis of an ethanol-based whole-crop refinery system in China. Chin J Chem Eng 24:1609–1618

64. Balat M (2011) Production of bioethanol from lignocellulosic materials via the biochemical pathway: a review. Energy Convers Manag 52:858–875

65. Ferreira JA, Mahboubi A, Lennartsson PR, Taherzadeh MJ (2016) Waste biorefineries using filamentous Ascomycetes fungi: present status and future prospects. Bioresour Technol 215:334–345

66. Taghizadeh-Alisaraeia A, Hosseinia SH, Ghobadianb B, Motevalic A (2017) Biofuel production from citrus wastes: a feasibility study in Iran. Renew Sustain Energy Rev 69:1100–1112

67. Sánchez AS, Silva YL, Kalid RA et al (2017) Waste bio-refineries for the cassava starch industry: new trends and review of alternatives. Renew Sustain Energy Rev 73:1265–1275

68. Gumisiriza R, Hawumba JF, Okure M, Hensel O (2017) Biomass waste-to-energy valorisation technologies: a review case for banana processing in Uganda. Biotechnol Biofuels 10:11

69. Radhakumari M, Taha M, Shahsavari E et al (2017) *Pongamia pinnata* Seed residue – a low cost inedible resource for on-site/in-house lignocellulases and sustainable ethanol production. Renew Energy 103:682–687

70. Badhan A, McAllister T (2014) Designer plants for biofuels: a review. Curr Metabolomics 2:1–7

71. Loque D, Scheller HV, Pauly M (2015) Engineering of plant cell walls for enhanced biofuel production. Curr Opin Plant Biol 25:151–161

72. Budzianowski WM (2017) High-value low-volume bioproducts coupled to bioenergies with potential to enhance business development of sustainable biorefineries. Renew Sustain Energy Rev 70:793–804

73. Lau CS, Carrier DJ, Beitle RR et al (2005) A glycoside flavonoid in kudzu (*Pueraria lobate*). Appl Biochem Biotechnol 121–124:783–794

74. Telysheva G, Dizhbite T, Bikovens O et al (2011) Structure and antioxidant activity of diarylheptanoids extracted from bark of grey alder (*Alnusincana*) and potential of biorefinery-based bark processing of European trees. Holzforschung 65:623–629

75. Devappa RK, Rakshit SK, Dekker RFH (2015) Forest biorefinery: potential of poplar phytochemicals as value-added co-products. Biotechnol Adv 33:681–716

76. Fernández-Dacosta C, Posada JA, Kleerebezem R et al (2015) Microbial community-based polyhydroxyalkanoates (PHAs) production from wastewater: techno-economic analysis and ex-ante environmental assessment. Bioresour Technol 185:368–377

Discovery of a Novel Fungus with an Extraordinary β-Glucosidase and Potential for On-Site Production of High Value Products

Peter Stephensen Lübeck and Mette Lübeck

Abstract

Among cellulases, β-glucosidases play a key role in the final conversion of cellulose into glucose as well as they boost the performance of the other cellulases, in particular cellobiohydrolases in relieving product inhibition. This chapter serves as case example from screening for novel fungal cellulases focusing on β-glucosidases to identifying a gene encoding the key β-glucosidase in the fungus with highest activity. In the case example, the β-glucosidase-producing fungus showed to belong to an unknown fungal species, *Aspergillus saccharolyticus*, not previously described. The gene was expressed in *Trichoderma reesei*, which has low indigenous β-glucosidase activity, and the activity of the purified enzyme was assessed in hydrolysis of various pretreated lignocellulosic biomasses. The potential of using the natural producing strain for on-site production of β-glucosidases using lignocellulosic biorefinery waste streams as substrates is discussed. Finally, the potential of the fungus for consolidated bioprocessing of waste streams into valuable compounds, such as organic acids is highlighted.

Key words Novel fungal species, β-glucosidase, On-site enzyme production, Organic acid production, Consolidated bioprocessing

1 Introduction

Cellulose is the most abundant renewable biomass compounds on earth and is thereby an important resource. To fully utilize this resource, the conversion of cellulose into valuable products are necessary in future in order to minimize the use of fossil materials for the production of fuels and a range of other carbon based materials for daily life [1]. The bioconversion of cellulose in nature takes place through microbial cellulases, which are the key enzymes for breaking down cellulosic material in plants [2]. By using microbial cellulases for breaking down cellulose to glucose in industrial settings, the glucose can be utilized as carbon source in fermentation of different valuable products through microbial biological

Mette Lübeck (ed.), *Cellulases: Methods and Protocols*, Methods in Molecular Biology, vol. 1796, https://doi.org/10.1007/978-1-4939-7877-9_2, © Springer Science+Business Media, LLC, part of Springer Nature 2018

production. This can become a central part for replacing oil-based refineries with lignocellulosic-based biorefineries in future [2].

In order to break down cellulose efficiently within the biorefineries, it is important to have efficient cellulases if biorefineries shall be economically interesting for commercialization. Even though there has been an intensive hunt for novel cellulases, there is still a need to identify new cellulases that can meet the commercial demands of the future.

Among cellulases, β-glucosidase is a key enzyme for efficient conversion of cellobiose and short chain cellodextrins to glucose, which is the last step in cellulose conversion [3]. With the first generations of commercial cellulase cocktails, it was evident, that β-glucosidase was a show stopper if not working properly, resulting in accumulation of cellobiose which again had a negative influence on the action of cellobiohydrolases due to product inhibition [3, 4]. Similarly, β-glucosidases themselves often are prone to product inhibition, so the accumulation of glucose inhibits the enzymes. The initial commercial β-glucosidase product were derived from *Aspergillus niger* and branded with the name Novozyme 188, which had to be mixed with a cellulase cocktail derived from *Trichoderma reesei*, such as Celluclast, to hydrolyze cellulose (Novozymes A/S). Still, most β-glucosidase containing products produced by Novozymes A/S, including Novozyme 188 and the CellicCTec-line, are enzymes originating from *A. niger*. In general, the enzymes are expensive and hamper commercial exploitation especially in small biorefinery companies [5]. Therefore, it is interesting to look for new alternative β-glucosidases with the same capability for cheap on-site production or better performance than the currently known enzymes.

2   Screening for β-Glucosidases

We initiated a screening program for identifying new and better β-glucosidases among fungal strains from our strain collection and new isolates found at several places around Denmark [6] using *Aspergillus* strains belonging to different species from the strain collection of Professor Jens C. Frisvad, Technical University of Denmark, as reference strains. The new fungi from the screening was isolated from materials both outdoor and indoor, such as decomposed wood in a local swamp, treated hardwood, or from food sources such as rye bread. In total 86 strains from 19 different genera primarily within the Ascomycota phylum were tested for β-glucosidase activity [6]. The fungi were grown in liquid fermentation using wheat bran as substrate and the fermentation broth was assayed using pNPG (p-nitrophenyl-β-D-glucopyranoside) as substrate in a microplate format.

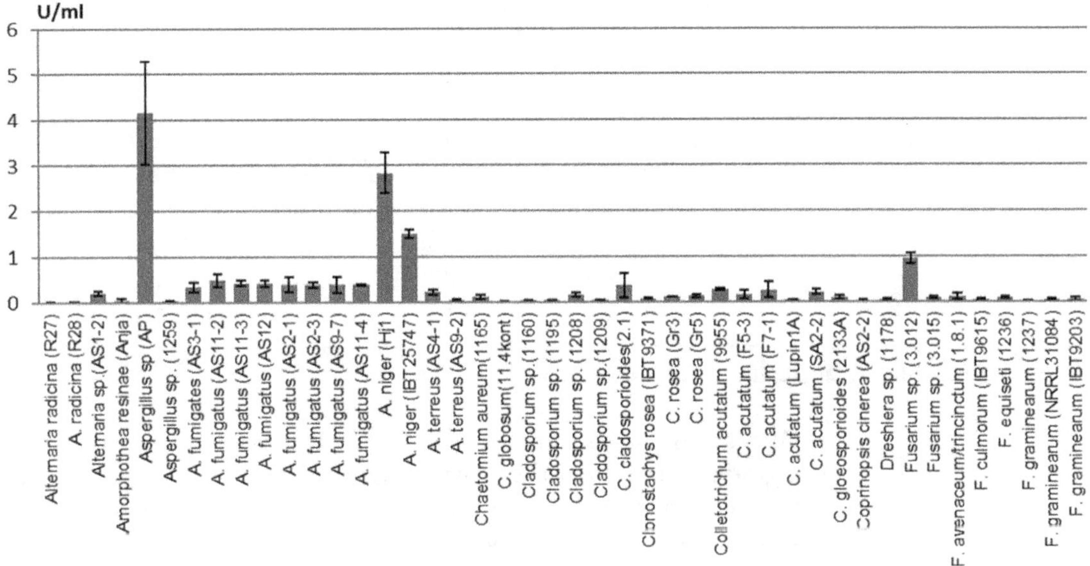

Fig. 1 Extracellular β-glucosidase activity of the fungi grown in simple submerged fermentation. The assay used was pNPG (p-nitrophenyl-β-D-glucopyranoside), with 1 unit (U) of enzyme activity defined as the amount of enzyme needed to hydrolyze 1 μmol of pNPG in 1 min

Among the fungal strains screened one fungus, an isolate (AP) obtained from a wooden toilet seat in Copenhagen, Denmark, showed significantly greater β-glucosidase activity (tenfold) than all other strains (Fig. 1), except one, an *A. niger* strain, that showed an activity about two-thirds of the AP isolate [6]. Following screening, the fungus was used for production of a fermentation broth containing the secreted β-glucosidases, which were compared with Novozyme 188 and Cellic CTec. It was shown that the AP enzymes had increased thermostability, as well as they were better to degrade cellodextrins and hydrolyze pretreated bagasse to monomeric sugars when combined with Celluclast 1.5 L [6].

3 Identification of a Novel Species

Since the β-glucosidase activity was very interesting, it was relevant to identify the strain and by using morphological observations, we initially found that it resembled *Aspergillus aculeatus* and *Aspergillus japonicus*. However, in order to test if the isolate was a new strain of these known *Aspergilli*, the isolate was subjected to morphological and biochemical analyzes as well as molecular methods [7]. No data were similar to known fungi and therefore the unknown isolate was identified as a new species, which was named *Aspergillus saccharolyticus* (*Aspergillus saccharolyticus* Sørensen, Lübeck et Frisvad sp. nov.), because of its efficient saccharification of cellulose [7]. *A. saccharolyticus* is a black Aspergillus belonging to the *Aspergillus* section *Nigri* group and the closest relatives to *A. saccharolyticus* are *Aspergillus homomorphus* and the recently identified *Aspergillus labruscus* placed in the *A. homomorphus* clade [8], not far from the clade with *A. aculeatus* and *A. japonicus*. The strain AP is the type strain, CBS 127449[T] (=IBT 28509[T]) and so far, only one other strain has since been identified in Australia, IBT 28231, which is very similar to the type strain (J.C. Frisvad, pers. comm.).

4   Isolation of the Superior β-Glucosidase

Due to the surprising capability of β-glucosidase activity, it was investigated whether the activity originated from one single enzyme or a collection of enzymes. *A. niger* has 11 β-glucosidases some of which are extracellular proteins [9]. Therefore, a growth experiment with *A. saccharolyticus* was made using solid-state fermentation followed by fractionating the culture broth by ion exchange chromatography in order to identify fractions with high β-glucosidase activity (Fig. 2) [10]. The fractions with high β-glucosidase activity were subjected to electrophoresis using SDS-PAGE. One dominating band was identified with a relevant size for typical β-glucosidases. This band was isolated from the gel and subjected

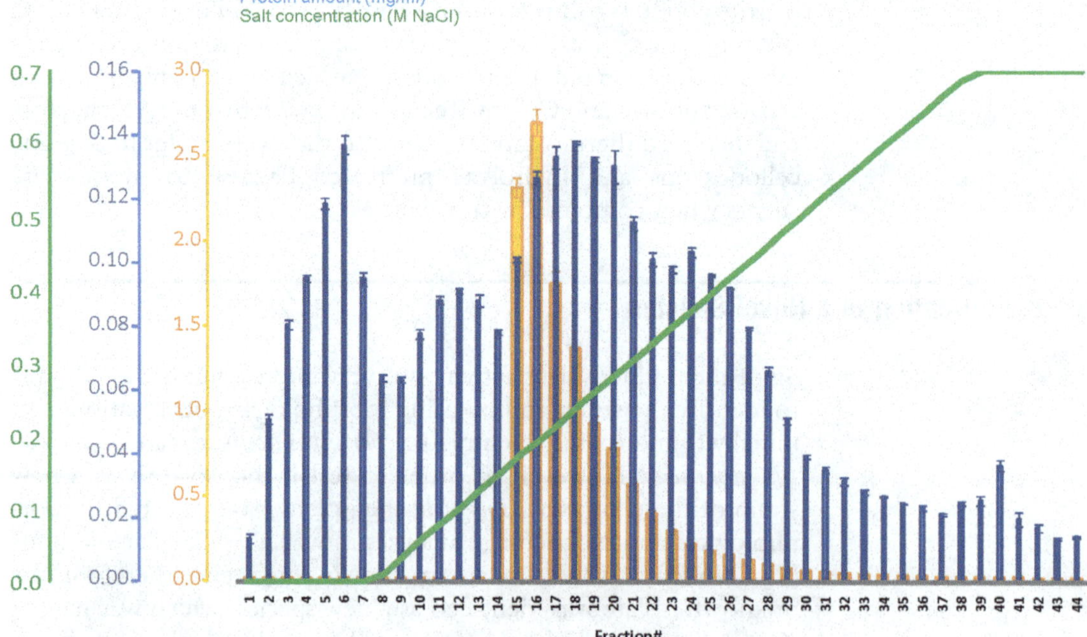

Fig. 2 β-glucosidase activity of the fractions of the crude culture broth of *A. saccharolyticus* obtained by IEX chromatography showed high activity in fractions 15–17

to amino acid sequencing. These data showed that the dominating band was a β-glucosidase (named *bgl1*), similar to β-glucosidases from other fungi especially β-glucosidases from 4 other Aspergilli, *A. aculatus*, *A. niger*, *Aspergillus terreus*, and *Aspergillus fumigatus* in the GH family 3 in the CAZY database [10].

In order to characterize the specific β-glucosidase, a genome walking experiment was carried out using PCR with primers design based on the amino acid sequence, and the full-length clone obtained was inserted in a *T. reesei* strain (QM6a) that has limited indigenous β-glucosidase activity in glucose-containing media. For being able to purify the protein for characterization, the gene inserted in *T. reesei* had a His-tag sequence attached for binding and purifying the tagged enzyme using an ÅKTA purifier system. The purified protein was then characterized for its thermostability, pH optimum, temperature range, ability to degrade cellodextrins, and kinetic studies. Moreover, a homology model of the catalytic module was made. All together, the data obtained showed that the enzyme (BGL1) was the main responsible enzyme for β-glucosidase activity in this fungus, that it has a high thermostability and that its hydrolytic activity is very efficient [10]. The enzyme was different from any known β-glucosidases in terms of amino acid sequence and as it also was found in a previously unknown fungus, the β-glucosidase was patented [11].

5   *A. saccharolyticus* as a Potential On-Site β-Glucosidase Producer

Some filamentous fungi produce high amounts of endoglucanases and cellobiohydrolases, e.g., strains of *Trichoderma* species while other fungi produces high amounts of β-glucosidases, e.g., strains of *Aspergillus* species [12]. Since endoglucanases, cellobiohydrolases and β-glucosidases all are important for efficient degradation of biomass cellulose due to their synergistic action, it is therefore known that by combining fungi that expresses different kinds of cellulases in their enzyme extracts, it is an efficient way to degrade cellulose. In a series of studies, enzyme extracts from *T. reesei* RUT C30 and *A. saccharolyticus* (in house) was tested alone, in combination or in combination with commercially derived enzyme products (Novozyme 188, Celluclast, and CellicCtec2) and used for hydrolysis of pretreated wheat bran, corn stover and loblolly pine biomass [13, 14]. The outcome of the enzyme treatment experiments in terms of glucose production was that the most efficient glucose production was obtained when the in house produced enzymes was combined with commercially produced enzymes. In the study by Kolasa et al. [13], *T. reesei* RUT C30 was combined with *A. saccharolyticus*, *Aspergillus carbonarius*, or *A. niger* in cocultivation experiments in order to study the effect on degradation of pretreated wheat straw when combining fungal strains with different biomass cell wall degrading enzyme profiles. Of the different combinations, the enzyme broth from *T. reesei* RUT C30 and *A. saccharolyticus* mixture showed the best competitive behavior to Celluclast and Novozyme 188 compared to the combinations with *A. carbonarius* and *A. niger* [13]. It was further revealed, that cocultivation with the two fungi did not function in submerged fermentation but in solid-state fermentation, which likely is due to the nature of the solids where each fungus can find niches for growing. Furthermore, *T. reesei* has to be inoculated 2 days earlier than the *Aspergilli* strains due to its slower growth.

These experiments indicates clearly that there is an economical advantage to produce biomass degrading enzymes on-site by combining fungal strains with a different enzyme profile. Such on-site enzyme broths can be combined with minor amounts of commercially produced enzymes, and optimized to a very efficient enzyme product for degradation of pretreated biomasses. This would still be a cheaper solution than commercially produced enzymes.

In a biorefinery concept, where it is possible to produce different kind of products it is very important that all processes are as cheap as possible and therefore each step has to be evaluated for its efficiency potential as well as the economy. *A. saccharolyticus* has through the described experiments above shown its potential as a good and efficient enzyme producer of plant cell wall-degrading enzymes for degradation of biomasses in biorefineries and especially if the fungus either is grown together with a fungus with a different enzyme profile or its enzyme broth is supplemented with broths from other fungi.

Utilization of cheap agricultural waste streams from the biorefinery as substrates for on-site enzyme production will aid in reducing costs. In order to identify, which carbon sources that support growth and stimulate *A. saccharolyticus* to produce β-glucosidases, carbon sources, ranging from monomer sugars to complex lignocellulosic biomasses, including pretreated and hydrolyzed corn stover fractions, were studied in a micro titer plate experimental setup for fast screening for β-glucosidase activity [15]. The greatest activity was found when *A. saccharolyticus* was cultivated on media containing xylose, xylan, wheat bran, and pretreated corn stover. Thus, in a biorefinery, on-site β-glucosidase production by *A. saccharolyticus* could be based on the hemicelluloses from lignocellulosic biomasses. The advantage of this is that glucose from cellulose then is reserved to fermentation, e.g., with conventional yeast.

For further developing *A. saccharolyticus* into an industrial enzyme producing strain, a proteomics study was carried out (Lübeck et al. unpublished data) in order to identify and in detail characterize the enzymes involved in lignocellulosic biomass degradation for comprehensive evaluation. This would help in understanding of the properties of the enzyme system and will give background knowledge of value for the further genetic improvement of the strain. The study was carried out through a time course protein secretome analysis using wheat bran, CMC, and Avicel as growth substrates, and the data collected is still under investigation. Many interesting enzymes have been identified belonging to many different CAZY families and among the enzymes identified are also several accessory enzymes belonging to the AA9 group of oxidoreductases, which are enzymes shown to boost the activity of "traditional" cellulases [16]. These results are interesting since the knowledge of the complete enzyme apparatus for plant cell wall degradation is important to gain an understanding of the needed enzymes involved and to design improved biocatalysts for biomass degradation.

6 Potential of *A. saccharolyticus* for Production of Other Compounds

An important aspect is to study whether the novel fungus *A. saccharolyticus* has other interesting abilities such as production of interesting compounds. The extrolite production showed that it did not produce any known mycotoxins [7], but is capable of producing secalonic acid D, neoxaline, aculene A and B, and homomorphosins [8]. The fungus was also studied for organic acid production, since several *Aspergilli* is highly capable of production of organic acids [17], which can be used as building blocks for different kinds of biochemicals. In a recent study, it was found that *A. saccharolyticus* naturally produces malic acid and succinic acid in sugar rich media [18]. In the study, the fungus produces more malic acid (17.6 g/L) than succinic acid (3.8 g/L) but because of its ability to produce both acids, it was studied whether the production of succinic acid production could be increased by genetic engineering. The production of malic acid, fumaric acid, and succinic acid is primarily carried out in the mitochondria but some malic acid is also produced in the cytosol. In filamentous fungi, there is a flow between fumaric acid and malic acid in the cytosol, which is in a balance with the TCA production in the mitochondria, but production of succinic acid is not possible in the cytosol because of the lack of a fumarate reductase. Therefore, the produced succinic acid must be derived from the mitochondria and sent to the cytosol before it is exported to the environment. In order to be able to produce more succinic acid, a codon-optimized fumarate reductase gene from *Trypanosoma brucei* was inserted and expressed in *A. saccharolyticus* [18]. The result showed that the fungus produced much more succinic acid (16.2 g/L) and less malic acid (13.1 g/L) than before, so it seems that by introducing a fumarate reductase the flow from fumaric acid to malic acid is partly changed to a flow into succinic acid in the cytosol instead [18]. Moreover, it was shown that *A. saccharolyticus* is a good producer of both malic acid and succinic acid in a biomass hydrolysate from wheat straw (Yang, L., Lübeck, M., Lübeck, P. S. unpublished data).

7   Potential of *A. saccharolyticus* in Consolidated Bioprocessing

Due to the different capabilities, the fungus has a very high potential for the use in a consolidated approach in a biorefinery concept. In most cases, enzyme decomposition of pretreated biomass to sugar and production of valuable products from the sugar platform derived from enzyme degradation are two separate processes [19]. This is due to that the biocatalysts used in the two processes normally has potential for either being good enzyme producers that can degrade cell wall material or good producers of for example ethanol or organic acids that can be converted into valuable bioproducts. An approach could be to use the same biocatalytic organism for both biomass decomposition and for bioproduct production in one process [19]. This requires a biocatalyst that is able to at the same time degrading the cell wall material to sugar and utilizing the sugar for producing one or more compounds. In theory, *A. saccharolyticus* has this capability. It degrades plant biomass efficiently, and it produces high amounts of organic acids naturally and is capable of growing and producing using pretreated biomass hydrolysate where there are several inhibitors known to limit the growth of many microorganisms. Therefore *A. saccharolyticus* would be relevant choice for a consolidated bioprocess in a biorefinery concept either as the sole biocatalyst or in combination with another biocatalyst, e.g., *T. reesei*.

8   Conclusions

This case example shows an example of screening and selection of a novel fungus with a superior β-glucosidase with potential for industrial production. In addition to a potential for an on-site production of β-glucosidases within biorefineries, *A. saccharolyticus* also proved to have a production capacity of industrial building blocks and thereby has different potentials as a novel fungal cell factory.

References[1]

1. Klemm D, Heublein B, Fink HP, Bohn A (2005) Cellulose: fascinating biopolymer and sustainable raw material. Angew Chem Int Ed 44:3358–3393.   https://doi.org/10.1002/anie.200460587

2. Perez J, Munoz-Dorado J, de la Rubia T, Martinez J (2002) Biodegradation and biological treatments of cellulose, hemicellulose and lignin: an overview. Int Microbiol 5:53–56

3. Sørensen A, Lübeck M, Lübeck PS, Ahring BK (2013) Fungal beta-glucosidases: a bottleneck in industrial us of lignocellulosic materials. Biomol Ther 3:612–631

4. Xiao ZZ, Zhang X, Gregg DJ, Saddler JN (2004) Effects of sugar inhibition on cellulases and beta-glucosidase during enzymatic hydrolysis of softwood substrates. Appl Biochem Biotechnol 113–116:1115–1126

---

[1] Cellulose: Fascinating Biopolymer and Sustainable Raw Material
 Dieter Klemm,* Brigitte Heublein, Hans-Peter Fink,* and Andreas Bohn

5. Lau MW, Bals BD, Chundawat SPS et al (2012) An integrated paradigm for cellulosic biorefineries: utilization of lignocellulosic biomass as self-sufficient feedstocks for fuel, food precursors and saccharolytic enzyme production. Energy Environ Sci 5:7100–7110

6. Sørensen A, Lübeck PS, Lübeck M et al (2011) β-glucosidases from a new *Aspergillus* species can substitute commercial β-glucosidases for saccharification of lignocellulosic biomass. Can J Microbiol 57:638–650

7. Sørensen A, Lübeck PS, Lübeck M et al (2011) *Aspergillus saccharolyticus* sp. nov., a new black *Aspergillus* species isolated from treated oak wood in Denmark. Int J Sys Evolution Microbiol 61:3077–3083

8. Fungaro MHP, Ferranti LS, Massi FP et al (2017) *Aspergillus labruscus* sp. nov., a new species of *Aspergillus* section *Nigri* discovered in Brazil. Sci Rep 7(1):6203. https://doi.org/10.1038/s41598-017-06589-y

9. Pel HJ, de Winde JH, Archer DB et al (2007) Genome sequencing and analysis of the versatile cell factory *Aspergillus niger* CBS 513.88. Nat Biotechnol 25(2):221–231. https://doi.org/10.1038/nbt1282

10. Sørensen A, Ahring BK, Lübeck M et al (2012) Identifying and characterizing the most significant β-glucosidase of the novel species *Aspergillus saccharolyticus*. Can J Microbiol 58:1035–1046

11. Ahring BK, Lübeck PS, Sørensen A, Teller P (2010) *Aspergillus* containing beta-glucosidase, beta-glucosidases and nucleic acids encoding the same. WO 2012013197:A3

12. Gutierrez-Correa M, Portal L, Moreno P, Tengerdy RP (1999) Mixed culture solid substrate fermentation of *Trichoderma reesei* with *Aspergillus niger* on sugar cane bagasse. Bioresour Technol 68:173–178. https://doi.org/10.1016/S0960-8524(98)00139-4

13. Kolasa M, Ahring BK, Lübeck PS, Lübeck M (2014) Co-cultivation of *Trichoderma reesei* RutC30 with three black *Aspergillus* strains facilitates efficient hydrolysis of pretreated wheat straw and shows promises for on-site enzyme production. Bioresour Technol 169:143–148

14. Rana V, Eckard AD, Ahring BK (2014) Comparison of SHF and SSF of wet exploded corn Stover and loblolly pine using in-house enzymes produced from *T. reesei* RUT C30 and *A. saccharolyticus*. Spring 3:516–528

15. Sørensen A, Andersen JJ, Ahring BK et al (2014) Screening of carbon sources for beta-glucosidase production by *Aspergillus saccharolyticus*. Int Biodeter Biodegrad 93:78–83. https://doi.org/10.1016/j.ibiod.2014.05.011

16. Vaaje-Kolstad G, Westereng B, Horn SJ et al (2010) An oxidative enzyme boosting the enzymatic conversion of recalcitrant polysaccharides. Science 330:219–222. https://doi.org/10.1126/science.1192231

17. Yang L, Lübeck M, Lübeck PS (2017) *Aspergillus* as a versatile cell factory for organic acid production. Fungal Biol Rev 31:33–49. https://doi.org/10.1016/j.fbr.2016.11.001

18. Yang L, Lübeck M, Ahring BK, Lübeck PS (2016) Enhanced succinic acid production in *Aspergillus saccharolyticus* by heterologous expression of fumarate reductase from *Trypanosoma brucei*. Appl Microbiol Biotech 100:1799–1809

19. Zoglowek M, Hansen GH, Lübeck PS, Lübeck M (2015) Fungal consortia for conversion of lignocellulose into bioproducts. In: Silva RN (ed) Fungal biotechnology for biofuels. Mycology: current and future developments, vol 1. Bentham eBooks, Bentham Science Publishers, Sharjah, pp 329–365. https://doi.org/10.2174/9781681080741115010

Part II

Protocols

Chapter 3

Isolation and Screening of Cellulolytic Filamentous Fungi

Mette Lübeck and Peter Stephensen Lübeck

Abstract

Filamentous fungi are among the microorganisms that most efficiently are able to degrade plant biomass by secreting cell wall-degrading enzymes and they are therefore used extensively in the industry as workhorses for the production of enzymes, including cellulases for the use in second-generation biorefinery concepts. Fungi are therefore of interest both as resources for the search of novel cellulolytic enzymes and for production of enzymes and enzyme cocktails, which also can be carried out on-site using cheap lignocellulosic substrates for growth and enzyme production. Fungi can be isolated from different environmental niches, such as soil, compost, decaying wood, decaying plant material, building materials, and different foodstuffs. Selective isolation can be carried out using simple cellulosic or complex plant material in the media. In this chapter, methods used for the isolation and screening of cellulolytic fungi isolated from different ecological niches are presented. The screening assay presented in the chapter is an easy semiquantitative high-throughput agar plate screening method using azurine-cross-linked (AZCL) cellulose substrates.

Key words Cellulolytic fungi, Qualitative methods, AZCL-cellulose, Decaying plant material, Compost

1 Introduction

The amount of plant biomass on earth is enormous, and the lignocellulosic part of the biomasses will constitute an important resource in future in the second-generation biorefineries as a substitution for oil-based refineries [1]. In order to utilize the lignocellulose as a resource for biofuels, biochemicals, and other carbon-based compounds, the lignocellulose has to be broken down to small molecules that can be utilized as backbone for biomaterials of interest. Cellulose is a very important part of the lignocellulose and in order to break down this material, the lignocellulose has to go through the process steps of pretreatment and enzymatic hydrolysis before it can be used in fermentation for production of valuable products [2]. Most pretreatment methods facilitate the removal of lignin, partly dissolve hemicelluloses but cannot break down the recalcitrant cellulose fibrils, which therefore is the main target for

Mette Lübeck (ed.), *Cellulases: Methods and Protocols*, Methods in Molecular Biology, vol. 1796,
https://doi.org/10.1007/978-1-4939-7877-9_3, © Springer Science+Business Media, LLC, part of Springer Nature 2018

enzyme hydrolysis. Although the enzyme industry has spent plenty of resources into identification of efficient cellulases and developed commercial preparations of cellulolytic enzyme cocktails, the hunt for novel and better cellulases still is relevant [3]. The desired properties of cellulases depend on the exact biorefining purpose. To mention a few examples: pretreatment most often is carried out at elevated temperatures, and enzyme reactions in general work faster at higher temperatures, which has resulted in a focus on thermostable enzymes. Since cellulases also are prone to product inhibition [4], it is also of interest to identify enzymes with less sensitivity to inhibition. However, in processes where saccharification is carried out simultaneous with fermentation (SSF), the optimal working temperature of cellulases may be different [5].

Filamentous fungi have attracted great attention as efficient enzyme producers, especially regarding enzymes of interest for lignocellulosic biomass deconstruction. Fungi have adapted to very different environmental niches [6], and their adaptation is possible due to a diversity of fungal enzymes that are secreted extracellularly and synergistically used for deconstructing various insoluble plant substrates into soluble sugar nutrients. The ability of fungi to utilize solid substrates at low water content results in high concentration of secreted enzymes and high biocatalysis efficiency [7]. Furthermore, fungi penetrate the solid plant materials by hyphal extensions and secrete the enzymes from the hyphal tips. Due to their versatility, a great variety of fungal species can be isolated from very different environmental niches, including extreme temperatures and pH, although fungal growth is restricted at temperatures beyond 65 °C [8, 9].

In order to discover new efficient hydrolytic enzymes for lignocellulose conversion, screening of fungal isolates from environmental samples can be carried out using different isolation techniques. One of the main problems in the isolation and screening techniques is that the conditions that are set up in the laboratory on the one hand narrow down the amount of isolates to screen and on the other hand limits the discovery of novel enzymes. Perhaps the fungi contain an efficient cellulase that is not expressed to a high degree in the laboratory conditions. Mainly, the fungi that grow well at laboratory conditions also will be among those with high enzyme activity as the activity is also related to growth. Since the qualitative screening assays are nonspecific, the measured activity may result from a synergistic array of enzymes, rather than a single very efficient enzyme. This can be put down to the sentence "you get what you screen for" [10]. From the different genomic projects, it is evident that fungi possess a great variety of cellulose-degrading enzymes in their genomes [11, 12], and proteomic studies also show that many of these are secreted and present when the fungi grow on lignocellulosic substrates [13, 14]. However, the discovery of novel enzymes involves as its main aim identifying enzymes with superior catalytic abilities and industrial applications.

In this chapter, isolation methods based on the use of selective petri dishes with either simple cellulosic material (Avicel, carboxymethyl cellulose (CMC)) or complex lignocellulosic substrates such as wheat bran, wheat straw, other agricultural residues, or waste streams from biorefineries are presented, together with a simple AZCL screening method for identifying efficient cellulolytic enzymes [15]. Wheat bran is a very rich and nutritious lignocellulosic substrate that has been widely used for fungal enzyme production [16]. The isolated fungi can also be cultivated on the selective media or screened using the methods described in Chapter 4.

2 Materials

2.1 Isolation and Collection of Fungal Samples

1. Soil samples.

2. Compost samples.

3. Decaying wood samples.

4. Decaying plant samples (e.g., leaves from defoliating trees).

5. Diseased plant samples.

6. Building materials with visible fungal growth.

7. Foodstuff (bread, vegetables, fruit).

8. Petri dishes.

9. Plastic container.

10. Inoculation loops.

11. Test tubes: 15-mL and 50-mL Falcon tubes, microcentrifuge tubes, 100-mL Erlenmeyer flasks.

12. Shaking incubator.

13. Thermoshaker.

14. Microcentrifuge.

15. Laminar flow bench (optional).

16. Magnet and magnet stirrer.

17. Vacuum filter.

18. Czapek (Cz) liquid medium without a carbon source: 3 g/L $NaNO_3$, 1 g/L K_2HPO_4, 0.5 g/L KCl, 0.5 g/L $MgSO_4 \cdot H_2O$, 0.01 g/L $FeSO_4 \cdot 7H_2O$. Adjust pH to 4.8. Add 1 mL trace metal solution (TMS: 0.01 g/L $ZnSO_4 \cdot 7H_2O$, 0.05 g/L $CuSO_4 \cdot 7H_2O$ in 100 mL water) to the medium just before autoclaving at 121 °C for 20 min [17].

19. Cz agar medium: Add 15 g/L agar and eventually 200 μg/mL Triton X-100 before autoclaving (*see* **Note 1**).

20. Cz medium with a carbon source: Add 2% of a selected carbon source, e.g., Avicel, carboxymethyl cellulose, or lignocellulosic biomass before autoclaving (*see* **Note 2**).

21. Cz medium with antibiotics: Add 50 µg/mL tetracycline and 50 µg/mL chloramphenicol after autoclaving and cooling to approx 60 °C.

22. Solid-state medium: Use 60 g sphagnum peat, 100 g wheat bran (or another selected cellulose component), and 230 mL H_2O and mix these three components well. Then add 10 g of the mixture to a 100 mL Erlenmeyer flask and close the flask using wadding and aluminum foil. Autoclave the medium twice at 121 °C for 60 min (may take 2 days).

23. Britten Robinson Universal buffer: Make 400 mL of 0.04 M H_3BO_3, 0.04 M H_3PO_4, and 0.04 M CH_3COOH in H_2O.

24. AZCL-HE-Cellulose (I-AZCL, Megazymes).

---

3   Methods

3.1   Preparation of Media for Isolation of Fungi

Choose a strategy for selective media for isolation of fungi, e.g., using media containing either simple cellulosic or complex lignocellulosic plant material. Prepare the selected Cz medium with agar, autoclave and add antibiotics after cooling to below 60 °C according to the description above and distribute the medium into petri dishes (approx. 15 mL in each plate). The inclusion of antibiotics in the media prevents the growth of bacteria. Especially when isolating fungi from soil or compost samples, it is recommended to add Triton X-100 to restrict colony spreading on the plates [18]. This facilitates isolation of a greater variation of colonies of fungi and prevents the rapidly growing fungi from overgrowing the more slowly growing fungi on the petri dish.

3.2   Isolation and Collection of Samples

3.2.1   Isolation from Soil or Compost Samples

1. Add 100 mg of soil or compost to 10 mL of sterile water or Cz medium in a 15 mL Falcon tube.

2. Make a dilution series by adding (after whirl mixing) 1 mL of the solution to 9 mL in another Falcon tube three times, as shown in Fig. 1.

3. From each dilution, and the undiluted sample, take out 100 µL solution and plate on the petri dishes containing the selected medium (*see* **Note 1**) using a Drigalski spatula, in triplicate preferably carried out in a laminar flow bench to prevent contamination (*see* **Note 3**).

4. Incubate the media at room temperature (RT) for some days, up to 1 week (*see* **Note 4**).

5. Follow the development of colonies and make pure cultivations of the colonies before they grow into each other.

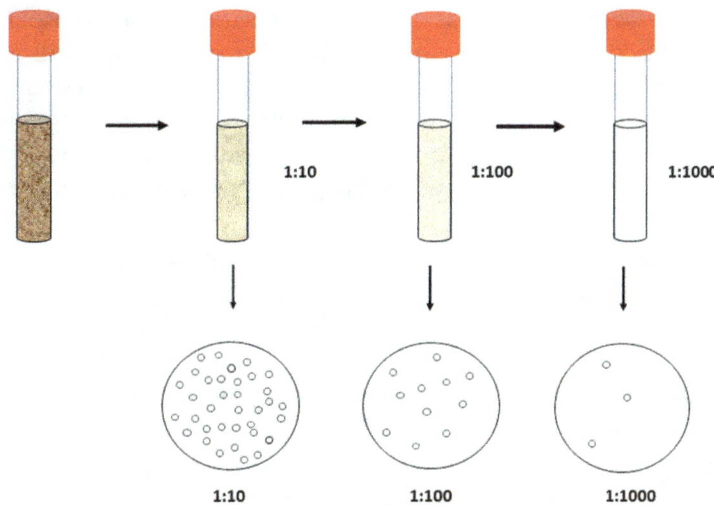

Fig. 1 Scheme showing a dilution series and plating on selective petri dishes

6. For pure cultivations, transfer spores or a small piece of hyphae using an inoculation loop to a new plate and place it in the middle of the plate or streak the inoculum out on the entire plate (*see* **Note 5**).

7. Incubate at the same temperature as before and follow the growth. If the colony is not clean (e.g., has a different morphology or color or is recognized as more than one colony), you may have to do a second or several rounds of pure cultivation.

8. Store the pure cultures by either harvest spores on the plates and add them to 1 mL 10% glycerol in a cryo vial, or add 1–4 agar plugs to 1 mL 10% glycerol in a cryo vial, and keep at −80 °C.

3.2.2 Isolation from Different Types of Plant Material

Several techniques can be used to isolate fungi from different plant materials such as decaying wood, defoliating leaves, and diseased plants, or from foodstuff:

1. Place the material in water, whirl-mix, and make a dilution series as described in Subheading 3.2.1.

2. Use an inoculation loop (if there is visible spores) as described in Subheading 3.2.3.

3. Place the plant material in the middle of a petri dish containing the selected medium, followed by incubation for up to a week.

4. Preincubate the plant material on wetted paper in a plastic container with a lid and incubate for a couple of days for the fungi to develop.

5. For all the above methods, follow **steps 4–7** of Subheading 3.2.1.

3.2.3 Isolation from
Building Materials
with Visible Fungal Growth

1. Use an inoculation loop to isolate fungi from houses with visible fungal growth, e.g., wall-paper in humid buildings or in wooden building materials such as window frames, by touching or scraping the visible fungal material and streak out on a petri dish containing selection medium with antibiotics.

2. Follow **steps 4–7** of Subheading 3.2.1.

3.3 Extraction of Enzymes from Pure Cultures

Screening of pure cultures can be performed either directly using the fungal material growing on the cellulosic substrate, or using the broth after liquid cultivation in the substrate without agar.

3.3.1 Extraction
of Enzymes from Agar
Plates

1. Use 7-day-old fungal cultures grown on selected cellulose medium.

2. Cut a 1 cm^2 block out of the culture including agar with a scalpel.

3. Cut the block into as many small cubes as possible, and transfer them to 1 mL sterile H_2O in a microcentrifuge tube and incubate overnight (ON) at 30 °C with shaking.

4. Centrifuge for 2 min at 13,000 rpm (= 17,900 × g in a microcentrifuge to pellet cell debris and agar.

5. Transfer the supernatant with enzymes to a new microcentrifuge tube (*see* **Note 6**).

3.3.2 Liquid Cultivation
and Extraction of Enzymes

1. Harvest spores from 7-day-old culture and add 1 × 10^6 spores/mL into a Falcon tube containing 5 mL liquid medium with the selected carbon source (*see* **Notes 7–10**).

2. Cultivate the culture at 180 rpm at RT for up to 7 days (*see* **Notes 5** and **10**).

3. Centrifuge the culture in the Falcon tube for 10 min at 2000 rpm (= 650 × g) to pellet cell debris and transfer the broth containing the secreted enzymes to microcentrifuge tubes (*see* **Note 6**).

3.3.3 Solid State
Cultivation and Extraction
of Enzymes

1. Harvest spores from a 7-day-old culture and add 10^6 spores into a 100 mL Erlenmeyer flask containing 10 mg solid medium including the selected carbon source (*see* **Notes 7** and **8**).

2. Incubate at RT for up to 10 days and shake manually once or twice per day to mix the medium with the culture (*see* **Note 5**).

3. Add 10 mL of H_2O and incubate for a couple of hours, and shake manually a couple of times during incubation.

4. Extract as much liquid containing the secreted enzymes eventually using a vacuum filter and transfer the enzyme extracts into 15-mL Falcon tubes (*see* **Note 6**).

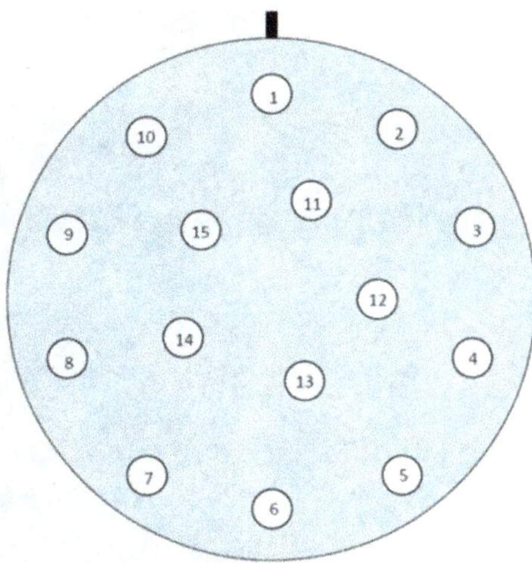

Fig. 2 Template for making holes in the AZCL-agarose in the petri dish

3.4 Screening of Isolated Pure Fungal Cultures Using AZCL-Cellulose Plates

3.4.1 Preparation of AZCL-Cellulose Plates

1. Use 72 mL stock solution of the Britten Robison universal buffer for a 250 mL final solution.

2. Add 100 mL of H_2O and adjust pH to 4.5 with 5 M sodium hydroxide.

3. Add H_2O to 250 mL.

4. Add 2.5 g agarose powder to a final concentration of 1% and melt it in a microwave oven.

5. Cool the solution down to 65 °C (use preferably a water bath).

6. Add 0.25 g AZCL-HE-cellulose to a final concentration of 0.1% while stirring (*see* **Note 11**).

7. Add 15 mL to petri dishes and solidify.

8. Punch out 15 holes/plate using a template (Fig. 2).

3.4.2 Screening for Cellulase Activity

1. Pipet 15 μL of enzyme solution into each hole of the AZCL agarose plate.

2. Incubate ON at 30 °C.

3. Inspect the plates and measure the radius of the developed halo around each of the holes with visible enzyme reaction (Fig. 3).

4 Notes

1. Addition of Triton X-100 to the medium is highly recommended, especially for the isolation of fungi from soil and compost.

2. Addition of a lignocellulosic carbon source can be wheat bran, known to be a very good fungal substrate [15], different agricultural wastes (wheat straw, rice straw, etc.) or a biorefinery

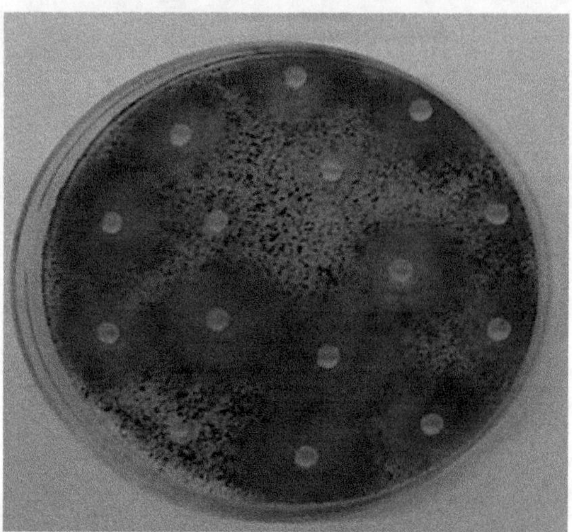

Fig. 3 Example of enzyme reactions on AZCL-cellulose agarose using fungal extracts from *Aspergillus*, *Fusarium*, *Penicillium*, and *Trichoderma* cultivated in different ways

waste (corn fiber, bagasse, filter cake after hydrolysis, etc.). If the aim with the screening is to identify an on-site fungal candidate with an efficient enzyme cocktail, the choice of lignocellulosic material may be the same, as one would wish to decompose using the produced enzyme cocktail. It is important to add a magnet and use magnetic stirrer with the insoluble carbon source in order to distribute it as evenly as possible.

3. The work can be performed at the table as the laminar flow bench is not absolutely needed since the soil or the compost is not sterile.

4. The same procedure can be followed with incubation at different temperatures, e.g., for isolation of psychrophilic fungi (4 °C) or thermophilic fungi (40–45 °C). However, incubation at higher temperatures >50 °C will require a different solidifying material than agar, as it starts to liquefy at high temperatures.

5. Use eventually small petri dishes to save media, and do include antibiotics in the medium at this stage, provided that you do the pure cultivation in the laminar flow bench.

6. The enzyme supernatant can be stored at −20 °C until use.

7. For harvesting spores, add 1–5 mL sterile H_2O or sterile liquid Cz medium to the petri dish with the culture, and use an inoculation loop, and Drigalski spatula or a scalpel to gently scrape off the spores and pipet them into a 15 mL Falcon tube. Count the spores using a hemacytometer.

8. Alternatively, it is possible to add 2–3 pieces of agar containing the culture instead of spores. This can be the best option if the culture is a nonsporulating fungus.

9. Use 1/10 volume of medium at maximum to the size of the Falcon tube. If Erlenmeyer flasks are used instead, do not use more than 1/5 volume of medium. This ensures proper aeration.

10. It is not recommended to close the lid of the Falcon tubes too tight; instead use tape to ensure that it stays on. Furthermore, it is recommended to tilt the Falcon tubes in the incubator so that the surface is larger. Both will ensure proper aeration.

11. It may be an advantage to add 1 mL of H_2O to the AZCL powder and make a slurry before adding it to the melted agarose solution.

References

1. Cherubini F (2010) The biorefinery concept: using biomass instead of oil for producing energy and chemicals. Energy Conver Manag 51:1412–1421

2. Lynd LR, Weimer PJ, van Zyl WH, Pretorius IS (2002) Microbial cellulose utilization: fundamentals and biotechnology. Microbiol Mol Biol Rev 66:506–577

3. Escamilla-Alvarado C, Pérez-Pimienta JA, Ponce-Noyolac T, Poggi-Varaldo HM (2017) An overview of the enzyme potential in bioenergy-producing biorefineries. Chem Technol Biotechnol 92:906–924

4. Murphy L, Bohlin C, Baumann MJ et al (2013) Product inhibition of five *Hypocrea jecorina* cellulases. Enzym Microb Technol 52:163–169

5. Gladis A, Bondesson PM, Galbe M, Zacchi G (2015) Influence of different SSF conditions on ethanol production from corn Stover at high solids loadings. Energy Sci Eng 3:481–489. https://doi.org/10.1002/ese3.83

6. Blackwell M (2011) The fungi: 1, 2, 3… 5.1 million species? Am J Bot 98:426–438

7. Raimbault M (1998) General and microbiological aspects of solid substrate fermentation. Electron J Biotechnol 1:26–27

8. Selbmann L, Egidi E, Isola D et al (2013) Biodiversity, evolution and adaptation of fungi in extreme environments. Plant BioSys 147:237–246

9. Maheshwari R, Bharadwaj G, Bhat MK (2000) Thermophilic fungi: their physiology and enzymes. Microbiol Mol Biol Rev 64:461–488

10. Schmidt-Dannert C, Arnold FH (1999) Directed evolution of industrial enzymes. Trends Biotechnol 1:135–136

11. Kubicek CP, Herrera-Estrella A, Seidl V et al (2011) Comparative genome sequence analysis underscores mycoparasitism as the ancestral life style of *Trichoderma*. Genome Biol 12:R40. https://doi.org/10.1186/gb-2011-12-4-r40

12. Benocci T, Aguilar-Pontes MV, Zhou M et al (2017) Regulators of plant biomass degradation in ascomycetous fungi. Biotechnol Biofuels 10:152. https://doi.org/10.1186/s13068-017-0841-x

13. Song W, Han X, Qian Y et al (2016) Proteomic analysis of the biomass hydrolytic potentials of *Penicillium oxalicum* lignocellulolytic enzyme system. Biotechnol Biofuels 9:68

14. Schneider WDH, Gonçalves TA, Uchima CA et al (2016) *Penicillium echinulatum* secretome analysis reveals the fungi potential for degradation of lignocellulosic biomass. Biotechnol Biofuels 9:66

15. Pedersen M, Hollensted M, Lange L, Andersen B (2009) Screening for cellulose and hemicellulose degrading enzymes from the fungal genus *Ulocladium*. Int Biodeter Biodegrad 63:484–489. https://doi.org/10.1016/j.ibiod.2009.01.006

16. Javed MM, Zahoor S, Shafaat S et al (2012) Wheat bran as a brown gold: nutritious value and its biotechnological applications. African J Microbiol Res 6:724–733

17. Samson R, Horkstra ES, Frisvad J, Filtenborg O (2002). Introduction to Food and Airborne fungi, Sixth edition. Centraalbureau Voor Schimmelcultures Utrecht. 1–282

18. Beuchat LR, De Dasa MST (1992) Evaluation of chemicals for restricting colony spreading by a xerophilic mold, *Eurotium amstelodami*, on Dichloran-18% glycerol agar. Appl Environ Microbiol 58:2093–2095

Chapter 4

Isolation and Screening of Cellulose-Degrading Microorganisms from Different Ecological Niches

Ayyappa Kumar Sista Kameshwar and Wensheng Qin

Abstract

Increased interest in developing cellulose-based ethanol over the last few years was the main reason behind inflated research to find cellulose-degrading microorganisms. Several methods have been developed in the past for efficient isolation and characterization of cellulolytic microorganisms. However, it is critical to choose a specific method from a list of qualitative methods for the characterization of cellulose degrading microorganisms. In this chapter, we have extensively listed various qualitative methods used for the isolation and characterization of the cellulolytic microorganisms isolated from different ecological niches such as soil, decaying wood, gut, and rumen.

Key words Cellulosic ethanol, Cellulolytic microorganisms, Qualitative methods, Soil, Decaying wood, Rumen, Gut

1 Introduction

Cellulose is the most abundant organic polysaccharide (linear chain of glucose units joined by β $(1\rightarrow4)$ glycosidic bonds) on the earth's surface, and thus cellulolytic microorganisms play a key role in maintenance of the global carbon cycle [1]. Studies conducted in the past have revealed that majority of cellulolytic bacteria belongs to Actinobacteria (aerobic bacteria) and Firmicutes (anaerobic bacteria), while the entire kingdom of filamentous fungi including Ascomycetes, Basidiomycetes, Deuteromycetes (aerobic fungi) and Chytridiomycetes (anaerobic fungi) contain larger groups of cellulolytic fungal species [1, 2]. Cellulose-degrading microorganisms majorly exhibit two types of cellulose hydrolysis patterns such as synchronized extracellular synergistic system, e.g., *Trichoderma* species, cellulosomes-based cellulose hydrolysis, e.g., *Clostridium* species and Neocallimastigomycetes fungi (anaerobic microbes) [3, 4].

In the past, several studies have reported a wide range of methods and assays for the efficient isolation and characterization

Mette Lübeck (ed.), *Cellulases: Methods and Protocols*, Methods in Molecular Biology, vol. 1796, https://doi.org/10.1007/978-1-4939-7877-9_4, © Springer Science+Business Media, LLC, part of Springer Nature 2018

of cellulolytic microorganisms. Traditionally, cellulose degrading microorganisms are isolated and characterized using cellulose (insoluble in growth media), which is highly reliable and carboxymethyl cellulose (CMC) (soluble in growth media) degraded by many microorganisms which can produce endoglucanase, will give a positive result. Several reports have used filter paper, ball-milled cellulose, acid-treated cellulose, dewaxed cotton string, bacterial cellulose, Avicel, carboxymethyl cellulose, cellobiose, xylan, wood fractions, plant biomass and wood pulp as cellulose substrates for testing the cellulolytic ability of microorganisms [1]. In this chapter, we have outlined the detailed protocols of qualitative and semiquantitative assays for isolation and characterization of cellulose degrading microorganisms, developed over the last few years, and discussed the advantages and disadvantages of the methods, as shown in Table 1. This chapter can act as a consolidated (one-stop) document harboring the qualitative assays for the isolation of cellulose-degrading microorganisms from different habitats such as soil, decaying wood, gut, and rumen.

Table 1
Advantages and drawbacks of different quantitative methods used for the estimation of cellulases

Quantitative method	Drawbacks and advantages
Filter paper degradation	The ambiguity caused due to visual interpretation of filter paper degradation is a major disadvantage of this assay, and the data obtained by this method cannot be used for comparative studies [7]
Cellulose agar plate	Recording the cellulose clearance zones in the growth medium is difficult, especially with fungal strains with dense and dark hyphae [8–10]
Esculin agar plate	Formation of black color in the esculin agar medium clearly indicates the breakdown of cellobiose to glucose by β-glucosidase. Dark hyphal growth by test fungal strain might ambiguity in the results [7]
CMC-Congo red	Compared to other qualitative methods, the CMC-Congo red agar plate method is easy and clear for screening cellulase producing microbial strains. The Congo red staining method can create ambiguity when used for screening fungal strains with dark hyphal growth [6, 7]
CMC-1% HAB	The clearance zones obtained from CMC-1% HAB are hazy and is costly and time expensive study [6]
CMC-Gram's iodine	When compared to CMC-Congo red and CMC-1% HAB methods CMC-Gram's iodine method is easier, cheap and rapid. Clear and sharp clearance zones were developed by this method [6, 7]
Dye diffusion method	Though this method is time-consuming it can be simultaneously used for both cellulolysis and lignin modifying activities. This study must be performed carefully as there are good chances of contamination [6, 7]

2   Materials

Analytical grade reagents must be used, with all the solutions prepared using the ultrapure water. All the reagents and solutions can be stored at room temperature, until and unless specified otherwise in the text.

2.1   Isolation and Collection of Samples

2.1.1   From Insect Gut Samples

1. 70% ethanol solution.
2. Sterilized distilled water.
3. Phosphate buffer solution ($1\times$PBS): 1.44 g Na_2HPO_4, 0.2 g KCl, 8 g NaCl, and 0.24 g KH_2PO_4, dissolve all the salts in 800 mL, followed by adjusting the pH to 7.4 using HCl further water is added appropriately to make the total volume to 1 L.
4. 1.5 mL microcentrifuge tube.
5. Sonicator, vortex.
6. 250 mL Erlenmeyer flasks, plastic pestle.
7. Tryptic soy broth: 3 g of tryptic soy broth, 15 g of agar, 1 L of distilled water with pH maintained at 7.0.
8. Luria–Bertani broth: 10 g Bacto tryptone, 5 g Bacto yeast extract, 5 g NaCl with pH maintained at 7.0.

2.1.2   From Rumen Samples

1. Collecting bottle with 250 mL capacity.
2. Cheesecloth.
3. −86 °C freezer.
4. Two types of growth mediums can be used for the isolation of cellulose-degrading microorganisms from the rumen:
5. Growth medium A
   (a) *15 mL Mineral Solution I:* 3.0 g KH_2PO_4; 6.0 g $(NH_4)_2SO_4$; 6.0 g NaCl; 0.6 g $MgSO_4$; 0.795 g $CaCl_2 \cdot 2H_2O$ per L.
   (b) *15 mL Mineral Solution II:* 3 g K_2HPO_4, 0.25 g yeast extract, 1 g tryptone, 0.1 mL 0.1% resazurin, 0.2 mL 0.05% hemin, 0.5 g microcrystalline cellulose, 0.1 g cellobiose, 0.4 g sodium carbonate, 20 mL clear rumen fluid, 50 mL distilled water, and 50 mg cysteine hydrochloride.
6. Growth medium B
   (a) Scott and Dehority agar growth medium supplemented with clarified rumen fluid 5% (v/v), 0–3% (w/v) glucose, cellobiose, and 0–6% (w/v) starch can be used for the culture of microbes.
7. The cultures must be incubated in an anaerobic chamber, which can maintain 90% of carbon hydroxide and 10% hydrogen.

8. Growth medium must be freed from oxygen gas by bubbling carbondioxide gas into the anaerobic culture/Hungate tubes.

9. Please refer the cited state-of-the-art review on preparation of culture mediums for isolating the anaerobic microorganisms [1].

2.1.3 Collection of Soil and Degrading Wood Samples

1. Soil samples.

2. Decaying wood samples.

3. Test tubes, 250-mL Erlenmeyer flasks.

4. Phosphate buffer saline solution ($1\times$PBS) composition is same as above.

5. Shaking incubator.

6. Potato dextrose broth: potato infusion 4 g and dextrose 20 g dissolved in 1 L of distilled water.

7. Nutrient broth: peptone 10 g, beef extract 10 g, NaCl 5 g, dissolved in 1 L of distilled water.

2.2 Methods for Isolation of Cellulase Producing Bacteria and Fungi

To achieve good results, it is necessary to employ uniform inoculation procedures. The test microbial strains (fungi) must be inoculated on minimal salt medium supplemented with 0.4% w/v glucose and 1.6% w/v agar. The test strains inoculated on minimal salt mediums can be used for the qualitative assays, as it does not carry over nutrients and does not interfere with result interpretations. Single agar disc with the test strains can be used further for inoculating the assay medium; the test strains can also be cultured using the cellulose basal medium, see below.

2.2.1 Filter Paper Degradation Method

1. Cellulose basal medium (CBM): 5 g of cellulose, 1 g KH_2PO_4, 0.5 g $MgSO_4.7\ H_2O$, 0.1 g yeast extract, and 0.001 g $CaCl_2.2H_2O$ are dissolved in 1 L distilled water.

2. 200-mL or 250-mL Erlenmeyer flasks.

3. Whatman's No. 1 filter paper (25×5 mm) or sterile filter paper (almost 100% cellulose).

2.2.2 Cellulose Agar Plate Method

1. 4% (w/v) cellulose is supplemented to the cellulose basal medium along with 1.6% (w/v) agar.

2. Autoclave at 121 °C for 15 min.

3. Petri plates.

2.2.3 Esculin-Iron Agar Plate Method

1. Cellulose basal medium is supplemented with 0.5% (w/v) esculin and 1.6% (w/v) agar.

2. 1 mL of 2% (w/v) aqueous ferric sulfate solution for 100 mL CBM growth medium.

3. Petri plates.

<table>
<tr><td>

2.2.4 Cellulose Agar Staining Methods

Method a (CMC-Congo Red)

</td><td>

1. Composition of carboxymethyl cellulose sodium: 0.05% K_2HPO_4, 0.025% $MgSO_4$, 0.188% CMC sodium salt, 0.02% Congo Red, 1.5% agar, 0.2% gelatin.

2. Petri plates.

3. For Congo-Red staining method: use the CMC agar plates same composition as mentioned above except Congo Red.

4. 1 M NaCl, distilled water.

</td></tr>
<tr><td>

Method B (Hexadecyltrimethyl Ammonium Bromide)

</td><td>

1. 2% (w/v) carboxymethyl cellulose (CMC), 0.2% $NaNO_3$, 0.1% K_2HPO_4, 0.05% KCl, 0.05% $MgSO_4$, 0.02% peptone, and 1.7% (w/v) agar.

2. 1% hexadecyltrimethyl ammonium bromide (HAB).

3. 1 M NaCl and distilled water.

4. Petri plates.

</td></tr>
<tr><td>

Method C (Gram's Iodine Staining)

</td><td>

1. 2% (w/v) carboxymethylcellulose, 0.2% $NaNO_3$, 0.1% K_2HPO_4, 0.05% KCl, 0.05% $MgSO_4$, 0.02% peptone, and 1.7% (w/v) agar.

2. Gram's Iodine complex (2.0 g KI and 1.0 g iodine dissolved in 300 mL distilled water).

3. Petri plates.

</td></tr>
<tr><td>

Method D (Dye Diffusion Method):

</td><td>

1. Cellulose basal medium, 1% (w/v) cellulose azure (or) cellulose dyed with Remazol Brilliant Blue R and Brilliant Blue R, 1.6% (w/v) agar.

2. 10-mL glass culture bottle.

3. Distilled water.

</td></tr>
</table>

---

3   Methods

<table>
<tr><td>

3.1 Isolation and Collection of Samples

3.1.1 Isolation of Gut from Insect Samples

</td><td>

1. Sterilize the surface of the test insect samples using 70% ethanol for 1 min.

2. Rinse the insect samples using sterile distilled water.

3. Following sterilization, suspend the insect samples in 10 mM sterile phosphate buffer saline ($1 \times$ PBS) solution.

4. Dissect the above insect samples inside a sterile laminar air flow chamber using a dissection scissors and fine tipped forceps.

5. Using dissection scissors, detach the head and last abdominal segments of the larva.

6. To release the gut, apply pressure on the anterior region of the crop.

</td></tr>
</table>

7. Held the thorax portion of the insect with the help of forceps, and pull the head from the thorax to stretch the gut out from the insect crop.

8. Separate the gut from the insect samples by detaching the extremities in a drop of sterilized PBS solution.

9. To avoid possible contamination from other tissues, rinse the gut samples using sterile PBS solution.

10. Gut samples can be either pooled or transferred individually to 1.5 mL micro centrifuge tube containing PBS solution (50, 100, or 500 μL).

11. Finally, sonicate the gut samples for 30 sec at 50/60 Hz, 117 V, 1.0 Amps, macerate with a plastic pestle and vortex at medium speed for 10 s for the separation of bacterial cells from the insect gut wall.

3.1.2 Isolation of Rumen Samples

1. Collect the rumen fluid from the ruminating animals, e.g., from the local slaughterhouse.

2. Place the rumen fluid immediately on dry ice and store further at $-50\,°C$.

3. Clarify the obtained rumen samples initially by straining rumen fluid using two-layered cheesecloth.

4. Autoclave the filtered rumen samples using the standard conditions followed by centrifuging the samples at $6000 \times g$ [5] (*see* **Notes 1** and **2**).

5. Later the rumen samples are ice thawed and used for further screening experiments.

3.2 Methods for Characterization of Cellulolytic Microorganisms

All the experiments must be carried out at room temperature unless the temperature conditions are specified otherwise.

3.2.1 Filter Paper Degradation

1. Prepare cellulose basal medium (CBM) and transfer 10 mL of the CBM solution into glass bottles followed by autoclaving at $121\,°C$ and 15 psi conditions (*see* **Notes 1** and **3**).

2. Transfer aseptically 25×5 mm sterile filter paper strips to the above glass bottles.

3. Inoculate the test strain into the glass bottles and incubate at corresponding temperatures (fungi $28\,°C$ and bacteria $37\,°C$).

4. Use uninoculated glass bottles as control.

5. Assess the degradation of filter paper based on the physical degradation and increased opacity in the inoculated samples.

6. Compare inoculated samples with the uninoculated samples to observe the difference.

3.2.2 Cellulose Agar Plate

1. Prepare growth medium with CBM supplemented with 4% (w/v) cellulose and 1.6% (w/v) agar (*see* **Note 4**).

2. After standard sterilization (gently mix the contents of the medium for uniform distribution) transfer appropriate amounts of medium to petri plates (*see* **Note 3**).

3. In a laminar air flow inoculate the test strain (fungi/bacteria) and incubate the plates at 27 °C and 37 °C respectively.

4. Observe the plates for the development of clearance zones around the test colonies on the opaque agar.

3.2.3 Esculin Agar Plate Method

1. Prepare growth medium with CBM supplemented with 0.5% (w/v) esculin and 1.6% (w/v) agar dissolved in 1 L distilled water, and add 1 mL of 2% (w/v) aqueous ferric sulfate for every 100 mL of CBM solution added to the sterilized medium (*see* **Note 1**).

2. Gently mix the medium contents, until uniform distribution is attained.

3. Aseptically transfer appropriate amounts of growth medium to sterile petri plates.

4. Inoculate the test strain on the petri plates under laminar airflow cabinet, and uninoculated plates can be used as control.

5. Incubate these petri plates at 27 °C in darkness, and monitor the plates for the development of black color, which indicates the production of β-glucosidase.

3.2.4 Dye Diffusion Based Methods

Method a (CMC-Congo Red Plate)

1. For Congo-Red staining method, prepare CMC-Congo Red agar growth medium based on the abovementioned composition dissolved in 1 L of distilled water and autoclave at standard conditions (*see* **Notes 1** and **5**).

2. Aseptically transfer the growth medium into sterile petri plates.

3. Inoculate 5 μL of test microbial strain on to the center of the plate, uninoculated plates are used as controls.

4. Incubate the inoculated plates at 27 °C for fungi and 37 °C for bacteria respectively.

5. Monitor daily for the development of clearance zones around the growing colony.

6. For Congo-Red staining method: flood the CMC agar plates with 0.1% Congo Red and incubate for 15–20 min, followed by 1 M NaCl for 15–20 min (*see* **Note 5**).

7. Observe the development of clearance zones around the growing colonies (Fig. 1).

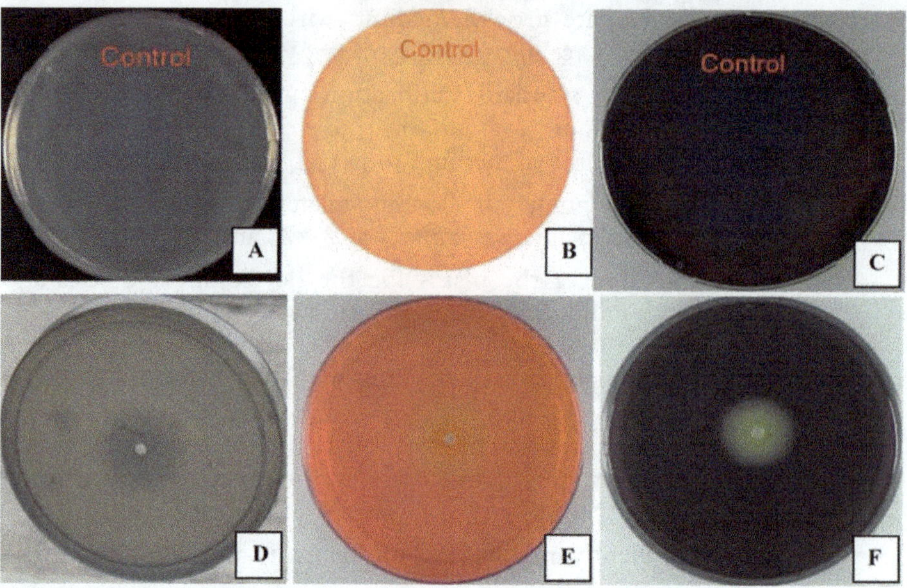

Fig. 1 Qualitative assays for the estimation of cellulases using the methods CMC-1% HAB (**a**) control, (**d**) test strain, CMC-Congo Red (**b**) control, (**e**) test strain and CMC-Gram's Iodine (**c**) control, (**f**) test strain (Reprinted with permission from [6])

Method B (CMC-HAB Method)

1. Prepare CMC agar medium based on the abovementioned compositions and autoclave at standard conditions (*see* **Notes 1** and **5**).

2. Aseptically transfer the growth medium into sterile petri plates.

3. Inoculate 5 µL of test microbial strain on to the center of the plate, uninoculated plates are used as controls.

4. Incubate the inoculated plates at 27 °C for fungi and 37 °C for bacteria respectively.

5. Plates are flooded with 1% HAB for 30–40 min (*see* **Note 5**).

6. Observe the development of clearance zones around the growing colonies (Fig. 1).

Method C (CMC-Iodine Stain)

1. Prepare CMC agar medium based on the abovementioned compositions and autoclave at standard conditions (*see* **Notes 1** and **5**).

2. Aseptically transfer the growth medium into sterile petri plates.

3. Inoculate 5 µL of test microbial strain on to the center of the plate, uninoculated plates are used as controls.

4. Incubate the inoculated plates at 27 °C for fungi and 37 °C for bacteria respectively.

5. Prepare Gram's Iodine solution by adding 2.0 g KI and 1.0 g iodine in 300 mL of distilled water (*see* **Note 5**).

6. Flood the plates with Gram's Iodine solution and incubate the plates for 3–5 min.

7. Observe the development of clearance zones around the growing colonies (Fig. 1).

Method D (Dye Diffusion Method)

1. Prepare CBM growth medium supplemented with 1.6% (w/v) agar dissolved in distilled water and autoclaved at standard conditions (*see* **Notes 1** and **3**).

2. Prepare 1% (w/v) cellulose azure supplemented with CBM and 1.6% (w/v) agar dissolved in distilled water and autoclaved at standard conditions (*see* **Note 1**).

3. Transfer the CBM agar solution prepared in the **step 1** to 10 mL to glass culture bottles.

4. Mix the cellulose azure agar prepared in **step 2** gently to attain uniform distribution and transfer 0.1 mL aseptically on to the surface (as an overlay) of the solidified CBM agar.

5. Uninoculated glass culture bottles can be used as controls.

6. Incubate the inoculated glass culture bottles at 27 °C and 37 °C for fungi and bacterial strains respectively.

7. Monitor the cultures carefully for the migration of dye from upper layer (cellulose azure agar) to lower layer (CBM agar), which indicates cellulolysis. The subsequent dye degradation of color in the upper layer indicates the presence of lignin-modifying activity.

4 Notes

1. Please refer to the appropriate laboratory manual before working with autoclave and other standard microbiology lab equipment mentioned in our chapter.

2. Centrifugation is performed to remove the particulate material present in the ruminal fluid. Be careful when you are using the autoclave, please read and understand the functioning of the autoclave.

3. Mix CMC thoroughly and ensure that no lumps are formed.

4. The cellulose source used in this assay can be ball-milled cellulose, acid-swollen and microcrystalline cellulose thus making the medium appear opaque. Cellulolysis leads to development of clearance zones around the colonies.

5. Avoid physical contact (by wearing gloves, lab coat, and eye gear) while using synthetic chemical dyes with Congo Red, Iodine solution, and hexadecyltrimethyl ammonium bromide.

5 Summary

In the past, several studies have reported a wide range of methods and assays for the efficient isolation and characterization of cellulolytic microorganisms. In this chapter, we have discussed the detailed protocols of qualitative and semi-quantitative assays for isolation and characterization of cellulose degrading microorganisms, developed in the last few years. This chapter can act as a consolidated (one-stop) document harboring the protocols using qualitative and semi-quantitative assays for the isolation of cellulose-degrading microorganisms from different habitats such as soil, decaying wood, gut, and rumen samples.

References

1. McDonald JE, Rooks DJ, McCarthy AJ (2012) Methods for the isolation of cellulose-degrading microorganisms. Methods Enzymol 510:349–374

2. Lynd LR, Weimer PJ, Van Zyl WH, Pretorius IS (2002) Microbial cellulose utilization: fundamentals and biotechnology. Microbiol Mol Biol Rev 66:506–577

3. Saloheimo M, Pakula TM (2012) The cargo and the transport system: secreted proteins and protein secretion in *Trichoderma reesei* (*Hypocrea jecorina*). Microbiology 158:46–57

4. Gilbert HJ (2007) Cellulosomes: microbial nanomachines that display plasticity in quaternary structure. Mol Microbiol 63:1568–1576

5. Yanke L, Cheng K (1998) A method for the selective enumeration and isolation of ruminal Lactobacillus and Streptococcus. Lett Appl Microbiol 26:248–252

6. Kasana RC, Salwan R, Dhar H et al (2008) A rapid and easy method for the detection of microbial cellulases on agar plates using Gram's iodine. Curr Microbiol 57:503–507

7. Pointing SB (1999) Qualitative methods for the determination of lignocellulolytic enzyme production by tropical fungi. Fungal Div 2:17–33

8. Rautela GS, Cowling EB (1966) Simple cultural test for relative cellulolytic activity of fungi. Appl Microbiol 14:892–898

9. Egger KN (1986) Substrate hydrolysis patterns of post-fire ascomycetes (Pezizales). Mycologia 78:771–780

10. Rohrmann S, Molitoris H-P (1992) Screening for wood-degrading enzymes in marine fungi. Can J Bot 70:2116–2123

Chapter 5

Isolation of Cellulolytic Bacteria from the Rumen

Makoto Mitsumori

Abstract

To isolate strictly anaerobic rumen bacteria capable of degrading cellulose, environmental and nutritional conditions similar to the rumen environment should be simulated in vitro. One of the most useful techniques for isolating rumen bacteria is the roll-tube technique. In this chapter, the roll-tube technique for isolating cellulolytic rumen bacteria is briefly outlined.

Key words Rumen bacteria, Strictly anaerobic bacteria, Roll-tube technique, CO_2-free from O_2

1 Introduction

Ruminants, major herbivores, have the rumen, in which the rumen microbial ecosystem convert dietary plant materials to short-chain fatty acids and microbial cells, which are energy sources and body constituents, respectively, for ruminants [1, 2]. Studies on the rumen microbial ecosystem are extensively carried out in terms of both animal production and interest for cellulolysis. Most of rumen bacteria are obligate anaerobic bacteria, because they inhabit strictly anaerobic rumen conditions. Such conditions complicate culturing of rumen bacteria, which requires strict anaerobic conditions in vitro. Hungate attempted to culture cellulolytic rumen bacteria using a liquid medium that provided environmental and nutritional conditions simulated to the rumen [3], and subsequently developed a roll-tube method, which consisted of some components including anaerobic gassing system, for isolating cellulolytic bacteria from the rumen [4, 5]. The roll-tube method can maintain strict anaerobic conditions in vitro; the method is used for culture of strictly anaerobic bacteria obtained not only from the digestive tracts but also from various environments [6–8].

Mette Lübeck (ed.), *Cellulases: Methods and Protocols*, Methods in Molecular Biology, vol. 1796,
https://doi.org/10.1007/978-1-4939-7877-9_5, © Springer Science+Business Media, LLC, part of Springer Nature 2018

2 Materials

2.1 Dilution Solutions

1. Mineral solution D-I: 0.6% K_2HPO_4 [9]. Store at 4 °C.
2. Mineral solution D-II: 1.2% NaCl, 1.2% $(NH_4)_2SO_4$, 0.6% KH_2PO_4, 0.12% $CaCl_2$, 0.25% $MgSO_4 \cdot 7H_2O$. Store at 4 °C.
3. Resazurin: 0.1% solution in water.
4. Dilution solution (1 L): Mix 75 mL of mineral solution D-I, 75 mL of mineral solution D-II, 1 mL of 0.1% resazurin, and 0.5 g of agar. Then, go to Subheading 2.3.

2.2 Media

1. Clarified rumen fluid: Rumen contents are collected from the rumen by through rumen fistula. The rumen contents are then autoclaved (121 °C, 20 min) and cooled down at room temperature. Squeeze the autoclaved rumen contents through metal mesh (5-mm mesh) and collect squeezed rumen fluid (SRF). Centrifuge the SRF at 30,000 × g, 15 min. The supernatant is stored at 4 °C or −20 °C until used as clarified rumen fluid [10].
2. Mineral solution R-I: 0.45% KH_2PO_4 [11]. Store at 4 °C.
3. Mineral solution R-II: 0.45% NaCl, 0.45% $(NH_4)_2SO_4$, 0.025% $CaCl_2$, 0.025% $MgSO_4$, 0.01% $MnSO_4 \cdot H_2O$, 0.01% $FeSO_4 \cdot 7H_2O$, 0.01% $ZnSO_4 \cdot 7H_2O$, 0.001% $CoCl_2 \cdot 6H_2O$. Store at 4 °C.
4. Vitamin stock solution: Weigh 0.5 g folic acid, 0.5 g d-biotin, and 0.05 g vitamin B_{12}. Make up to 100 mL with water. Store an aliquot (1 mL) of the solution at −20 °C.
5. Vitamin solution: Weigh 0.2 g pyridoxal hydrochloride, 0.2 g vitamin B_2, 0.2 g vitamin B_1 hydrochloride, 0.2 g nicotinamide, 0.2 g d-pantothenic acid calcium, and 0.01 g p-aminobenzoic acid. Add water to a volume of 900 mL. Add an aliquot (1 mL) of the vitamin stock solution and make up to 1 L with water. Store at 4 °C.
6. Volatile fatty acids (VFAs) solution: Mix 20 mL acetic acid, 1 mL iso-butyric acid, 1.2 mL iso-valeric acid, 1.2 mL valeric acid and 1.2 mL 2-methyl butyric acid. Make up to 1 L with water. Store at 4 °C.
7. Resazurin: 0.1% solution in water.
8. Hemin: 0.05% solution in water.
9. Cellulose medium (1 L): Mix 75 mL mineral solution D-I, 75 mL mineral solution D-II, 300 mL clarified rumen fluid, 2.0 g trypticase peptone, 1.2 g yeast extract, 0.2 g cellobiose, 1 mL 0.1% resazurin, and 700 mL water. Then, go to Subheading 2.3. In Subheading 2.3, **item 7**, a test tube containing a piece of Whatman No. 1 filter paper (1 × 3 cm) is served for 10 mL medium [10].

10. Rumen fluid-glucose-cellobiose agar (RGCA) medium (1 L): Mix 200 mL mineral solution R-I, 200 mL mineral solution R-II, 400 mL clarified rumen fluid, 66.7 mL VFA solution, 40 mL vitamin solution, 2.0 g casamino acids (low NaCl), 1 mL 0.05% hemin, 1 mL 0.1% resazurin, 0.25 g glucose, 0.25 g cellobiose, and 18 g of agar. Then, go to Subheading 2.3.

2.3 Basic Procedure for Preparation of Anaerobic Media and the Dilution Solution

1. Add water to the mixture to be a 90% volume of the final volume.

2. Adjust pH to 6.8 with NaOH solution.

3. Add 8% Na_2CO_3 and make up 100% volume with water.

4. Heat the medium in boiling water under CO_2 gas for 15–30 min. Check change of violet resazurin to red, pink, and colorless. Alternatively, the medium is heated at 105 °C for 5 min in an autoclave to remove air from the medium (*see* **Note 1**). After autoclaving, a bottle of the medium is gently taken from the autoclave (*see* **Note 2**). Then, insert a nozzle into the bottle and vent CO_2 gas. If the medium contains agar, the medium should be keep at 55 °C in a water bath.

5. Bubble with CO_2 gas for 15–30 min (*see* **Note 3**). Then, add L-cysteine–HCl.

6. Dispense into test tubes under a stream of CO_2 gas (*see* Subheading 3). Cap a test tube with a butyl rubber stopper (Fig. 1).

7. Place test tubes in a cage and press the top (Fig. 2).

8. Autoclave at 121 °C for 15 min.

9. Cool at room temperature or, if tubes are prepared for an agar slant medium, at 55 °C in a water bath. Take out test tubes from the cage (*see* **Note 4**). Store the test tubes at room temperature.

3 Methods

Prepare the dilution solution and media using distilled water and special grade reagents. Be careful about hot materials including media, rumen fluid, water, and the dexoygenator column heater.

3.1 Anaerobic Gassing System

Cultures of typical rumen bacteria require obligate anaerobic condition. The anaerobic gassing system, which consists of a gas tank, a dexoygenator, and an outlet, prepares strict anaerobic gasses. A dexoygenator removes traces of oxygen contained in commercial gasses by a copper column for supply gasses free from oxygen. Because you handle high-pressure gasses and dangerous gasses (CO_2 and H_2), watch over the line of the gassing system all the time. Make sure CO_2 concentration in air always using a CO_2 alarm (Fig. 3).

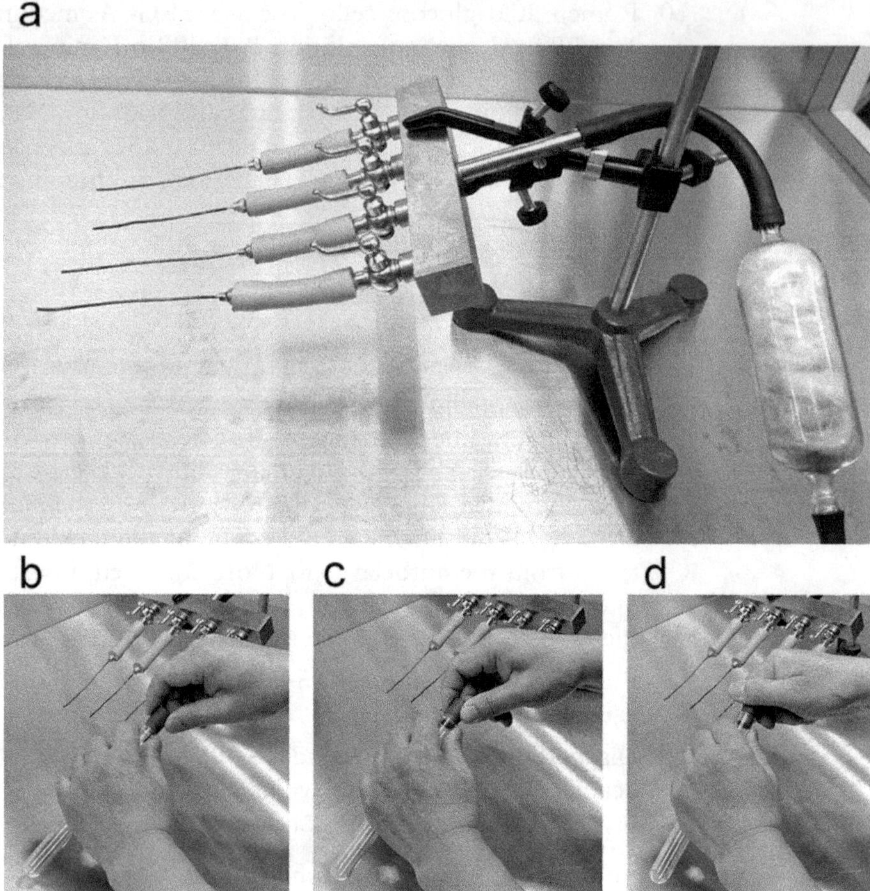

Fig. 1 Flushing needles (**a**) and closing a tube with a butyl rubber stopper (**b–d**). The rubber stopper slide on a needle (**b**), obstructing a tube (**c**), and tightly closing a tube (**d**). Because a tube is occasionally broken in this procedure, grasp a top of the tube with a hand to prevent cut the other hand with the edge of a broken tube

1. Check inlets and outlets of the gas lines are closed.

2. Open taps of outlet. To flush air from the lines, loosen the valve of a CO_2 cylinder in a short time. Then, all the taps on the gas lines are closed.

3. Switch on the dexoygenator column heater and wait for about 30 min for the heater to heat up. Do not touch the hot heater.

4. Look at the copper column inside the heater and blackening copper, which indicates that copper is oxidized.

5. Open taps of outlet. To flush air from the lines again, loosen the valve of a CO_2 cylinder in a short time. Do not close the outlet.

6. Switch the CO_2 line to the H_2 line. To reduce oxidized copper, loosen the valve of a H_2 cylinder in a short time (ca. 10 s) and close the valve. Wait the pressure on the gauge has dropped

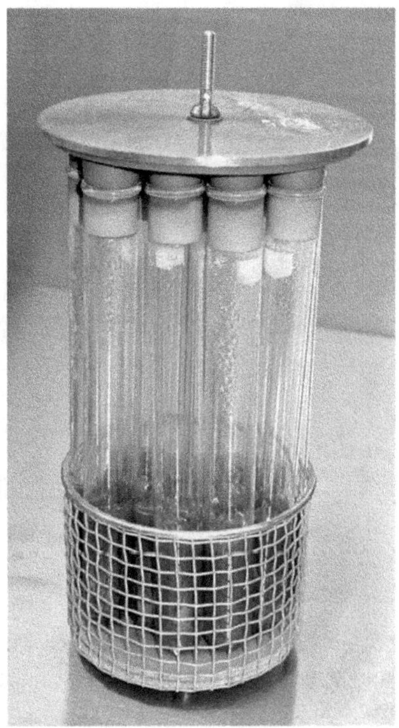

Fig. 2 Tubes are placed in to a metal cage and covered with a metal plate to prevent a stopper falling from a tube during autoclave processing because of high gas pressure in a test tube

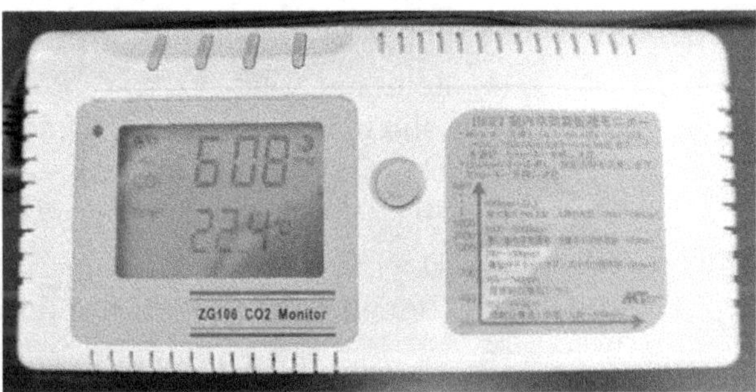

Fig. 3 CO_2 monitor. This monitor sounds at 5000 ppm of CO_2 concentration

    zero. Look at the copper in the column has changed to orange (Fig. 4). If the copper still blackens, flush H_2 gas again.

7. Switch the H_2 line to the CO_2 line. Open the CO_2 line and flow CO_2 gas at the minimum flow rate until starting an experiment.

8. After preparation of media or an experiment, close the inlet flow and look at the pressure on the gauge is zero. Then, close the outlet lines.

9. Turn off the heater on the deoxygenator.

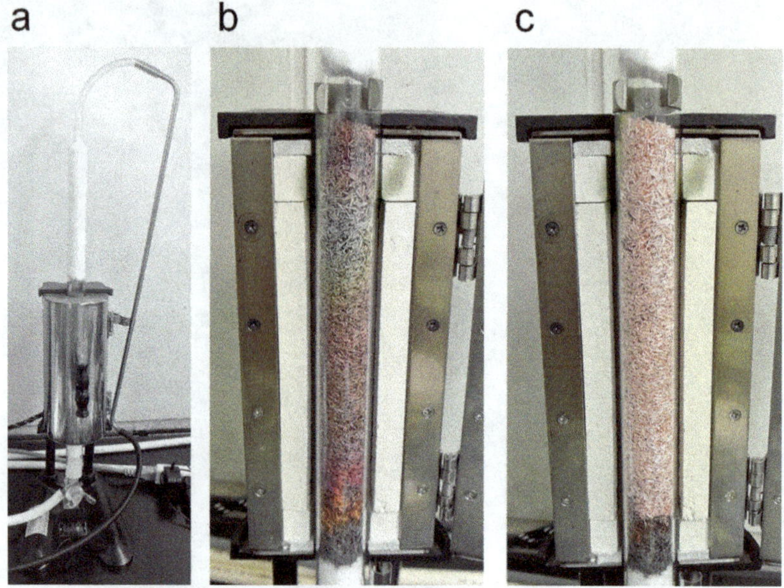

Fig. 4 Dexoygenator column heater. Panel (**a**) shows the whole apparatus. Gases flow from the lower pipe and are discharged from the upper pipe. Panel (**b**) shows oxidized copper colored red and black. Panel (**c**) shows reducted copper colored brilliant orange

3.2 Dilution of a Rumen Sample

1. Before going to cattle shed, run the anaerobic gassing system.

2. At the cattle shed, collect rumen contents (ca. 1000 g) through a rumen fistula and transport to the laboratory immediately.

3. Squeeze rumen contents by gloved hands and separate rumen fluid and solid materials. Weight rumen fluid (ca. 600 g) and solid materials (ca. 150 g). Place them in a sterile blender cup of a Waring blender under CO_2 gas stream. Blend the mixture for 30 s at high speed in the Waring blender.

4. The mixture (1 mL) is anaerobically transferred into a test tube containing the dilution solution (9 mL). Vortex the test tube.

5. Transfer this 10^{-1} dilution (1.0 mL) to a test tube with dilution solution (9.0 mL) and vortex (10^{-2} dilution). Serial10-fold dilutions are prepared up to 10^{-8} dilutions.

3.3 Cellulose-Enrichment Culture and Isolation of Cellulolytic Bacteria

1. Inoculate each (0.2 mL) of the mixture of rumen sample and/or diluted rumen samples into the cellulose medium (10 mL).

2. Incubate at 39 °C for 24–96 h.

3. Vortex the tube to collapse the filter paper and dilute the medium up to 10^{-8} dilution.

a b

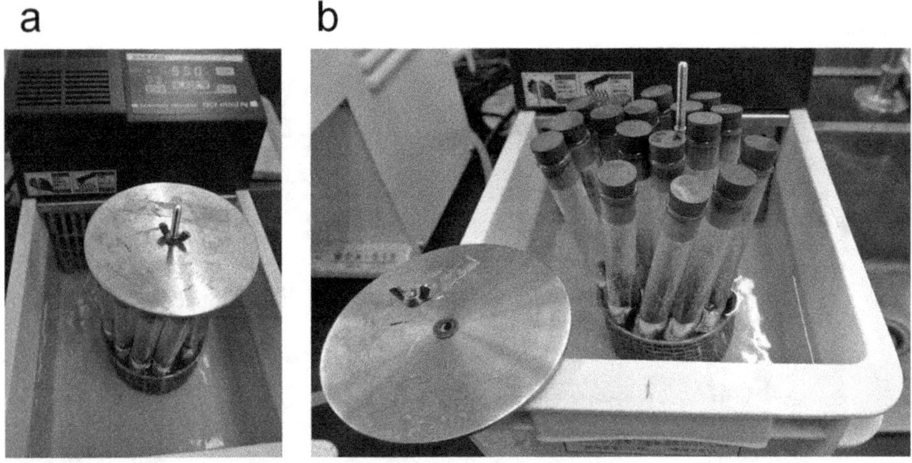

Fig. 5 Cool tubes containing an agar medium in a 55 °C water bath. Remove a metal plate before use

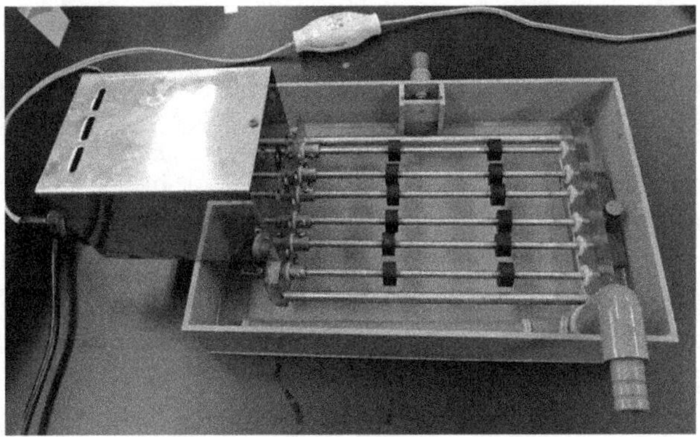

Fig. 6 A roll-tube maker. Fill the tray with ice-cold water. Tube rollers on the tray are rotated

4. Inoculate the dilution (0.2 mL) into the RGCA medium (10 mL), which is melted and kept at 55 °C in a water bath (Fig. 5).

5. After closing the tube with a butyl rubber stopper, the medium is gently shaken upside down. Place the tube on a tube roller filled with ice-cold water to form roll-tube cultures [9] (Figs. 6 and 7).

6. Incubate at 39 °C for 1–3 days.

7. Pick colonies from the roll-tube cultures (*see* **Note 5**) and transfer each colony to a slant medium of RGCA.

8. Incubate at 39 °C for 1–3 days.

9. Place the tube in a −80 °C freezer to store the culture.

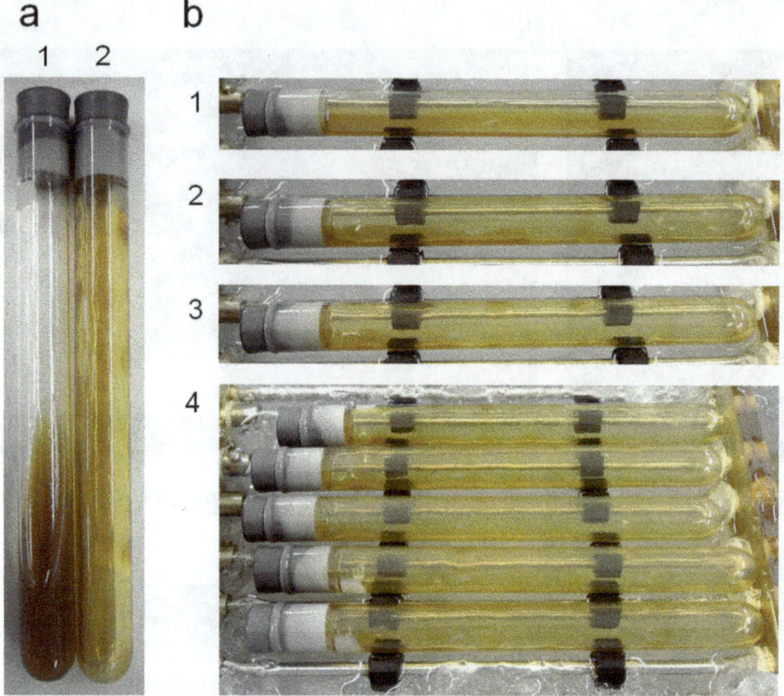

Fig. 7 Take a tube containing an agar medium from a water bath (Fig. 5b), add an inoculum to a tube and close a tube with a butyl rubber stopper (**a1**). Shake the tube gently upside down and place on a tube roller (**b1**). Forms a thin layer of the agar medium inside the tube by rotation (**b2–4**). Take a tube from the tube roller (**a2**)

4 Notes

1. A compact autoclave is recommended, because temperature regulation is easy. We use an autoclave (ES-215, Tomy, Tokyo, Japan). http://www.digital-biology.co.jp/man ufactured/products/es/.

2. To avoiding bumping of the medium, do not take out the bottle from the autoclave, when the temperature is higher than 85 °C.

3. Sometimes, the color of resazurin is still pink, especially in preparation of the dilution solution. It is no problem to go to the next step. In many cases, the color of resazurin is colorless after autoclaving.

4. The internal gas pressure of test tubes is higher than 1 atm, when the temperature of a test tube is higher than at room temperature.

5. Some modifications are applicable to increase diversity of isolates [12].

References

1. Russell JB, Rychlik JL (2001) Factors that alter rumen microbial ecology. Science 292:1119–1122
2. Hobson PN, Stewart CS (eds) (2012) The rumen microbial ecosystem. Springer Science & Business Media, New York
3. Hungate RE (1947) Studies on cellulose fermentation: III. The culture and isolation for cellulose-decomposing Bacteria from the rumen of cattle 1. J Bacteriol 53:631–645
4. Hungate RE (1950) The anaerobic mesophilic cellulolytic bacteria. Bacteriol Rev 14:1–49
5. Hungate RE, Macy J (1973) The roll-tube method for cultivation of strict anaerobes. Bull Ecological Res Commit 17:123–126
6. Holdeman LV, Cato EP, Moore WEC (eds) (1977) Anaerobe laboratory manual. Virginia Polytechnic Institute and State University, Blacksburg
7. Akasaka H, Izawa T, Ueki K, Ueki A (2003) Phylogeny of numerically abundant culturable anaerobic bacteria associated with degradation of rice plant residue in Japanese paddy field soil. FEMS Microbiol Ecol 43:149–161
8. Mikucki JA, Liu Y, Delwiche M et al (2003) Isolation of a methanogen from deep marine sediments that contain methane hydrates, and description of *Methanoculleus submarinus* sp. nov. Appl Environ Microbiol 69:3311–3316
9. Bryant MP, Burkey LA (1953) Cultural methods and some characteristics of some of the more numerous groups of bacteria in the bovine rumen. J Dairy Sci 36:205–217
10. Kudo H, Imai S, Arakaki C et al (1991) Practical laboratory manual for ruminology. Japan International Research Center for Agricultural Sciences (JIRCAS, Japan) and Instituto Nacional de Tecnología Agropecuaria (INTA, Argentina), Tsukuba, Japan
11. Scott HW, Dehority BA (1965) Vitamin requirements of several cellulolytic rumen bacteria. J Bacteriol 89:1169–1175
12. Nyonyo T, Shinkai T, Mitsumori M (2014) Improved culturability of cellulolytic rumen bacteria and phylogenetic diversity of culturable cellulolytic and xylanolytic bacteria newly isolated from the bovine rumen. FEMS Microbiol Ecol 88:528–537

Chapter 6

Methods for Discovery of Novel Cellulosomal Cellulases Using Genomics and Biochemical Tools

Yonit Ben-David, Bareket Dassa, Lizi Bensoussan, Edward A. Bayer, and Sarah Moraïs

Abstract

Cell wall degradation by cellulases is extensively explored owing to its potential contribution to biofuel production. The cellulosome is an extracellular multienzyme complex that can degrade the plant cell wall very efficiently, and cellulosomal enzymes are therefore of great interest. The cellulosomal cellulases are defined as enzymes that contain a dockerin module, which can interact with a cohesin module contained in multiple copies in a noncatalytic protein, termed scaffoldin. The assembly of the cellulosomal cellulases into the cellulosomal complex occurs via specific protein–protein interactions. Cellulosome systems have been described initially only in several anaerobic cellulolytic bacteria. However, owing to ongoing genome sequencing and metagenomic projects, the discovery of novel cellulosome-producing bacteria and the description of their cellulosomal genes have dramatically increased in the recent years. In this chapter, methods for discovery of novel cellulosomal cellulases from a DNA sequence by bioinformatics and biochemical tools are described. Their biochemical characterization is also described, including both the enzymatic activity of the putative cellulases and their assembly into mature designer cellulosomes.

Key words Glycoside hydrolases, Cellulosomes, Cohesin, Dockerin, Cellulases, Biomass degradation, Biofuels, Scaffoldin, Carbohydrate-binding module (CBM)

1 Introduction

A few decades ago, when cellulolytic activity was observed in bacterial or fungal cultures, one of the empirical methods to reveal cellulases was through their isolation, purification, and amino-acid sequencing [1]. Subsequently, additional cellulase genes could be revealed by genome walking [2], since cellulase genes are often observed in clusters [3–6].

Owing to bioinformatics, we can now establish novel, faster and more systematic methods for the discovery of novel cellulases directly from the genome sequence. Bioinformatics is based on accumulated knowledge and therefore is continuously evolving. One of the major tools that bioinformatics has provided is the

Mette Lübeck (ed.), *Cellulases: Methods and Protocols*, Methods in Molecular Biology, vol. 1796,
https://doi.org/10.1007/978-1-4939-7877-9_6, © Springer Science+Business Media, LLC, part of Springer Nature 2018

collection of databases, such as CAZy [7] or MG-RAST [8], which contain thousands of cellulase sequences, either presumed (owing to sequence conservation) or thoroughly characterized. Glycoside hydrolases (GHs) have been classified thus far in more than 130 different families [7] on the basis of their sequence homologies. Novel cellulase families, such as the GH family 124 (GH124) [9], are occasionally added, and a great many are yet to be discovered. The classification of the latter enzyme type as a cellulase was achieved only after strict biochemical analyses, combined with crystallographic structure determination [9]. We cannot rely currently only on bioinformatic tools to determine cellulolytic activity and initial bioinformatic indications must be validated by biochemical characterization. For example, bioinformatics alone cannot discriminate between possible xyloglucanase and cellulase activity in GH9 [10].

Here, we detail the discovery of ten cellulosomal cellulases from *Ruminococcus champanellensis,* a cellulosome-producer that inhabits the human gut [11]. First, we describe the methodologies that allowed detection of 65 dockerin-containing proteins in the *R. champanellensis* genome [12], using a database of known dockerin modules. The presence of a dockerin in a protein strongly indicates its cellulosomal nature. After biochemical characterization of the binding activity of these dockerins, either by ELISA or microarray, the parent dockerin-containing proteins are carefully screened for glycoside hydrolase modules. Out of these 65 proteins, 25 could be assigned bioinformatically to glycoside hydrolase families [13]. We then focused only on specific GH families that are known to exhibit cellulolytic activities. We describe the biochemical characterization of these putative cellulases using cellulosic substrates and the DNS method to determine reducing sugar production.

2   Materials

2.1 Cellulose Microarray

1. Tris-buffered saline (TBS): Prepare 1 L of $10\times$ stock solution. Weigh 8 g NaCl, 2 g KCl and 30 g Tris (hydroxymethyl) amino methane. Transfer to a glass beaker and add 900 mL double-distilled water. Mix and adjust pH to 7.4 with HCl. Bring to 1 L with double-distilled water. Store at 4 °C.

2. Calcium chloride ($CaCl_2$): prepare 50 mL of 2 M stock solution. Weigh 11.1 g $CaCl_2$, add double-distilled water up to 50 mL, mix well and store at 4 °C.

3. Washing Buffer: TBS, supplemented with 10 mM $CaCl_2$ and 0.05% Tween 20. For 0.5 L, add 50 mL $10\times$ TBS, 2.5 mL of 2 M $CaCl_2$, and 250 µL of Tween 20. Make up to 0.5 L with double-distilled water. Store at 4 °C.

4. Blocking Buffer: TBS, supplemented with 10 mM $CaCl_2$, 0.05% Tween 20 and 2% BSA. For 50 mL add 5 mL of TBS 10×, 250 µL of 2 M $CaCl_2$ solution, 25 µL of Tween 20, and 1 g of BSA into a 50 mL falcon tube. Make up to 50 mL. Mix vigorously, and store at 4 °C prepare fresh blocking buffer for every experiment.

5. Anti-xylanase Cy-3, anti-CBM Cy-5: The labeling of the fluorescent antibodies was performed using GE Healthcare's (Uppsala, Sweden), N-hydroxysuccinimide-ester-activated Cy-5 dye and Cy-3 kits. The dyes are resuspended in 0.1 M sodium carbonate buffer, pH 9, and mixed with the antibody (1 mg in 1 mL), according to the manufacturer's instructions. Free dye was removed by overnight dialysis against 1 L TBS buffer with two buffer changes. The fluorescence-labeled antibody was stored in 50% glycerol at −20 °C.

2.2 Cellulase Activities

1. Tris-buffered saline (TBS): Prepare 1 L of 10× stock solution *see* Subheading 2.1, **item 1**.

2. Calcium chloride: 2 M stock solution *see* Subheaidng 2.1 **item 2**.

3. EDTA (ethylenediaminetetraacetic acid): prepare 50 mL of stock solution at 0.5 M; weigh 9.3 g EDTA and 1 g NaOH (pellets), add 30 mL double-distilled water (*see* **Note 1**). Stir, heat until it dissolves, and adjust pH to 8 with 10 M NaOH. Bring to 50 mL with double-distilled water.

4. Acetate buffer 0.5 M: prepare 50 mL of 0.5 M sodium acetate; weigh 2.05 g sodium acetate and add 30 mL double-distilled water. Adjust pH to 5 with 0.5 M acetic acid. Bring to 50 mL with double-distilled water.

5. Citrate buffer 0.5 M: prepare 50 mL of 0.5 M sodium citrate; weigh 7.35 g sodium citrate and add 30 mL double-distilled water. Adjust pH to 6 with 0.5 M citric acid. Bring to 50 mL with double-distilled water.

6. 2% carboxymethyl cellulose (CMC) (VWR International Ltd., Lutterworth, England): weigh 2 g and dissolve in 100 mL double-distilled water under heating. Store at 4 °C (*see* **Note 2**).

7. Phosphoric acid swollen cellulose (PASC) suspension at 6 g/L. In a 3-L beaker add 6 g of microcrystalline cellulose (Avicel, Sigma-Aldrich, St. Louis, MO), a stirrer and a small amount of double-distilled water to get a homogeneous suspension. Add 300 mL phosphoric acid and stir 3 h at room temperature in the chemical hood. Add double-distilled water to 1.5 L while stirring (should produce an amorphous cellulose gel). Centrifuge at $14,000 \times g$ for 30 min, discard supernatant and wash three times with double-distilled water. Centrifuge and resuspend in 800 mL, and titrate with 10 M NaOH to a pH of 7. Use at a final concentration of 0.75% (w/v).

8. 10% Avicel (Sigma-Aldrich): weigh 1 g and dissolve in 10 mL double-distilled water. Store at 4 °C.

9. pNP-β-D-glucuronide (pNPG) (Sigma-Aldrich) at 50 mM. Store at −20 °C.

10. 3,5-Dinitro-salicylic acid (DNS) solution: In chemical hood, Weigh 40 g DNS (Sigma-Aldrich), 8 g phenol, 2 g sodium sulfite (Na_2SO_3) and 800 g Na-K tartrate (Sigma-Aldrich). Add all the above to 2 L of 2% NaOH. Mix with a magnetic stirrer overnight at room temperature, cover with aluminum foil. Complete with double-distilled water to a final volume of 4 L. Store in a dark bottle at 4 °C.

3   Methods

3.1 Bioinformatic-Based Discovery of Cellulosomal Cellulases

The wealth and availability of genome sequencing projects have enable the bioinformatic discovery of putative cellulases, and, more broadly, cellulosome-related carbohydrate active enzymes (CAZymes), which bear a dockerin module. An elaborate reservoir of dockerin-containing cellulases was observed in numerous glycoside hydrolases (GHs), carbohydrate esterases (CEs), polysaccharide lyases (PLs), and in proteins bearing carbohydrate-binding modules (CBMs). Thus, an important feature of cellulosomal cellulases is the presence of a dockerin module, which mediates the incorporation of the cellulase into the cellulosomal scaffoldin, via a specific cohesin–dockerin interaction. Therefore, by using bioinformatics-based identification of the dockerin module it is possible to fish out and identify its cellulase or related host protein.

As detailed below, the bioinformatic identification of cellulosomal modules is based on sequence similarity to previously characterized motifs or structural folds of the dockerin-bearing protein. It also takes into account the functional classification of the protein sequence, such as additional domains on the host protein, which are involved in polysaccharide degradation. The search must also allow for variations and diversity from canonical and conserved sequence motifs, in order to detect newly identified modules within genomes.

The dockerin module, as initially characterized in *C. thermocellum* [14], is typically a protein fold of ~70 amino acids long, that resides within carbohydrate-degrading enzymes, and serves to anchor the enzyme into the cellulosome by direct interaction with cohesin modules on the scaffoldin. Dockerins are comprised of two conserved repeats of calcium-binding loops followed by an "F helix," and connected by a linker region, variable in length and in composition [15] (Fig. 1). Moreover, Asp and Asn are usually highly conserved at the calcium-coordinating positions (1, 3, 5, 9, and 12), together with a conserved N-terminal Gly

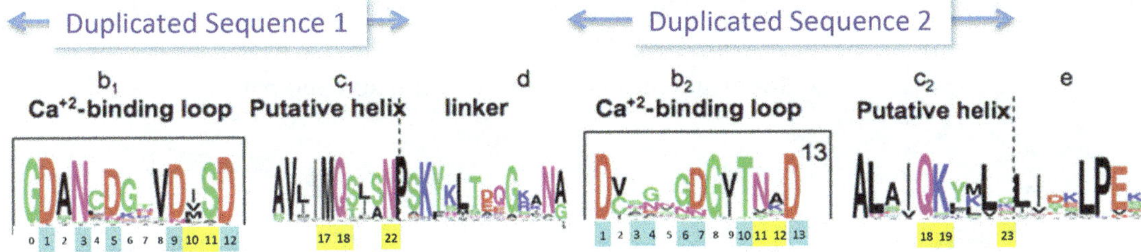

Fig. 1 Conserved sequence features of dockerin modules in cellulases, demonstrated by group 1 dockerins from *R. flavefaciens* FD-1. Dockerins are composed of two conserved repeats of calcium-binding loops followed by an "F helix" and a variable linker region. In the conventional dockerin, the calcium-coordinating residues (highlighted in cyan), typically Asp and Asn, appear at positions 1, 3, 5, 9, and 12 of the calcium-binding loop. Likewise, suspected specificity residues (highlighted in yellow) appear at positions 10, 11, 17, 18, and 22. This is true for Duplicated Sequence 1 of *R. flavefaciens* FD-1 group 1 dockerins shown in the figure but not for Duplicated Sequence 2, where an extra residue in the calcium-binding loop causes a displacement of the norm and an unconventional second repeat. This illustrates the variations and diversity from canonical and conserved sequence motifs that must be taken into account when searching for cellulosome-related modules. For construction of dockerin logos, multiple sequence alignments were created using the CLUSTAL server at the EBI (http:/www.ebi.ac.uk/Tools/msa/clustalw2/). Amino acid sequence logos were performed using the WebLogo application, version 2.8.2 [7]. The figure was adapted from Rincon et al. [19]

residue. In addition, biochemical and structural studies suggested that positions 10, 11, 17, 18, and 22 within the calcium-binding repeat are specificity determinants of the dockerin domain, which allow a species-specific recognition of the dockerin with its cohesin mate [16–18]. The dockerin module can be further categorized to type according to conserved and significant sequence features. Specifically, dockerins of Ruminococcal strains were further classified to at least 6 groups and 11 subgroups, based on residues of the Ca^{2+}-binding repeats or in their flanking regions, in references [19] and [20]. Recently, the unique cohesin–dockerin interaction for each of the dockerin groups was demonstration by Israeli-Ruimy and colleagues [21].

3.1.1 Genome-Wide Identification and Categorization of Dockerins

The above conserved features in the dockerin sequence allows the identification of dockerin modules by sequence similarity, using the BLASTP algorithm [22]. Here we detail the search for dockerins in the *R. champanellensis* 18P13 genome as an example.

1. Sequence search is performed with BLAST algorithm (https://blast.ncbi.nlm.nih.gov/Blast.cgi) using a large number of representative dockerin sequences as query (http://www.weizmann.ac.il/Biomolecular_Sciences/Bayer/sites/Biomolecular_Sciences.Bayer/files/uploads/dockerins_setup.txt), and *R. champanellensis* 18P13 genome or its deduced amino acid sequences as a subject.

2. Take the significant hit sequences below an E-value of 10^{-4}.

72   Yonit Ben-David et al.

3. Perform multiple sequence alignment with Clustal Omega
   (http://www.ebi.ac.uk/Tools/msa/clustalo/).

4. Identify conserved residues and specific patterns (Fig. 2):

Group 1

```
>DocJ  GDVNMDGRVIVADAVLVLEACARMAAGEPCPLSDVQQKLGDLVPDGYIKVSDAVELLGIIARGIAG
>DocF  GDADGNGTVEVADAVEVLQYCAEAAVGQTPDQSYAWLCGADATENGTVEVADAVAILQYCARTLVD
>DocG  GDVDLDGDVDVADAVAVLQASAEQMVTGENPLSKDARFGADVNDDSRVDVSDAVLILQYSSMKIAN
>DocH  GDVNMDGALTVADAVTVLQACAQVTAGGESPLTDQQKKLADMNQDGNVSVGDAVDILVTIAQSMVG
       **.:  :*  :  ***** **: .*.   .        :        .*   :. :.*.*** :*   :  :..
```

Group 2

```
>GH5A   GDVNDDGEVTIADVVRLCRYVAQDDTLAPAMTAAQVANADVNGDAIVDSSDITVLARFLAHLVETV.
>GH44   GDVNSDGEVTIADVVRLCRYVAEDTELTPAMTEEQLPCADVNGDSIVDSSDITLIARYLAHLVDAL.
>GH9A   GDVDGNGKVEVNDLVRLARYVAQDQELTPALTAQQVTNADVNCDGTVDASDITMIARALARLTSLED
>GH9F   GDVDCNGKIEINDIVLLSRYVAQDATAK-APTAQGLINADCEYNGSIDAGDITAIARYLAHLCEASE
>GH48   GDVDCNGKVEINDVVLLSRYVAQDSTIPNPPSAQGLKNADCKYDGTIDADDITAIARYLAHLIEQSE
>GH9D   GDVDCNGVIEINDIVLLCRYVAQDSTLPNPPTEYGLANADCVRDNSIDSGDITAISRYLAHLIEESD
>GH5C   GDVDCNGVVEINDIVLLARYIAQDQGIPA-ISAEGLANANCVQDTNVDSSDLTQIARYLAHLITAEE
>GH10A  GDVDCNGVVEINDAVLLARYVAQDDTVK--ITVQGVANGDYNQDGSVDSTDITAVCRQLAHLTDEDA
>GH74   GDVNADGIVDIADTVLLARYIAQDPAVK--LTKQGLQNADCDPDGNIDSSDLTAIARYLAHLVDKLG
>GH26B  GDVNADGIVDIADTVLLARYIAQDNAVK---VSEAGLQNADCVLDGNIDASDLTAIARYLAHLTDADQ
        ***: :* : : * * *.**:*:*        :     :   .:   :    :*: *:* :.* **:*
```

Group 3

```
>GH43B  GDVNGDGMVNGMDLARMRRALLG-QSLNRN--ALRRADLNGTGSVEIADAVVLTKFLLGGSKN.
>GH43C  GDLNEDGRIDGMDLALIRRGLTVGFADSYA---ELAADVNQDTQITVADAVLIQNYLIGRIREFPV
>GH10B  GDLNFDGKINGADLALERRGLIKGFTDNYA---KLAADVNQDGDVIVADAVLIQSFLIGKIKEFPI
>GH9B   GDVNNDKTINAFDLVLARRGLTTGFDNQSA---EQAADVNQDGKVIVADIVVMQQYLLGKIKAFSG
>GH98   GDVNLDGVINGFDLILAKRGMANG-FTDTRS--FDAADVDKDEKLEKDDITQLQNFIKLSAKEFTE
>GH30   GDINLDGVIDGFDLAAAKRGLAAGGFDNVFS--AKVADVNKSTVFDAEDVQELSSFLLKKTTAFTD
        **:* *   ::. **    :*.:      .    *    **::     .    *     :  .::
```

Group 4

```
>GH26A  GDINLDGKLSVSDVVLLQKYLIKSG--KLTADQAKRADLNGDGVLNAYDLALLKHTMLGK.
>GH43A  GDVNADGSFNIADAVLLQKWLLAVPNTELADWKA--GDFCEDDRLTVFDLCLMKRALIQNK.
>GH8    GDVNQDGACTVADVIALQKWLLHTPDASLADWKA--GDLCPDGKLNAMDLCLLKRELLNLA.
>GH43D  GDVNADGVCSVADVVALQKYLIKQTD-TLADSKA--GDYNGDGVLTGMDLVRMKRALLG.
>GH11   GDVDGSGTVDAKDVKALQNYLTRK-ASTLANAEA--ADLDGNGVINAMDLALLKRSLLGNQ
>GH9G   GDANADGKVQIADVVTLNKYLVGAG--KLTARGAKNADMDDNGRLDAVDAVLLKRIFVS.
>GH9E   GDVDCNGKVNVQDYVLLKQFMVKKA--EITAKGGVNADMNQDRALTVSDCVLLARKLLS.
        **  : .*      *.   *:::*        *:   *  .*   :. :   *    :*: ::
```

Fig. 2 *R. champanellensis* dockerin alignment groups. Selected dockerin sequences of *R. champanellensis* scaffoldins and glycoside hydrolases were divided into four groups, using bioinformatics-based criteria. Positions of calcium-binding residues are highlighted in cyan, and putative recognition residues are highlighted in yellow. Note: The character of the latter recognition residues is largely similar within the different groups

- Dockerins are usually about 70 amino acids long. Shorter sequences (<50 amino acids) or longer sequences (>90 amino acids) should be manually inspected.

- Check if the sequence contains the characteristic conserved duplicated segments of the Ca^{2+}-binding repeats, with a linker between them.

- Look for the GDxDxDxxxDxxD motif in the first Ca^{2+}-binding segment, x being any amino acid (*see* **Note 3**).

- Look for the DxDxDxxxDxxD motif in the second Ca^{2+}-binding segment, x being any amino acid (*see* **Note 4**).

- Look for conserved amino acids in positions 10, 11, 17, 18, and 22 of each segment, which are putative specificity or cohesin-recognition residues.

5. Reorganize your sequences based on the amino acids that define the dockerin, as mentioned above, and group them by similarity (*see* **Note 5**).

Using the above detailed bioinformatic analysis, it is possible to identify sequences of dockerin-containing proteins within genomes and metagenomes from all three domains of life [23], although they are more common in bacteria. Comprehensive resources for genome sequences of bacteria are publically available, at the National Center for Biotechnology Information (NCBI) (http://www.ncbi.nlm.nih.gov/) and at the Genome OnLine Database (JGI-GOLD) (gold.jgi.doe.gov).

Cellulolytic enzymes which harbor a dockerin module can be assembled into the cellulosome via tight interaction with a cohesin module, contained in noncatalytic proteins termed scaffoldins [24]. Thus, the discovery of cellulosomal cellulases is intrinsically linked with the discovery of cohesin and scaffoldins within the same genome. Therefore, it is important to also identify cohesin-containing proteins in the bacterial genome, in order to verify the cellulosomal function of the enzyme and to determine the binding specificity of each cellulosomal enzyme (see below).

3.1.2 Genome-Wide Identification and Categorization of Cohesins

The scaffoldin usually contains multiple copies of cohesin modules, which are often very similar, flanked by a linker sequence. However, other types of monovalent scaffoldins have been reported with functional roles in cellulosome structure, such as the *R. flavefaciens* ScaC adaptor scaffoldin [20] and cell-surface ScaE-like anchoring proteins [25].

Cohesin modules are categorized into distinct types (I, II, and III), based on their sequence characteristics, structural topology and specificity of their interaction with dockerins [20, 26, 27]. They have an overall structure of jelly-roll topology that folds into a nine-stranded β-sandwich [28, 29]. Although the topology of cohesins of different types is analogous, type II cohesins contain two sequence insertions which are absent in the type I cohesin, which fold into two "β-flaps" (f4 and f8 as first defined in the structure of *Bacteroides cellulosolvens* scaffoldin).

While types I and II cohesins were mainly discovered in *C. thermocellum* and other cellulosome-producing Clostridia, more diverse examples of type III cohesins were discovered in elaborated *R. flavefaciens* cellulosomes [21, 30]. Identification of cohesin modules in the context of the scaffoldin host protein is accomplished by sequence similarity. Here we detail the search for cohesins in the *R. champanellensis* 18P13 genome as an example.

1. Sequence search of cohesins is performed with BLAST (https://blast.ncbi.nlm.nih.gov/Blast.cgi) using a large number of representative cohesin sequences that includes previously identified and confirmed type I, II, and III cohesin modules as query (http://www.weizmann.ac.il/Biomolecular_Sciences/ Bayer/sites/Biomolecular_Sciences.Bayer/files/uploads/ cohesins_setup.txt), and the *R. champanellensis* 18P13 genome or its deduced amino acid sequences as a subject.

2. Select the significant hit sequences below an E-value of 10^{-4}. Type I cohesins are typically ~140 amino acids long, type II cohesins are ~170 residues, and type III cohesins vary between ~130 to 170 residues in length. Shorter sequences (<130 amino acids) or longer sequences (>170 amino acids) should be manually inspected.

3. Identify multiple cohesin modules, which reside within a single scaffoldin protein. Preferably, name the cohesins according to their order of appearance on the scaffoldin. Conserved linker sequences (~10 residues), which appear adjacent to either the N- and/or C-terminus of the cohesin modules, may also be included in the analysis.

4. Perform multiple sequence alignment of the cohesin modules with Clustal Omega (http://www.ebi.ac.uk/Tools/msa/ clustalo/). The N- and C-termini of the cohesin modules may be less conserved in species with elaborate cellulosomes.

5. Define the type of cohesin, based on its sequence similarity to representative cohesin sequences. Construct a sequence-based dendrogram of cohesin modules to estimate the relationship between the cohesins, using the PhyML algorithms (with LG substitution model, and default parameters of the Approximate Likelihood-Ratio test) [31], and visualize using Interactive Tree Of Life (iTOL) online tool (http://itol.embl.de) [32] (Fig. 3). Significant bootstrap values in the dendrogram should be greater than 0.8.

6. Reorganize your sequences based on their amino acids similarity and define the borders of the cohesins.

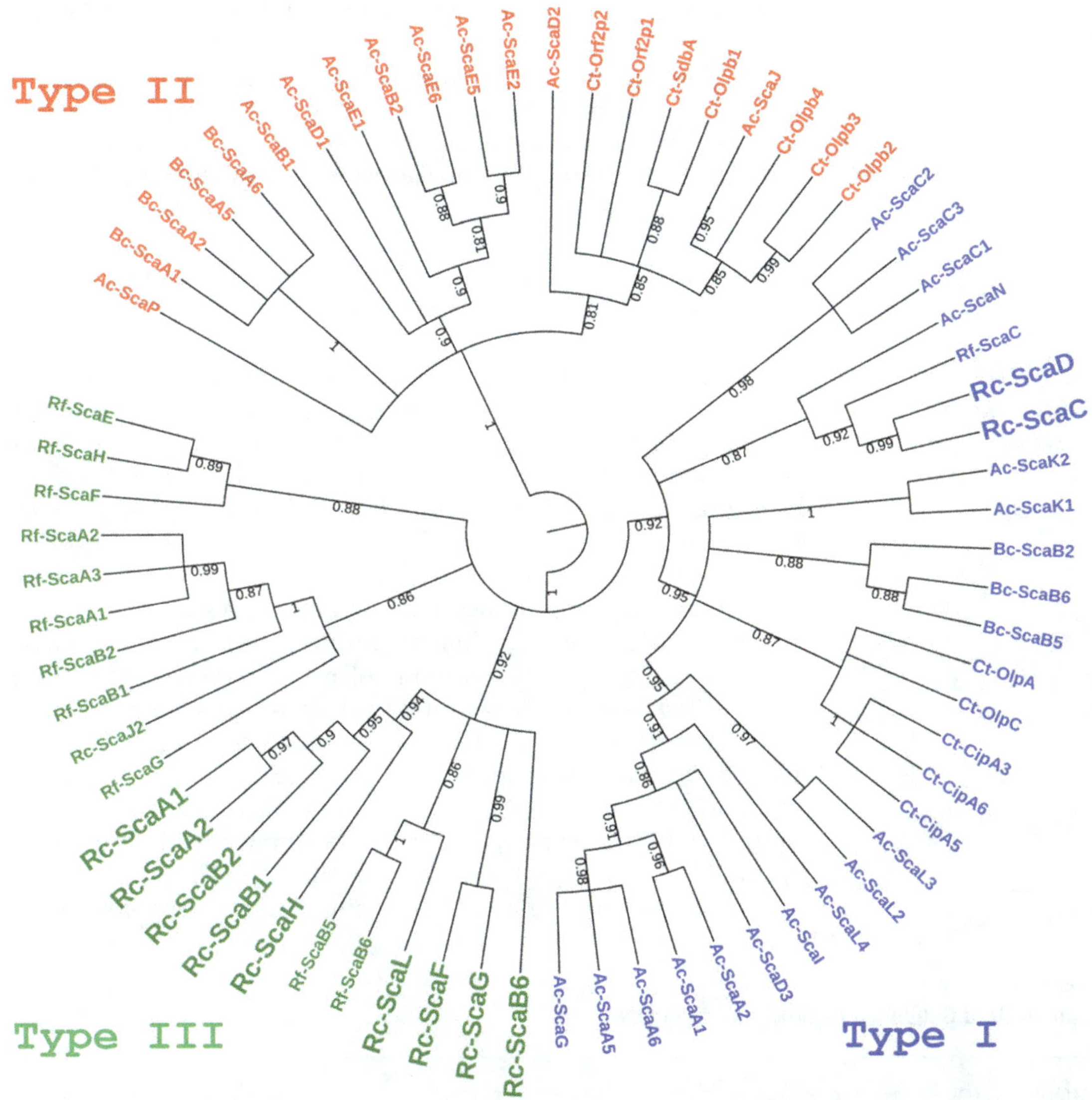

Fig. 3 Phylogenetic relationship of the *R. champanellensis* cohesins. *R. champanellensis* cohesins (Rc) are shown in larger, bold font, and selected known cohesins from other cellulosome-producing bacteria [*R. flavefaciens* (Rf), *C. thermocellum* (Ct) *B. cellulosolvens* (Bc), and *A. cellulolyticus* (Ac) define the distribution into the three cohesin types. The resultant dendrogram of the type I (blue), II (red), and III (green) cohesin modules indicates that the majority of *R. champanellensis* cohesins map together with the type III cohesins of *R. flavefaciens*, except for *R. champanellensis* ScaC and ScaD which map as type I. Bootstrapping confidence values higher than 0.8 are shown in black

3.2 Functional Characterization of Dockerin-Containing Cellulases

A comprehensive resource for Carbohydrate Active Enzymes is the CAZy database [7], which identifies and classifies the catalytic modules into family types, such as glycoside hydrolases, carbohydrate esterases, polysaccharide lyases, carbohydrate-binding modules and glycosyl transferases. The identification of modules is based on sequence comparison using BLAST or using hidden Markov models (HMMs) produced for each family.

Additional annotation for conserved protein domains can be analyzed using the CD-search website (http://www.ncbi.nlm.nih.gov/Structure/cdd/wrpsb.cgi) and the Pfam database [33]. Of note is that traditionally, dockerins appear within cellulosomal Carbohydrate Active Enzymes, yet dockerins may also appear in noncellulosomal contexts in noncellulolytic microorganisms [23] or in dockerin-containing proteins which were not annotated with any known carbohydrate-degrading domain [21].

3.2.1 Biochemical Characterization of Putative Dockerins

Cohesin–dockerin interactions are considered to be species-specific [16, 34]. However, subtyping of interaction can be found within the same species [12, 21], which cannot be predicted only by bioinformatics analysis. In general, for determining the specificity of a given cohesin–dockerin interaction, biochemical approaches must be applied. Two main methods are used to test the interaction between specific cohesin–dockerin pairs, ELISA [35] and cellulose microarray. Both approaches are based on expressing the cohesin and dockerin modules as chimeric fusion proteins, e.g., a cohesin fused (usually) at the C-terminus of a CBM module (from the *Clostridium thermocellum* scaffoldin) and a dockerin (usually) fused at the C-terminus of a xylanase module (xylanase T6 from *Geobacillus stearothermophilus*) [36, 37] (see amino-acid sequences in Table 1), for both overexpression of the fused protein and detection of the subsequent interaction via specific antibodies against the CBM or xylanase tag (*see* **Note 6**). The two methods are semiquantitative. The ELISA method is more sensitive but is

Table 1
Amino-Acid Sequence of CBM and Xylanase Tags

Tag	Sequence
CBM—CBM3a from *C. thermocellum*	MANTPVSGNLKVEFYNSNPSDTTNSINPQFKVTNTGSSAIDLSKL TLRYYYTVDGQKDQTFWCDHAAIIGSNGSYNGITSNVKGTF VKMSSSTNNADTYLEISFTGGTLEPGAHVQIQGRFAKNDWSN YTQSNDYSFKSASQFVEWDQVTAYLNGVLVWGKEPGGSVVP STQPVTTPPATTKPPATTIPPSDDPNA
Xylanase XynT6 from *G. stearothermophilus* (includes a his-tag at N-terminus)	MSHHHHHHKNADSYAKKPHISALNAPQLDQRYKNEFTIGAA VEPYQLQNEKDVQMLKRHFNSIVAENVMKPISI QPEEGKFNFEQADRIVKFAKANGMDIRFHTLVWHSQVP QWFFLDKEGKPMVNETDPVKREQNKQLLLKRLETHIKTIVE RYKDDIKYWDVVNEVVGDDGKLRNSPWYQIAGIDYIKVAF QAARKYGGDNIKLYMNDYNTEVEPKRTALYNLVKQLKEEG VPIDGIGHQSHIQIGWPSEAEIEKTINMFAALGLDNQITELD VSMYGWPPRAYPTYDAIPKQKFLDQAARYDRLFKLYEKLSDKI SNVTFWGIADNHTWLDSRADVYYDANGNVVVDPNAPYAK VEKGKGKDAPFVFGPDYKVKPAYWAIIDHKV

appropriate only for a small-scale screen, whereas the cellulose-based microarray allows for simultaneous screening of many cohesins against a specific dockerin.

3.2.2 Cellulose Microarray

1. Dilute the CBM-fused cohesin samples in TBS to a final concentration of 9 μM, and apply a volume of 70 μM to 96-well plates in 4–5 successive threefold dilutions (9, 3, 1, 0.3, and 0.1 μM).

2. Print spots of each sample dilution in triplicate on a cellulose-coated glass slide, either by manual or robotic spotter (*see* **Note** 7). Incubate the printed microarray slides in Blocking Buffer at room temperature for 30 min.

3. Incubate the slides at room temperature with the desired xylanase fused dockerin at a concentration of 3 nM in blocking buffer for 30 min.

4. Wash the slides three times (10 min each) with Wash Buffer.

5. Fluorescent staining: add Cy3-labeled anti-xylanase antibody and Cy5-labeled anti-CBM *C. thermocellum* antibody (diluted 1:1000) in Blocking Buffer. Incubate slides for 30 min.

6. Wash slides three times (10 min each) with Wash Buffer.

7. Remove the slides from the Wash buffer and air-dry them by air-dried or leave for 30 min at room temperature.

8. Scan slides for fluorescence signals using a fluorescent image analyzer (e.g., Typhoon 9400 Imager (GE Healthcare Bio-Sciences AB, Uppsala, Sweden)).

The cellulose microarray in Fig. 4 illustrates the specificity of the interactions: one of the dockerins (*R. champanellensis* DocJ) interacts with a single cohesin (*R. champanellensis* ScaE) while a second dockerin (*R. champanellensis* DocC) interacts with two different cohesins (strong interaction with *R. champanellensis* DocH but weakly with DocI) (*see* **Note 8**).

3.2.3 Biochemical Characterization of Cellulase Activities

Cellulases act on the cellulolytic substrate in three distinct ways: (1) Exoglucanases hydrolyze cellulose chains at one of its free termini (either reducing or nonreducing, depending on the specific mode of action of the enzyme), and then degrades the chains in a processive manner. (2) Endoglucanases hydrolyze the glycosidic bond internally at any site along the cellulose chain. (3) -β-glucosidases hydrolyze short cellodextrins, notably the disaccharide cellobiose, into glucose units. These three types of cellulolytic activities in dockerin-containing proteins were observed in thirteen different glycoside hydrolase families in anaerobic cellulosome-producing bacteria (see Table 2 for the list of GH families and their possible enzymatic activities). As can be seen in Table 2, *Ruminococcus champanellensis* (*18P13*) has putative cellulases from GH

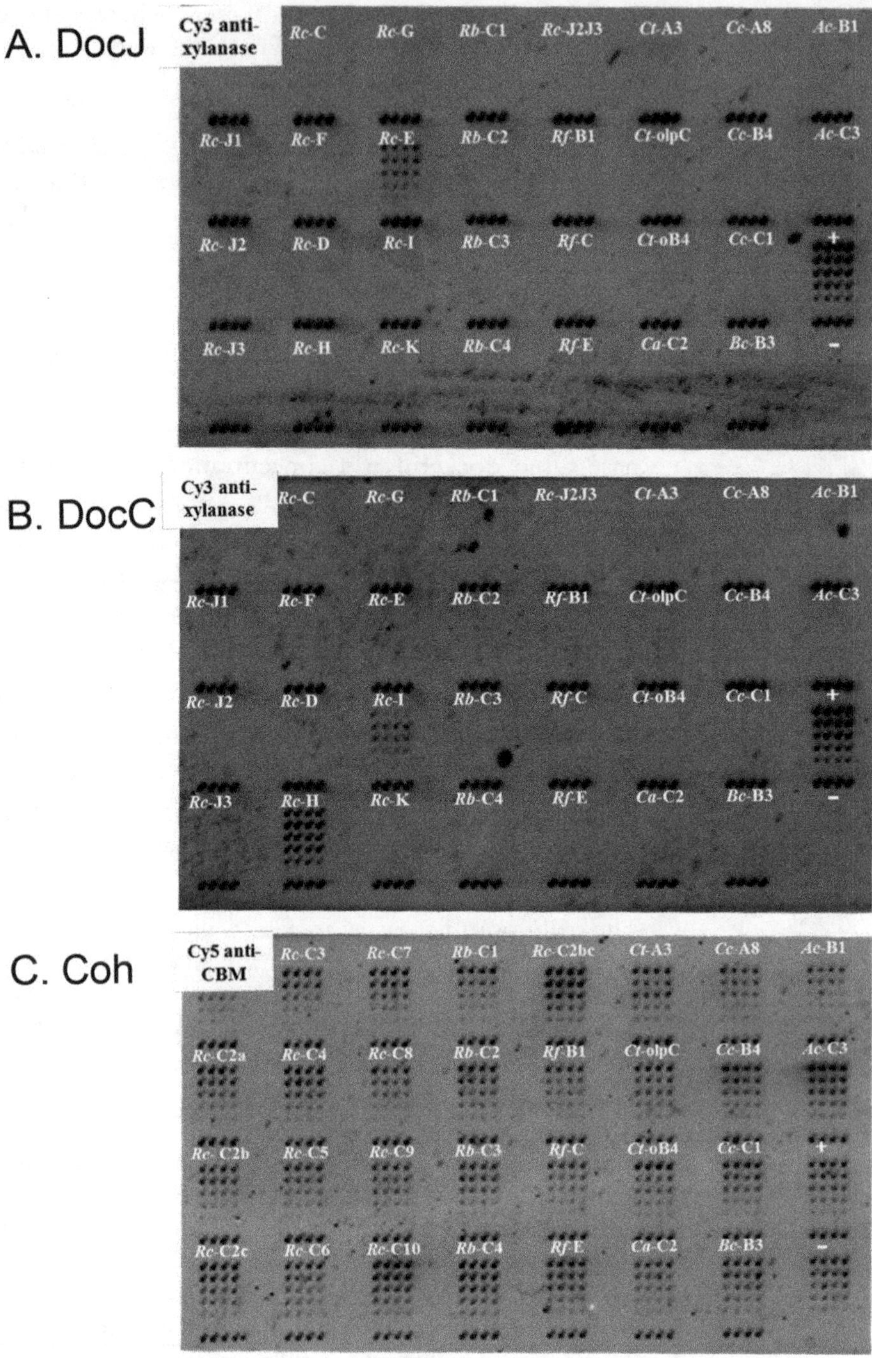

Fig. 4 Cellulose microarray of cohesin–dockerin recognition. (**a**) Interaction of the *R. champanellensis* dockerin J (DocJ as xylanase–dockerin fusion protein) with *R. champanellensis* ScaE cohesin (*Rc*-E as CBM-fused cohesin). (**b**) Preferential interaction of *R. champanellensis* dockerin C (DocC as xylanase–dockerin fusion protein) with *R. champanellensis* ScaH and (weakly) ScaI cohesins (*Rc*-H and *Rc*-I as CBM-fused cohesins). (**a** and **b**) Fluorescence scan showing Cy3-conjugated anti-xylanase antibody, indicating cohesin–dockerin binding. (**c**) Scan showing Cy5-conjugated anti-CBM antibody, indicating the relative amount of the different

families 5, 8, 9, 16, 26, 30, 44, 48, and 74. These putative cellulases were expressed, purified (*see* **Note 9**) and examined for cellulolytic activities with cellulosic substrates as described below.

3.2.4 Cellulase Activities   All assays must be performed at least twice in triplicate

1. Prepare 200 μL reactions containing: the protein at concentrations ranging from 0.1 to 1 μM, 50 mM acetate buffer, pH 5, 12 mM CaCl₂ (*see* **Note 10**), 2 mM EDTA, and the tested substrate (2% CMC/7.5 g/L PASC/10% Avicel or 12.5 mM pNPG (*see* **Note 11**).

2. Incubate the reactions at 37 °C (*see* **Note 12**) in a vertical shaking incubator for:

   - 1 h for CMC and PASC.
   - 24 h for Avicel.
   - 10–20 min for pNPG.

3. Terminate enzymatic reactions by transferring the tubes to an ice–water bath.

4. Centrifuge the tubes for 2 min at 14,000 rpm in the case of Avicel and PASC.

5. Enzymatic activity is then determined quantitatively by measuring the soluble reducing sugars released from the polysaccharide substrates by the DNS method [38, 39].

   - Add 150 μL DNS solution to 100 μL of reaction supernatant.
   - Prepare a glucose standard curve (using initial glucose solution at 20 mM diluted to 0.5 to 5 mM) for determination of sugar concentrations.
   - Boil the reaction mixture for 10 min.
   - Measure absorbance at 540 nm.

As can been seen in Fig. 5a, nine enzymes exhibited enzymatic activities on CMC substrate, ten enzymes exhibited activities on Avicel (Fig. 5b) with a synergistic effect observed between GH48 and GH5B or GH8 (*see* **Note 13**). None of the enzymes showed activity on pNPG (data not shown).

The lack of cellulolytic activity on CMC or Avicel of GH5C, GH16A, GH26A, GH26B, GH44A, and GH74A may well

---

Fig. 4 (continued) CBM-fused cohesin samples applied to the slide. Selected cohesins from other species *A. cellulolyticus (Ac)*, *B. cellulosolvens (Bc)*, *C. acetobutylicum (Ca)*, *C. cellulolyticum (Cc)*, *C. thermocellum (Ct)*, *R. bromii (Rb)*, and *R. flavefaciens (Rf)* were included as controls. A CBM-xylanase fusion-protein served as a positive control (+) and as a marker, which indicates the relative location of all samples on the cellulose slide

Table 2
Glycoside hydrolase families that include dockerin-containing enzymes with known cellulase activities. The information is organized according to CAZy classification [7]. Activities in boldface represent putative cellulolytic activities of the characterized family to date. The information is based on the genomes of *C. thermocellum* **DSM1313,** *C. cellulolyticum* **H10,** *C. papyrosolvens* **C7,** *C. cellulovorans* **743B,** *C. clariflavum* **DSM_19732,** *A. cellulolyticus* **CD2,** *B. cellulosolvens* **ATCC 35603,** *R. champanellensis* **18P13,** *R. albus* **strains SY3, 7 and 8,** *R. flavefaciens* **strains FD1, 17 and 007c,** *R. bromii* **L2–63 and** *C. alkalicellulosi* **Z-7026**

Glycoside hydrolase (GH) Family	Known enzymatic activities	Putative GHs in *R. champanellensis*	GI number
3	β-glucosidase, β-xylosidase, and exoglucanase	–	–
5	**Cellulase**, xylanase, β-mannanase, β-mannosidase, β-glucosidase, licheninase, galactanase, chitosanase	GH5A-doc GH5B-doc GH5C-doc	291,543,414 291,543,738 291,545,071
8	**Endoglucanase**, chitosanase, licheninase, xylanase	GH8A-doc	291,543,899
9	**Exo/endo/processive-glucanase**, lichenase, **β-glucosidase**, xyloglucanase, glucosaminidase	GH9A-CBM3c-doc CBM4-Fn3-GH9B-doc-GH16A GH9C-CBM3c-doc GH9D-doc GH9E-CBM3c-doc CBM4-Fn3-GH9F-doc GH9G-CBM3c-doc	291,543,282 291,543,673 291,543,938 291,544,445 291,544,574 291,544,575 291,545,280
16	**Endoglucanase**, licheninase, xyloglucanase, galactosidase, agarase	CBM4-Fn3-GH9B-doc-GH16A	291,543,673
26	Mannanase, xylanase, **endoglucanase**	CBM6-GH26A-CBM6-doc CBM6-GH26B-doc	291,544,512 291,545,037
30	Endoxylanase, β-xylosidase, **β-glucosidase**, β-glucuronidase, β-fuconidase, galactanase	GH30A-CBM4-doc-CE	291,544,794
44	**Endoglucanase**, xyloglucanase	GH44A-doc	291,543,699
48	**Exoglucanase, endoglucanase**, chitinase	GH48A-doc	291,544,207
74	Xyloglucanase, **endoglucanase**	GH74A-doc	291,543,413
81	Endoglucanase	–	–
116	**β-Glucosidase**	–	–
124	Endoglucanase	–	–

indicate that they are not cellulases, but act on other types of substrates (hemicelluloses), and exhibit an alternative activity consistent with the respective GH family listed in Table 2. The lack of

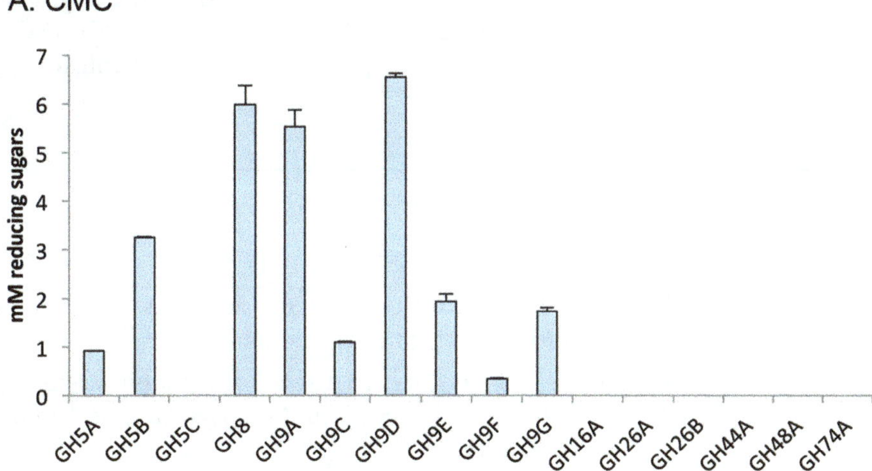

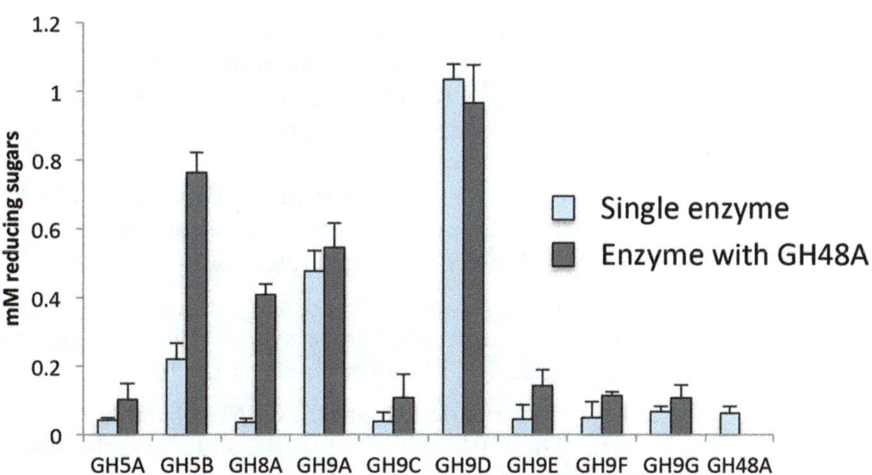

Fig. 5 Enzymatic activity of *R. champanellensis* cellulosomal cellulases. (**a**) Comparative enzymatic activity of the recombinant proteins on a soluble substrate at a concentration of 0.5 μM. Putative cellulases were tested for endoglucanase activity at pH 5 and 37 °C for 1 h with 2% carboxymethyl cellulose (CMC). (**b**) Comparative enzymatic activity of the recombinant proteins on an insoluble cellulose substrate at a protein concentration of 0.5 μM and synergism with the putative exoglucanase GH48A. The putative cellulases were tested at pH 5 and 37 °C for 24 h with 10% microcrystalline cellulose (Avicel). Reactions were performed at least twice in triplicate; standard deviations are indicated

CMC activity and low level of Avicel activity by the GH48 enzyme is typical of this type of enzyme. The synergistic boost in Avicel activity that it provides to GH5B and GH8A endoglucanases is indicative of its exoglucanase character. In addition, a protein that shows no activity may not have folded correctly, and further studies on additional substrates should be carried out.

Sugar analysis can also be performed on the degradation products after enzymatic degradation by HLPC or HPAEC, whereby the presence of short cellodextrins will confirm cellulolytic activities.

4   Notes

1. EDTA dissolves only close to pH 8.

2. Do not store CMC longer than a few weeks. Loss of viscosity in the solution is a sign of contamination.

3. Asn (N) can be observed instead of the amino acid Asp (D). In rare instances, Ser (S), Thr (T), Lys (K), and Glu (E) may sometimes replace D or N.

4. The second segment of the dockerin is often less conserved. The length of the calcium-binding loop may vary from the usual 12 residues.

5. Additional hits, such as truncated sequences with only a single calcium-binding loop, or less similar hits are to be examined individually, by searching for more characteristic sequence features. For example, the length of the second calcium-binding was shown to be atypically longer or shorter than 12 residues in specific group-1 dockerins of *R. flavefaciens* FD-1. Another example is the existence of internal sequence symmetry within group 4 dockerins in *R. flavefaciens* FD-1 [19].

6. CBM-fused cohesins are purified using the affinity of the CBM toward cellulose [40] while xylanase-fused dockerins are purified with Ni-NTA beads via a His-tag [41].

7. It is recommended to print a CBM-xylanase protein as a reference. The CBM will bind to the slide, and the fusion protein will be recognized by both the Cy3 anti-xylanase and Cy5 anti-CBM. It will serve both as a positive control and as a marker to locate the positions of the protein samples on the microarray.

8. In cases where no interaction is observed, it is sometimes recommended to express the protein with a longer linker of 5–10 amino acids (preferably at the N-terminus but the C-terminus can be also considered), and repeat the experiment. This is especially recommended when bioinformatics indicates that a module very similar in sequence—especially with identical or very similar putative recognition residues—would be expected to react with the intended counterpart.

9. For a rapid enzymatic activity screening, enzymatic assays can be performed on lysates of *E. coli* expressing cells as opposed to lysates of nonexpressing cells.

10. $CaCl_2$ is required for proper dockerin folding and should also be added during protein expression and purification.

11. For pNPG activity, add citrate buffer instead of acetate buffer.

12. The optimal conditions for each enzyme have to be calibrated (pH, temperature, and concentration, where activities versus time should be within the linear range).

13. To further confirm exoglucanase activity, synergistic activity on microcrystalline cellulose (Avicel) can be performed by combining the putative exoglucanase with a previously characterized endoglucanase. A synergy (the sum of the individual activities is lower than the activity of the combined enzymes) between the enzymes is often observed when exoglucanase and endoglucanase act in concert on the cellulosic substrate.

References

1. Stahlberg J, Johansson G, Pettersson G (1988) A binding-site-deficient, catalytically active, core protein of endoglucanase III from the culture filtrate of *Trichoderma reesei*. Eur J Biochem 173(1):179–183

2. Tamaru Y, Karita S, Ibrahim A et al (2000) A large gene cluster for the *Clostridium cellulovorans* cellulosome. J Bacteriol 182 (20):5906–5910

3. Bergquist PL, Gibbs MD, Morris DD et al (1999) Molecular diversity of thermophilic cellulolytic and hemicellulolytic bacteria. FEMS Microbiol Ecol 28(2):99–110

4. Centeno MS, Goyal A, Prates JA et al (2006) Novel modular enzymes encoded by a cellulase gene cluster in *Cellvibrio mixtus*. FEMS Microbiol Lett 265(1):26–34

5. Bagnara-Tardif C, Gaudin C, Belaich A et al (1992) Sequence analysis of a gene cluster encoding cellulases from *Clostridium cellulolyticum*. Gene 119(1):17–28

6. Druzhinina IS, Kopchinskiy AG, Kubicek EM, Kubicek CP (2016) A complete annotation of the chromosomes of the cellulase producer *Trichoderma reesei* provides insights in gene clusters, their expression and reveals genes required for fitness. Biotechnol Biofuels 9:75

7. Lombard V, Golaconda Ramulu H, Drula E et al (2014) The carbohydrate-active enzymes database (CAZy) in 2013. Nucleic Acids Res 42(Database issue):D490–D495

8. Meyer F, Paarman D, D'Souza M et al (2008) The metagenomics RAST server–a public resource for the automatic phylogenetic and functional analysis of metagenomes. BMC Bioinformatics 9:386

9. Bras JL, Cartmell A, Carvalho AL et al (2011) Structural insights into a unique cellulase fold and mechanism of cellulose hydrolysis. Proc Natl Acad Sci U S A 108(13):5237–5242

10. Ravachol J, Borne R, Tardif C et al (2014) Characterization of all family-9 glycoside hydrolases synthesized by the cellulosome-producing bacterium *Clostridium cellulolyticum*. J Biol Chem 289(11):7335–7348

11. Chassard C, Delmas E, Robert C et al (2012) *Ruminococcus champanellensis sp. nov.*, a cellulose-degrading bacterium from human gut microbiota. Int J Syst Evol Microbiol 62 (Pt1):138–143

12. Ben David Y, Dassa B, Borovok I et al (2015) *Ruminococcal* cellulosome systems from rumen to human. Environ Microbiol 17 (9):3407–3426

13. Morais S, Ben David Y, Bensoussan L et al (2016) Enzymatic profiling of cellulosomal enzymes from the human gut bacterium, *Ruminococcus champanellensis*, reveals a fine-tuned system for cohesin-dockerin recognition. Environ Microbiol 18(2):542–556

14. Tokatlidis K, Salamitou S, Béguin P et al (1991) Interaction of the duplicated segment carried by *Clostridium thermocellum* cellulases with cellulosome components. FEBS Lett 291 (2):185–188

15. Pages S, Bélaich A, Bélaich JP et al (1997) Species-specificity of the cohesin-dockerin interaction between *Clostridium thermocellum* and *Clostridium cellulolyticum*: prediction of specificity determinants of the dockerin domain. Proteins 29(4):517–527

16. Mechaly A, Yaron S, Lamed R et al (2000) Cohesin-dockerin recognition in cellulosome assembly: experiment versus hypothesis. Proteins 39(2):170–177

17. Schaeffer F, Matuschek M, Guglielmi G et al (2002) Duplicated dockerin subdomains of *Clostridium thermocellum* endoglucanase CelD bind to a cohesin domain of the scaffolding protein CipA with distinct thermodynamic

parameters and a negative cooperativity. Biochemistry 41(7):2106–2114

18. Pinheiro BA, Proctor MR, Martinez-Fleites C et al (2008) The *Clostridium cellulolyticum* dockerin displays a dual binding mode for its cohesin partner. J Biol Chem 283 (26):18422–18430

19. Rincon MT, Dassa B, Flint HJ et al (2010) Abundance and diversity of dockerin-containing proteins in the fiber-degrading rumen bacterium, *Ruminococcus flavefaciens* FD-1. PLoS One 5(8):e12476

20. Dassa B, Borovok I, Ruimy-Israeli V et al (2014) *Rumen cellulosomics:* divergent fiber-degrading strategies revealed by comparative genome-wide analysis of six ruminococcal strains. PLoS One 9(7):e99221

21. Israeli-Ruimy V, Bule P, Jindou S et al (2017) Complexity of the *Ruminococcus flavefaciens* FD-1 cellulosome reflects an expansion of family-related protein-protein interactions. Sci Rep 7:42355

22. Altschul SF, Madden TL, Schäffer AA et al (1997) Gapped BLAST and PSI-BLAST: a new generation of protein database search programs. Nucleic Acids Res 25(17):3389–3402

23. Peer A, Smith SP, Bayer EA et al (2009) Non-cellulosomal cohesin- and dockerin-like modules in the three domains of life. FEMS Microbiol Lett 291(1):1–16

24. Bayer EA, Belaich JP, Shoham Y, Lamed R (2004) The cellulosomes: multienzyme machines for degradation of plant cell wall polysaccharides. Annu Rev Microbiol 58:521–554

25. Rincon MT, Cepeljnik T, Martin JC et al (2005) Unconventional mode of attachment of the *Ruminococcus flavefaciens* cellulosome to the cell surface. J Bacteriol 187:7569–7578

26. Salamitou S, Tokatlidis K, Beguin P et al (1992) Involvement of separate domains of the cellulosomal protein S1 of *Clostridium thermocellum* in binding to cellulose and in anchoring of catalytic subunits to the cellulosome. FEBS Lett 304(1):89–92

27. Leibovitz E, Beguin P (1996) A new type of cohesin domain that specifically binds the dockerin domain of the clostridium thermocellum cellulosome-integrating protein CipA. J Bacteriol 178(11):3077–3084

28. Noach I, Frolow F, Jakoby H et al (2005) Crystal structure of a type-II cohesin module from the *Bacteroides cellulosolvens* cellulosome reveals novel and distinctive secondary structural elements. J Mol Biol 348(1):1–12

29. Shimon LJ, Bayer EA, Morag E et al (1997) A cohesin domain from *Clostridium thermocellum:* the crystal structure provides new insights into cellulosome assembly. Structure 5(3):381–390

30. Ding SY, Rincon MT, Lamed R et al (2001) Cellulosomal scaffoldin-like proteins from *Ruminococcus flavefaciens.* J Bacteriol 183 (6):1945–1953

31. Guindon S, Dufayard JF, Lefort V et al (2010) New algorithms and methods to estimate maximum-likelihood phylogenies: assessing the performance of PhyML 3.0. Syst Biol 59 (3):307–321

32. Letunic I, Bork P (2016) Interactive tree of life (iTOL) v3: an online tool for the display and annotation of phylogenetic and other trees. Nucleic Acids Res 44(W1):W242–W245

33. Bateman A, Coin L, Durbin R et al (2002) The Pfam protein families database. Nucleic Acids Res 30(1):276–280

34. Haimovitz R, Barak Y, Morag E et al (2008) Cohesin-dockerin microarray: diverse specificities between two complementary families of interacting protein modules. Proteomics 8 (5):968–979

35. Stern J, Artzi L, Moraïs S et al (2017) Carbohydrate depolymerization by intricate cellulosomal systems. In: Protein-carbohydrate interactions: methods and protocols, methods in molecular biology, D.W.A.A.A.L.v. Editor. Springer Science, Bueren, pp 93–116

36. Barak Y, Handelsman T, Nakar D et al (2005) Matching fusion protein systems for affinity analysis of two interacting families of proteins: the cohesin-dockerin interaction. J Mol Recognit 18(6):491–501

37. Lapidot A, Mechaly A, Shoham Y (1996) Overexpression and single-step purification of a thermostable xylanase from *Bacillus stearothermophilus* T-6. J Biotechnol 51(3):259–264

38. Miller GL (1959) Use of dinitrosalicylic acid reagent for determination of reducing sugar. Anal Biochem 31:426–428

39. Ghose TK (1987) Measurements of cellulase activity. Pure Appl Chem 59:257–268

40. Morais S, David YB, Bensoussan L et al (2015) Enzymatic profiling of cellulosomal enzymes from the human gut bacterium, *Ruminococcus champanellensis,* reveals a fine-tuned system for cohesin-dockerin recognition. Environ Microbiol 18(2):542–556

41. Caspi J, Irwin D, Lamed R et al (2006) *Thermobifida fusca* family-6 cellulases as potential designer cellulosome components. Biocatal Biotransformation 24:3–12

Metatranscriptomic Techniques for Identifying Cellulases in Termites and their Symbionts

Brittany F. Peterson and Michael E. Scharf

Abstract

Characterizing symbiotic communities, like that of the termite hindgut, is essential for understanding their functionality and capabilities. However, the same complexity that allows termites to digest wood so efficiently also makes them difficult to study. With the expansion in technology and sequencing strategies the feasibility of sequencing entire consortiums or microecosystems is now possible. Here we present an adapted library preparation strategy which allows for the detection and measurement of expressed genes from all three domains of life in a single sample simultaneously. This technique effectively captures the transcriptome contributions by the various members of the consortium regardless of their taxonomic identity, which can then be annotated using custom-built databases and reciprocal BLASTing. Joining the universality of this library prep strategy with the power of bioinformatics allows for the identification of cellulases and other genes encoding carbohydrate active enzymes from complex communities using metatranscriptomics.

Key words Metatranscriptome, Cellulases, Insect–microbe interactions, Symbionts, Termite

1   Introduction

Termites are the quintessential example of insects with complex gut microbial communities. The consortium of organisms working in hindguts of lower termites is hyperdiverse, with estimates of 12 species of protists, >4000 distinct bacterial OTUs, and a handful of archaeal OTUs per termite gut [1]. Next-generation sequencing technologies have allowed for increased sensitivity and sequencing depth to detect even relatively rare species compared to cloning-based strategies [2–4]. However, simply cataloging the diversity of species in the community is not sufficient to understand the relationships between the host and symbionts. Consortia like the termite hindgut microbiota are unique, complex, and often not found outside of this community. Reductionist strategies have approximated the prokaryotic contributions to termite cellulolytic activity [5], but the high level of interdependence among these microbes

Mette Lübeck (ed.), *Cellulases: Methods and Protocols*, Methods in Molecular Biology, vol. 1796,
https://doi.org/10.1007/978-1-4939-7877-9_7, © Springer Science+Business Media, LLC, part of Springer Nature 2018

and their host, along with complimentary physiological capabilities among consortium members, makes culturing them outside of the termite host nearly impossible reviewed in [6]. This makes characterizing the contributions of individual taxa exceedingly difficult.

Harnessing power from new sequencing platforms, transcriptomics has become a widely used tool to assess responses of organisms, and even individual cells, to experimental and environmental conditions. In whole organism studies, symbiotic microbes (pathogens, commensals, and mutualists) of termites are often detected to varying degrees [7–9]. However, depending on the library preparation and sequencing strategies employed, assumptions made based upon these data may be erroneous. Prokaryotic microbes, bacteria and archaea do not consistently polyadenylate mature messenger RNAs (mRNAs) due to the often-concurrent transcription and translation processes occurring in their cells [10]. Moreover, polyadenylation in prokaryotes is used to target mRNAs for degradation [11]. In the lower termite hindgut, this means that the eukaryotic members of the community, host and protists, are disproportionality studied for their cellulolytic potential reviewed in [12]. More recently the roles of prokaryotic members have been studied more directly, often by-passing transcriptomics entirely by analyzing the metagenomes of the consortium in an effort to identify genes with cellulolytic potential [13, 14]. This latter approach, however, poses significant assembly challenges due to genome similarities among many consortium members.

In this chapter, we present procedures developed in conjunction with our recent efforts to identify the contributions of prokaryotic members to the termite hindgut consortium [15]. By amending a commercially available method, we have created a custom library preparation protocol capable of reducing the rRNAs in our multidomain, multiorganism samples. We also describe the bioinformatic method by which cellulases, and other transcripts, can be annotated in metatranscriptome assemblies. Although the focus during the development of this procedure was the termite hindgut, these strategies have broad application for studies seeking to identify cellulases from various homogeneous or heterogeneous communities of prokaryotes and eukaryotes.

2 Materials

2.1 Termite Rearing and Maintenance in the Laboratory

1. Pine wood shims from hardware store (e.g., Nelson Wood Shims, Cohasset, MN, USA).

2. Georgia Pacific Envision™ brown paper towels (>50% postconsumer recycled content).

3. Tap water.

4. Round plastic containers (25 cm diameter) with lids that have pinholes to facilitate gas exchange (e.g., PWP Industries).

5. Spray bottle with mist capability.

2.2 Dissections and Gut Sample Collection

1. 3 × 1 in. glass slides.

2. Dissecting microscope with light source.

3. Fine Science Tools® Vannas spring scissors with 2 mm, straight cutting edge (Item #: 15000-03) or comparable fine-scale microdissection scissors.

4. Fine Science Tools® Moria ultra-fine forceps with straight tip (Item #: 11370–40) or comparable fine-tipped forceps.

5. ThermoFisher Scientific™ RNase Zap® decontamination solution (Item #: AM9780).

6. 70% Ethanol (in a spray or squirt bottle).

7. Small-sized Kimwipes™ delicate task wipers.

8. Sample collection buffer: 100 mM sodium phosphate buffer, pH 7.0

9. Sterile, nuclease-free 1.5 mL microcentrifuge tubes compatible with micropestles (e.g., Fisher Scientific Item #12-141-368).

10. RNA lysis buffer from the SV Total RNA Isolation Kit.

11. Ice bucket with ice.

12. Disposable gloves.

13. Petri dish or vessel to contain termites before dissection.

2.3 RNA Isolation

Materials associated with the SV Total RNA Isolation Kit (Promega, Madison, WI):

2.3.1 Materials Provided with the Kit

1. Spin column assemblies and elution tubs.

2. RNA lysis buffer.

3. β-mercaptoethanol (48.7%).

4. DNase I (lyophilized).

5. 0.09 M $MnCl_2$.

6. Yellow core buffer.

7. DNase stop solution (concentrated).

8. RNA wash buffer (concentrated).

9. Nuclease-free water.

2.3.2 Materials Provided by the User

1. Sterile micropestles compatible sterile microcentrifuge tubes (e.g., Fisher Scientific Item #12-141-368).

2. 95% ethanol (molecular grade).

3. Refrigerated, benchtop microcentrifuge (e.g., Eppendorf 5415R).

4. Digital dry bath (e.g., Bio-Rad Item #: 1660571).

5. ThermoFisher™ Scientific Nanodrop™ spectrophotometer (Item #: ND2000c).

6. ThermoFisher™ Scientific RNase Zap® decontamination solution (Item #: AM9780).

7. Disposable gloves.

8. Nuclease-free 1.5 mL microcentrifuge tubes.

2.4 Meta-transcriptome Library Preparation

Materials associated with the NuGen Ovation Complete Prokaryotic RNA-Seq DR Multiplex System 1–8 (San Carlos, CA):

2.4.1 Materials Provided with the Kit

1. First strand primer buffer (A1).

2. First strand buffer mix (A2).

3. First strand enzyme mix (A3).

4. Second strand buffer mix (B1).

5. Second strand enzyme mix (B2).

6. Second strand stop buffer (B3).

7. End repair buffer mix (ER1).

8. End repair enzyme mix (ER2).

9. End repair enhancer (ER3).

10. Ligation buffer mix (L1).

11. Ligation adapter mix (BC1–8).

12. Ligation enzyme mix (L3).

13. Strand selection buffer mix I (SS1).

14. Strand selection enzyme mix I (SS2).

15. Strand selection buffer mix II (SS3).

16. Strand selection enzyme mix II (SS4).

17. Strand selection reagent (SS5).

18. Strand selection enzyme mix III (SS6).

19. Ribosomal depletion enzyme mix (RD1).

20. Amplification buffer mix (P1).

21. Amplification primer mix (P2).

22. Amplification enzyme mix (P3).

23. Nuclease-free water.

24. Agencourt RNAClean XP beads.

<table>
<tr><td>

2.4.2 Materials Provided by the User

</td><td>

1. Ethanol (molecular grade).
2. Sterile, nuclease-free 1.5 mL microcentrifuge.
3. Sterile, nuclease-free 0.2 mL thin-walled PCR tubes.
4. Small-sized Kimwipes™ delicate task wipers.
5. Refrigerated, bench-top microcentrifuge (e.g., Eppendorf 5415R).
6. Vortex (e.g., VWR Genie 2 G560 Vortex Item #: VWR-G2).
7. Thermocycler with 0.2-mL tube heat block, heated lid, and 100 µL reaction capacity.
8. Covaris sonication system (E-series or S-series).
9. Covaris #520045 6 × 16 mm microtubes (or other appropriate sonication tubes).
10. Magnetic separation device (e.g., Agencourt SPRIPlate ring super magnet plate (Beckman Coulter Item #: A32782).
11. ThermoFisher™ Scientific RNase Zap® decontamination solution (Item #: AM9780).
12. Ice bucket with ice.
13. Disposable gloves.
14. Thermo Fisher™ NanoDrop™ spectrophotometer (Item #: ND2000c).

</td></tr>
<tr><td>

2.5 Quality Control, Library Sequencing, and Assembly

</td><td>

1. Agilent 2100 BioAnalyzer (or other materials/equipment for the electrophoresis and analysis of nucleic acids).
2. Illumina HiSeq 2500 System (or other high-throughput sequencing system).
3. Trinity RNA-Seq de novo assembly software (or other de novo assembly software) [16].
4. Access to high-performance, high-memory supercomputing node(s) (e.g., Node: 2 × 10-core Intel Xeon-E5 processors with 256 GB memory) running.

</td></tr>
<tr><td>

2.6 Cellulase Annotation Using Bioinformatics

</td><td>

1. Basic Local Alignment Search Tool (BLAST) [17].
2. Access to high-performance, high-memory supercomputing node(s) (e.g., Node: 2 x 10-core Intel Xeon-E5 processors with 256 GB memory) running
3. Species or system specific database(s) of RefSeq sequences downloaded from the Nation Center for Biotechnology Information (NCBI) file transfer protocol (FTP) website (ftp://ftp. ncbi.nlm.nih.gov/refseq/).

</td></tr>
</table>

---

3   Methods

3.1   Termite Rearing in the Laboratory

Termites were collected locally, in West Lafayette, Indiana, USA, and reared in the laboratory prior to use in experiments. The following protocol was used to collect and maintain *Reticulitermes flavipes* colonies.

1. Dampen corrugated cardboard rolls and place in preassembled, subterranean PVC pipe traps. After 24–48 h, collect cardboard rolls and bring into the lab.

2. Remove termites from the unrolled cardboard using paintbrushes and allow them to walk to a moistened brown paper towel on their own.

3. Premoisten pine wood shims to saturation and place inside the rearing container for nest material and to maintain colony humidity.

4. Add termites to the moistened nest material and move into the wood. Then cover nest material with a damp brown paper towel and close the container with a lid perforated for oxygen exchange.

5. Leave colonies undisturbed until needed for experimentation except for the regular addition of wood shims and rewetting of nest materials with a space bottle as needed.

6. To collect termites for experimentation, carefully deconstruct nests to access and remove workers. Take care to remove/damage as little of the colony as possible.

3.2   Dissections and Sample Collection

Following desired experimental conditions/treatments, termite gut tissue will be dissected out and harvested for further analysis. This process involves microdissection and will require mastery through practice.

1. Place the closed dish containing termites into a container of ice for 15–30 min prior to dissection to immobilize termites.

2. Prep the work area and instruments by surface-sterilizing with 70% ethanol on a Kimwipes.

3. Aliquot 175 µL of RNA lysis buffer into as many nuclease-free 1.5 mL microcentrifuge tubes, label appropriately, and keep on ice in anticipation of sample tissues.

4. Remove RNases from the work surface and instruments by spraying with RNase Zap and wiping with a Kimwipes.

5. Place a clean slide in the view area of the dissecting microscope and add a 50 µL droplet of sodium phosphate in the center of the viewing area.

6. Place an immobilized termite into this droplet using forceps.

7. While holding the termite steady with forceps carefully remove the head by cutting where the head and thorax meet with microscissors (*see* **Note 1**). Discard the head.

8. Continue to hold the termite steady, while also not squeezing it, and use the microscissors to remove the last posterior segment of the abdomen and discard.

9. Steady the termite, while not squeezing it, with one pair of forceps and use another to gently pull the rectal end of the gut through the opening at the posterior of the abdomen.

10. Continue gently pulling the gut until you see the foregut and salivary glands and it fully separates from the rest of the termite's body (*see* **Note 2**).

11. Use the droplet of buffer to wash any fat body or tissue from the gut.

12. Place tissue into designated, labeled microcentrifuge tube of RNA lysis buffer.

13. Repeat for all termites. Resterilize and RNase Zap treat workstation and instruments between condition/treatment types.

3.3 RNA Isolation

Total RNA is to be isolated from all samples. The following protocol is the manufacturer's protocol for the SV Total RNA Isolation Kit with slight modifications. All solutions should be prepared per manufacturer's recommendations prior to beginning.

1. Turn on the digital dry bath and set it to 70 °C. Transfer tubes of dissected tissue into a fresh bucket of ice. Turn on refrigerated centrifuge and set to 4 °C.

2. Using sterile micropestles, homogenize the fresh termite tissue in the RNA lysis buffer (175 μL) from dissection. Add 350 μL of the blue dilution buffer and mix by inverting 2–3 times to each tube.

3. Place all tubes in the digital dry bath and incubate at 70 °C for a maximum of 3 min. Monitor these samples closely and mix them by inverting every 30–45 s. After 3 min, place the samples into ice immediately.

4. Spin samples at 4 °C for 10 min at 14000 × g. While samples are spinning, label new tubes with sample identifiers and add 200 μL of 95% ethanol to each tube. Assemble and label spin columns.

5. Carefully aspirate supernatant from each sample without disturbing the pellet and add to the tube of ethanol prepared in the previous step. Pipet-mix supernatant and ethanol then add the mixture into the spin basket. Centrifuge samples in spin columns for 1 min at 14000 × g.

6. Decant flow-through and place spin basket back into the collection tube. Add 600 μL of wash buffer to the spin basket. Centrifuge samples in spin columns for 1 min at 14000 × g.

7. Decant flow-through and place spin basket back into the collection tube. Prepare DNase treatment per manufacturer's protocol (Per sample: 40 μL yellow core buffer, 5 μL 0.09 M manganese chloride, 5 μL DNase I; *see* **Note 3**). Gently pipet-mix DNase treatment solution, then apply 50 μL directly onto the membrane of the column. Incubate for 15 min at room temperature.

8. Add 200 μL of DNase stop solution to the column then centrifuge samples in spin columns for 1 min at 14000 × g.

9. Without decanting, add 600 μL of wash buffer to the column. Centrifuge samples in spin columns for 1 min at 14000 × g.

10. Decant flow-through and place spin basket back into the collection tube.

11. Apply 250 μL of wash buffer to the column for a final wash. Centrifuge samples in spin columns for 2 min at 14000 × g. During this spin, prepare and label a fresh, nuclease-free 1.5 mL microcentrifuge tube for final elution.

12. After removing samples from the centrifuge, discard collection tubes containing final flow-through. Place spin basket into the labeled tube for final elution.

13. Add 100 μL of nuclease-free water directly onto the membrane of the column. Incubate at room temperature for 2 min then spin for 1 min at 14000 × g to elute.

14. The quantity and quality of purified RNA can be measured by spectrophotometry using the ThermoFisher™ NanoDrop™ spectrophotometer. Additionally, gel electrophoresis can be used to assess the integrity of the total RNA in the samples.

15. Once isolated, quality RNA can be used as the template for library preparation. Spare RNA solution should be aliquoted and stored at −80 °C.

3.4 Metatranscriptome Library Preparation

Library preparation will synthesize cDNA from the total RNA isolated from samples, fragment cDNAs, purify cDNAs, deplete ribosomal RNAs, barcode samples, append necessary sequencing adapters for the Illumina sequencing platform, and amplify the library for sequencing. The following instructions are modified from the NuGen Ovation Complete Prokaryotic RNA-Seq DR Multiplex System 1–8 to accommodate custom oligos designed against the anticipated eukaryotic members of the termite gut community.

1. Before beginning with the manufacturer's protocol, prepare 500 nM mix of all custom oligos to be added to the kit.

2. Thaw reagents for first strand cDNA synthesis: A1, A2, and D1 at room temperature, vortex, and place on ice. Thaw and spin down enzyme mix A3; keep on ice. Prewarm a thermocycler to 65 °C.

3. Add 2 μL of A1 to a 0.2 mL, thin-walled PCR tube. Add 100–500 ng of total RNA to the tube containing primer mix and bring the total volume up to 7 μL with nuclease-free water. Pipet-mix, quick-spin, and place on ice.

4. Place tubes into the prewarmed thermocycler and incubate at 65 °C for 5 min for primer annealing. Following incubation, immediately snap-chill on ice. Adjust prewarmed thermocycler to 40 °C.

5. Prepare the first strand master mix (2.5 μL of A2 and 0.5 μL of A3 per sample) in a microcentrifuge tube. Add 3 μL of the first strand master mix to each tube containing RNA and primers. Pipet-mix, quick-spin, and keep on ice.

6. Place tubes into the prewarmed thermocycler and incubate at 40 °C for 30 min, followed by a 4 °C hold. Remove tubes from thermocycler, quick-spin, and place on ice. Adjust thermocycler to 16 °C.

7. Thaw tubes B1 and B3 at room temperature, vortex, quick-spin, and place on ice. Spin down enzyme mix B2 and keep on ice.

8. Prepare the second strand master mix (63 μL of B1 and 2 μL of B2 per sample) in a microcentrifuge tube. Add 65 μL of the second strand master mix to reaction tubes from first strand synthesis. Pipet-mix, quick-spin, and keep on ice.

9. Place tubes into the precooled thermocycler and incubate at 16 °C for 1 h, followed by a 4 °C hold. Remove tubes from thermocycler and quick-spin.

10. Add 45 μL of stop solution B3, pipet-mix, and quick-spin (*see* **Note 4**).

11. Transfer cDNA into appropriate Covaris microtubes (Fragment cDNA by sonication using a Covaris sonication E-series or S-series system (Table 1; *see* **Note 4**). Transfer fragmented cDNA from sonication microtubes into fresh 0.2 mL tubes.

12. Allow Agencourt RNAClean XP beads and nuclease-free water D1 to completely reach room temperature before proceeding to cDNA purification. Prepare fresh 70% ethanol wash solution from a new stock container (*see* **Note 5**).

13. Resuspend the beads by agitating and inverting the tube before adding 180 μL of magnetic beads to 100 μL of fragmented cDNA solution. Mix thoroughly by pipetting.

Table 1
Sonication parameters for Covaris sonication _E_-series or _S_-series systems

Parameter	_E_-series	_S_-series
Duty cycle	10%	
Intensity	Level 5	5%
Cycles per burst	200	
Time	180 s	
Water bath temperature	6–8 °C	
Power mode frequency	Sweeping	
Degassing mode	Continuous	
Sample volume	120 µL	
Water (fill/run)	Level 6	Level 12
Intensifier	Yes	

14. Split each sample into 2 × 140 µL aliquots in fresh 0.2 mL tubes. Incubate at room temperature for 10 min.

15. Transfer tubes to the magnet (Agencourt SPRIPlate) and allow for beads to completely clear from the solution (~5 min).

16. Remove 125 µL of the cleared binding buffer solution and take care not to aspirate beads. Magnet should prevent beads from dispersing. Bead loss at this point will affect yield.

17. With the tubes still on the magnet, add 200 µL of the freshly prepared 70% ethanol and allow to stand for 30 s. Aspirate ethanol with a pipette.

18. Repeat **step 16** for a second ethanol wash. Following the second wash remove as much ethanol as possible, it is recommended to do so in two pipetting steps allowing the ethanol to collect following the first step.

19. Air dry beads for 10 min. Before removing the tubes from the magnet, carefully inspect each tube and ensure that all ethanol has been removed/evaporated.

20. Add 12 µL of room temperature nuclease-free water D1 to the first aliquot of each sample. Mix thoroughly, resuspending all beads.

21. Transfer the first aliquot of resuspended beads to the second aliquot and mix thoroughly to resuspend the dried beads. Let stand at room temperature for 3 min.

22. Move tubes to the magnet and let stand for 3 min for beads to clear the eluate. Remove 10 µL of the eluate, taking care not to carry over magnetic beads, and transfer to a 0.2 mL PCR tubes. Places tubes on ice.

23. Thaw tube ER1 at room temperature, vortex, quick-spin, and place on ice. Spin down tubes ER2 and ER3; keep on ice. Prewarm thermocycler to 25 °C.

24. Prepare the end repair master mix (4 µL of ER1, 0.5 µL of ER2, and 0.5 µL of ER3 per sample) in a microcentrifuge tube. Add 5 µL of the end repair master mix to tubes of purified cDNA eluate. Pipet-mix, quick-spin, and keep on ice.

25. Place tubes into the prewarmed thermocycler and incubate at 25 °C for 30 min, followed by 10 min at 70 °C, and a 4 °C hold. Remove tubes from thermocycler, quick-spin, and place on ice. Adjust thermocycler to 25 °C.

26. Consult appendices B and C of the NuGen Ovation Complete Prokaryotic RNA-Seq DR Multiplex System 1–8 user guide to select barcode sequences for each sample. Thaw tubes L1, appropriate L2(s), and D1 at room temperature, vortex, quick-spin, and place on ice. Spin enzyme mix L3 and place on ice.

27. Add 3 µL of the selected L2 ligation adapter mix to each designated sample and mix thoroughly by pipetting.

28. Prepare the ligation master mix (4.5 µL of D1, 6 µL of L1 (*see* **Note 6**), and 1.5 µL of L3 per sample) in a microcentrifuge tube. Add 12 µL of the end repair master mix to end repair reaction tubes. Gently pipet-mix to avoid introducing bubbles, quick-spin, and keep on ice.

29. Place tubes into the prewarmed thermocycler and incubate at 25 °C for 30 min, followed by a 4 °C hold. Remove tubes from thermocycler, quick-spin, and place on ice. Adjust thermocycler to 72 °C.

30. Thaw tube SS1 at room temperature, vortex, quick-spin, and place on ice. Spin enzyme mix SS2 and place on ice.

31. Prepare the strand selection I master mix (69 µL of SS1 and 1 µL of SS2 per sample) in a microcentrifuge tube. Combine 70 µL of the strand selection I master mix to 30 µL of sample in a 0.2 mL tube. Gently pipet-mix to avoid introducing bubbles, quick-spin, and keep on ice.

32. Place tubes into the prewarmed thermocycler and incubate at 72 °C for 10 min, followed by a 4 °C hold. Remove tubes from thermocycler, quick-spin, and place on ice. Adjust thermocycler to 37 °C.

33. Ensure Agencourt RNAClean XP beads and nuclease-free water D1 at room temperature before proceeding to cDNA purification.

34. Resuspend the beads by agitating and inverting the tube before adding 80 µL of magnetic beads to strand selection I solution. Mix thoroughly by pipetting and incubate at room temperature for 10 min.

35. Transfer tubes to the magnet (Agencourt SPRIPlate) and allow for beads to completely clear from the solution (~5 min).

36. Remove 165 µL of the cleared binding buffer solution and take care not to aspirate beads. Magnet should prevent beads from dispersing. Bead loss at this point will affect yield.

37. With the tubes still on the magnet, add 200 µL of the freshly prepared 70% ethanol and allow to stand for 30 s. Aspirate ethanol with a pipette.

38. Repeat **step 37** for a second ethanol wash. Following the second wash remove as much ethanol as possible, it is recommended to do so in two pipetting steps allowing the ethanol to collect following the first step.

39. Air-dry beads for 10 min. Before removing the tubes from the magnet, carefully inspect each tube and ensure that all ethanol has been removed/evaporated.

40. Add 20 µL of room temperature nuclease-free water D1 to each sample. Mix thoroughly, resuspending all beads.

41. Move tubes to the magnet and let stand for 3 min for beads to clear the eluate. Remove 18 µL of the eluate, taking care not to carry over magnetic beads, and transfer to a 0.2 mL PCR tubes. Places tubes on ice.

42. Thaw tubes SS3, SS5, and the custom oligo mix at room temperature, vortex, quick-spin, and place on ice. Spin enzyme mixes SS4 and SS6; place on ice.

43. Prepare the custom strand selection reagent (CSS5) by combining 1 µL × number of samples plus one of SS5 with 0.5 µL of the custom oligo mix.

44. Prepare the strand selection II master mix (5 µL of SS3, 0.5 µL of SS4, 1 µL of CSS5, and 0.5 µL of SS6 per sample) in a microcentrifuge tube. Combine 7 µL of the strand selection II master mix to the 18 µL strand selection I eluate. Pipet-mix, quick-spin, and keep on ice.

45. Place tubes into the prewarmed thermocycler and run the following thermocycler program: 37 °C for 10 min, 95 °C for 2 min, 50 °C for 30 s, 65 °C for 5 min, and then a 4 °C hold. Remove tubes from thermocycler and quick-spin. Adjust thermocycler to 55 °C.

46. Spin enzyme mixes RD1 and place on ice.

47. Prepare the adapter cleavage master mix (17.5 µL of D1, 5 µL of SS3, and 2.5 µL of RD1 per sample) in a microcentrifuge tube. Add 25 µL of the adapter cleavage master mix to each sample. Pipet-mix, quick-spin, and keep on ice.

48. Place tubes into the prewarmed thermocycler and run the following thermocycler program: 55 °C for 5 min, 95 °C for 5 min, and then a 4 °C hold. Remove tubes from thermocycler and quick-spin. Adjust thermocycler to 95 °C.

49. Thaw tubes P1 and P2 at room temperature, vortex, quick-spin, and place on ice. Spin enzyme mix P3 and place on ice.

50. Prepare the amplification master mix (31.5 μL of D1, 10 μL of P1, 8 μL of P2, and 0.5 μL of P3 per sample) in a microcentrifuge tube. Gently pipet-mix taking care to avoid introducing bubbles, quick-spin, and keep on ice. On ice, add 50 μL of amplification master mix to each 50 μL of sample.

51. Place tubes into the prewarmed thermocycler and run the following thermocycler program: 95 °C for 2 minutes, 20 cycles (95 °C for 30 s and 60 °C for 90 s), and 65 °C for 5 min, and then a 4 °C hold. Remove tubes from thermocycler and quick-spin (*see* **Note 4**).

52. Ensure Agencourt RNAClean XP beads and nuclease-free water D1 at room temperature before proceeding to cDNA purification.

53. Resuspend the beads by agitating and inverting the tube before adding 100 μL of magnetic beads to each amplified library sample. Mix thoroughly by pipetting, it is recommended to do this with a multichannel pipette to minimize library-to-library variations. Incubate at room temperature for 10 min.

54. Transfer tubes to the magnet (Agencourt SPRIPlate) and allow for beads to completely clear from the solution (~5 min).

55. Remove 185 μL of the cleared binding buffer solution taking care not to aspirate beads. Magnet should prevent beads from dispersing, bead loss at this point will affect yield.

56. Remove tubes from the magnet, add 200 μL of the freshly prepared 70% ethanol, mix thoroughly by pipetting. Place tubes back on the magnet and allow to stand for 5 min. Aspirate ethanol with a pipette.

57. Repeat **step 56** for a second ethanol wash. Following the second wash remove as much ethanol as possible, it is recommended to do so in two pipetting steps allowing the ethanol to collect following the first step.

58. Air dry beads for 10 min. Before removing the tubes from the magnet, carefully inspect each tube and ensure that all ethanol has been removed/evaporated.

59. Remove tubes from the magnet and add 30 μL of room temperature nuclease-free water D1 to each sample. Mix thoroughly, resuspending all beads and then incubate at room temperature for 5 min.

60. Move tubes to the magnet and let stand for 2 min for beads to clear the eluate. Remove 25 µL of the eluate, taking care not to carry over magnetic beads, and transfer to a 0.2 mL PCR tubes. Places tubes on ice. Libraries are now ready for quality control checks.

3.5 Quality Control, Library Sequencing, and Assembly

Quality control and sequencing are commonly performed by core facilities, so the explanation of this is brief. Additionally, de novo transcriptome assembly is necessary given the complexity of metatranscriptome samples, a brief procedure is provided.

1. Prior to sequencing, sample quality and concentration should be validated using a sensitive measure such as the Agilent 2100 BioAnalyzer or similar equipment.

2. Sequence quality libraries on the Illumina HiSeq 2500 System or comparable platform to produce paired-end reads. Resulting raw reads should be filtered for quality, cleaned of rRNA contamination, and trimmed of adapter sequences prior to assembly.

3. Assemble the metatranscriptome into contigs using standard de novo assembly algorithms, like Trinity 2.1.1 [16, 18] (*see* **Note 7**).

3.6 Cellulase Annotation Using Bioinformatics

In order to identify putative cellulase genes from the metatranscriptome assembly, a number of bioinformatic analyses are performed. The strategy presented herein is a conservative approach, more liberal strategies may identify an increased number of putative, or even novel, cellulases.

1. Download appropriate species specific official gene sets for organisms with sequenced genomes or for closely related species to the species of interest (*Zootermopsis nevadensis* was used in this case) [19]. For those species that do not have sequenced genomes, and for which there are no closely related species to use a reference, download RefSeq databases from the NCBI FTP page (ftp://ftp.ncbi.nlm.nih.gov/refseq/).

2. Add a common identifier to the unique identifier given to each entry for each taxon (or taxon group, i.e., bacteria), this will enable sorting and shifting through large grouping more rapidly after annotation. Compile a local BLAST database consisting of these reference sequences (gene sets and RefSeq amino acid sequences).

3. Perform reciprocal BLAST jobs, tblastn and blastx, to generate two lists of best hits using format 6 output option (*see* **Note 8**). Using basic Unix commands, sort these lists and identify matches that occur in both lists. This this becomes a set of best Reciprocal Blast Hits (RBH).

4. Using the annotation given to the sequence from the compiled, custom database the new sequence from the metatranscriptome assembly now has an annotation based on a conservative, homology based strategy. Alternatively, a similar strategy can be performed using a motif search with a database of conserved cellulase domains/motifs. However, in a metatranscriptome of this size and complexity this strategy has the potential to generate a high frequency of type I errors in identification due the short length of motif sequences.

4 Notes

1. We recommend bracing/steading the termite using forceps in the nondominant hand and cutting/maneuvering the termite using the dominant hand.

2. The gut must be intact, not punctured, in order to accurately assess symbiont contributions and must retain salivary glands to accurately assess host termite contributions.

3. Upon reconstituting the DNase I enzyme, we recommend freezing aliquots of 2 reaction volumes worth per tube (10 µL) to will minimize freeze-thawing.

4. At this point, samples can be stored at $-20\,^{\circ}\mathrm{C}$ or the protocol can be carried forward to cDNA fragmentation.

5. At this point, there are no more stopping places until after library amplification, be aware that the remainder of the protocol will take several hours.

6. Ligation buffer L1 is very viscous, to ensure accurate volume measurement, pipet slowly or even remove a portion of the pipette tip with a razor blade to increase the opening diameter.

7. Based on the design of the experiment generating your samples and the complexity anticipated in the community, it may be impractical to assemble all libraries into a single reference. If this is the case, we recommend assembling the sample anticipated to have the highest sequence diversity based on your experimental design.

8. Depending on the computer power at your disposal, this will likely be an iterative process. We recommend breaking the queries into smaller manageable size files using a consistent, easily interpreted naming system for all files (query and output) and then consolidating them upon completion of all BLAST runs.

Acknowledgments

The authors acknowledge Dr. Zul Gulzar and technicians at NuGen Technologies for helpful consultation and design of the custom InDA-C oligos. We also thank Steve Kelley, Michael Gribskov, and Pete Pascuzzi for invaluable discussion, training, and guidance in the development of the annotation strategies detailed herein. Funding support for this work was provided by the O. Wayne Rollins-Orkin Endowment in the Department of Entomology at Purdue University, the Monsanto Research Grant from the Entomological Society of America, the Indiana Academy of Sciences Senior Research Grant (Grant #: 2014-13), and the Postdoctoral Excellence in Research and Teaching Fellowship administered by the Center for Insect Science at the University of Arizona funded through the NIH-IRACDA program (Grant #: K12GM000708-17).

References

1. Boucias DG, Cai Y, Sun Y et al (2013) The hindgut lumen prokaryotic microbiota of the termite *Reticulitermes flavipes* and its responses to dietary lignocellulose composition. Mol Ecol 22(7):1836–1853. doi: 10.111/mec.12230

2. Hongoh Y, Ohkuma M, Kudo T (2003) Molecular analysis of bacterial microbiota in the gut of the termite *Reticulitermes speratus*. FEMS Microbiol Ecol 44(2):231–242. https://doi.org/10.1016/S0168-6496(03)00026-6

3. Yang H, Schmitt-Wagner D, Stingl U, Brune A (2005) Niche heterogeneity determines bacterial community structure in the termite gut (*Reticulitermes santonensis*). Environ Microbiol 7(7):916.932. doi: 10.111/j.1462-2920.2005.00760.x

4. Fisher M, Miller D, Brewster C et al (2007) Diversity of gut bacteria in *Reticulitermes flavipes* as examined by 16S rRNA gene sequencing and amplified rDNA restriction analysis. Curr Microbiol 55(3):254–259. https://doi.org/10.1007/s00284-007-0136-8

5. Peterson BF, Stewart HL, Scharf ME (2015) Quantification of symbiotic contributions to lower termite lignocellulose digestion using antimicrobial treatments. Insect Biochem Mol Biol 59:80–88. https://doi.org/10.1016/j.ibmb.2015.02.009

6. Hongoh Y (2011) Toward the functional analysis of uncultivable, symbiotic microorganisms in the termite gut. Cell Mol Life Sci 68(8):1311–1325. https://doi.org/10.1007/s00018-011-0648-z

7. Scharf ME, Wu-Scharf D, Pittendrigh BR, Bennett GW (2003) Caste- and development-associated gene expression in a lower termite. Genome Biol 4:R62. https://doi.org/10.1186/gb-2003-4-10-r62

8. Scharf ME, Wu-Scharf D, Zhou X et al (2005) Gene expression profiles among immature and adult reproductive castes of the termite *Reticulitermes flavipes*. Insect Mol Biol 14:31–44. doi: 10.111/j.1365-2583.2004.00527.x

9. Steller MM, Kambhampati S, Caragea D (2010) Comparative analysis of expressed sequence tags from three castes and two life stages of the termite *Reticulitermes flavipes*. BMC Genomics 11:463. https://doi.org/10.1186/1471-2164-11-463

10. Gowrishankar J, Harinarayanan R (2004) Why is transcription coupled to translation in bacteria? Mol Microbiol 54(3):598–603. doi:10.111/j.1365-2958.2004.04289.x

11. Deutscher MP (2006) Degradation of RNA in bacteria: comparison of mRNA and stable RNA. Nucleic Acids Res 34(2):659–666. https://doi.org/10.1093/nar/gkj472

12. Scharf ME (2015) Omic research in termites: an overview and a roadmap. Front Genet 5:76. https://doi.org/10.3389/fgene.2015.00076

13. He S, Ivanova N, Kirton E et al (2013) Comparative metagenomic and metatranscriptomic analysis of hindgut paunch microbiota in wood- and dung-feeding higher termites.

PLoS One 8(4):e61126. https://doi.org/10. 1371/journal.pone.0061126

14. Do TH, Nguyen TT, Nguyen TN et al (2014) Mining biomass-degrading genes through Illumina-based *de novo* sequencing and metagenomic analysis of free-living bacteria in the gut of the lower termite *Coptotermes gestroi* harvested in Vietnam. J Biosci Bioeng 118 (6):665–371. https://doi.org/10.1016/j. jbiosc.2014.05.010

15. Peterson BF, Scharf ME (2016) Metatranscriptomic analysis reveals bacterial symbiont contributions to lower termite physiology and potential immune functions. BMC Genomics 17:772. https://doi.org/10.1186/s12864-016-3126-z

16. Grabherr MG, Haas BJ, Yassour M et al (2011) Full-length transcriptome assembly from RNA-seq data without a reference genome. Nat Biotechnol 29(7):644–652. https://doi. org/10.1038/nbt.1883

17. Altschul SF, Gish W, Miller W, Myers EW, Lipman DJ (1990) Basic local alignment search tool. J Mol Biol 215(3):403–410. https:// doi.org/10.1016/S0022-2836(05)80360-2

18. Haas BJ, Papanicolaou A, Yassour M et al (2013) De novo transcript sequence reconstruction from RNA-seq using the trinity platform for reference generation and analysis. Nat Protoc 8(8):1494–1512. https://doi.org/10. 1038/nprot.2013.084

19. Terrapon N, Li C, Robertson HM, Ji L et al (2014) Molecular traces of alternative social organization in a termite genome. Nat Commun 5:3636. https://doi.org/10.1038/ ncomms4636

Chapter 8

Discovery of Novel Cellulases Using Proteomic Strategies

Marta Zoglowek, Heather Brewer, and Angela Norbeck

Abstract

In order to develop cost-effective processes for conversion of lignocellulosic biomass, discovery of novel enzymes for enhanced lignocellulose hydrolysis is one of the main scientific and industrial goals. This could be achieved by applying proteomic strategies for identification of proteins secreted by filamentous fungi that are among the most powerful producers of biomass-degrading enzymes. Here a strategy for a comparative study of proteins differentially secreted on media inducing production of biomass-degrading enzymes (e.g., lignocellulosic biomass) and media repressing secretion of those enzymes (e.g., glucose) are presented. The protocols presented include preparation of samples for mass spectrometry and identification of cellulolytic and other carbohydrate-degrading enzymes using bioinformatics.

Key words Filamentous fungi, Cellulases, Sample preparation for mass spectrometry, Proteomics, Bioinformatics

1 Introduction

Filamentous fungi are known for high secretion of both native and heterologously expressed proteins such as cellulases which have significant value for the biotechnology industry [1]. Secretomics, identifying the whole catalog of extracellular proteins expressed and secreted under certain conditions, plays an important role in the discovery of new enzymes and determination of optimal and tailored enzyme mixtures for biomass conversion in biorefineries [2]. In addition, proteomic strategies can reveal relative abundances of the enzymes. The identification of proteins by secretomic methods is improved by increasing availability of sequenced and annotated fungal genomes. The proteins of interest might include various enzymes from CAZY families such as: cellulases, hemicellulases and auxiliary proteins for biomass conversion. The selection of growth media and conditions will influence the types of secreted proteins, and therefore they have to be carefully designed in order to obtain reliable results [2].

Mette Lübeck (ed.), *Cellulases: Methods and Protocols*, Methods in Molecular Biology, vol. 1796,
https://doi.org/10.1007/978-1-4939-7877-9_8, © Springer Science+Business Media, LLC, part of Springer Nature 2018

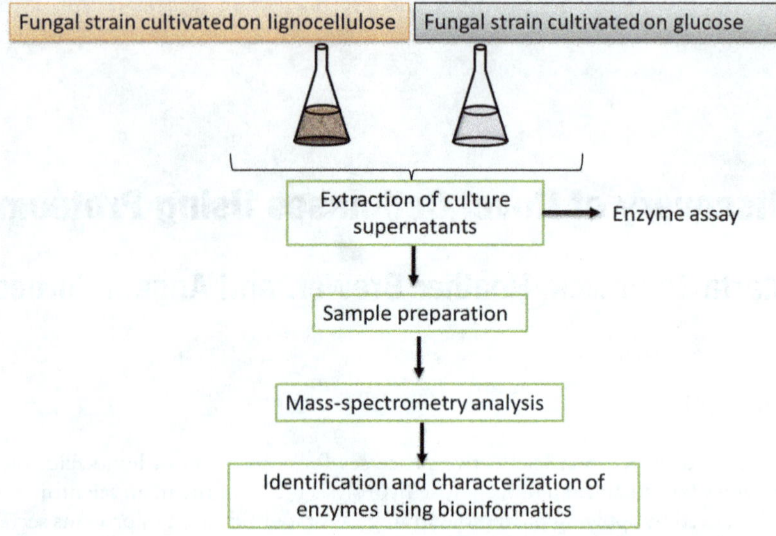

Fig. 1 Overview of the procedure for enzyme identification using secretomics

In this chapter, protocols for comparative studies of proteins differentially secreted on media inducing production of biomass-degrading enzymes (synthetic cellulosic media, lignocellulosic biomass) and medium repressing secretion of those enzymes (glucose) are presented. Also, methods for preparation of the samples for mass-spectrometry and further bioinformatics identification of proteins are described. The general procedure for the protocol is shown on Fig. 1.

2 Materials

1. Strain(s) of selected cellulolytic fungi.

2. Miracloth (Merck, item #: 475855).

3. Hemocytometer (Fuchs-Rosenthal, Assistent).

4. Thermoshaker.

5. Microcentrifuge.

6. Microtiter plates, e.g., Thermo Scientific™ Pierce™ 96-Well Plates, Product No. 15041.

7. Spectrophotometer with microplate reader.

8. Urea pellets

9. 30 K MWCO (molecular weight cutoff) filter (Amicon).

10. Ultracentrifuge.

11. Liquid N_2.

12. *p*-nitrophenol (pNP) (Sigma-Aldrich, item #: 1048)

13. Pierce BCA Protein Assay Kit (Thermo Scientific, item #: 23225).

14. Standard Ampules with Bovine Serum Albumine

2.1 Media

1. Potato dextrose agar (PDA) (Sigma-Aldrich, item #: 70139).

2. Potato dextrose broth (PDB) (Sigma-Aldrich, item #: P6685).

3. Trace metal solution: 1 g $ZnSO_4 \cdot 7H_2O$, 0.5 g $CuSO_4 \cdot 5H_2O$ in 100 mL water.

4. Basic Czapek medium: 3 g/L $NaNO_3$, 1 g/L K_2HPO_4, 0.5 g/L KCl, 0.5 g/L $MgSO_4 \cdot 7 H_2O$, 0.01 g/L $FeSO_4 \cdot 7 H_2O$, add water to 1 L [3]. Adjust pH to 4.8. Add 1 mL of trace metal solution (from **step 3**) per liter medium just before autoclaving at 121 °C for 20 min.

5. Crystalline cellulose medium: 2% (w/v) Avicel PH-101 (Sigma-Aldrich, item #: 11365) in basic Czapek medium.

6. Amorphous cellulose medium: 2% (w/v) carboxymethyl cellulose sodium salt (Sigma-Aldrich, item #: C5678) in basic Czapek medium.

7. Biomass medium: 2% (w/v) wheat bran (or other lignocellulosic biomass) in basic Czapek medium.

8. Noncellulosic medium: 2% (w/v) glucose in basic Czapek medium.

9. Agar medium with basic Czapek medium and either 2% Avicel or 2% carboxymethyl cellulose or 2% wheat bran or 2% glucose: prepare as above and include 1.5% agar prior to autoclaving.

2.2 Solutions

1. 0. 1% Tween 80 (Sigma-Aldrich, item #: P1754).

2. BCA Reagent A, 1000 mL (Thermo Scientific Pierce BCA Protein Assay Kit, item #: 23225) or 500 mL (in item #: 23227), containing sodium carbonate, sodium bicarbonate, bicinchoninic acid, and sodium tartrate in 0.1 M sodium hydroxide.

3. BCA Reagent B, 25 mL, containing 4% cupric sulfate.

4. Albumin Standard Ampules, 2 mg/mL, 10 × 1 mL ampules, containing bovine serum albumin (BSA) at 2 mg/mL in 0.9% saline and 0.05% sodium azide.

5. BCA working solution: mix BCA reagent A and BCA reagent B in the ratio 50:1.

6. Urea buffer: Prepare 8 M urea solution (480 mg/mL) in 100 mM NH_4HCO_3.

7. 50 mM *p*-nitrophenyl β- D-glucopyranoside (pNPG) (Sigma-Aldrich, item #: N7006).

8. Stop solution: 1 M Na_2CO_3.

9. Buffer solution (Sodium acetate buffer, 100 mM, pH 4.6): Add 6.0 g of glacial acetic acid (1.05 g/mL) to 800 mL of distilled water. Adjust the pH to 4.6 by the addition of 5 M (20 g/ 100 mL) sodium hydroxide solution. Approximately 50 mL is required. Adjust the volume to 1 L.

10. CM-Cellulose 4 M: Add 2 g of powdered substrate (Megazyme, Bray, Ireland) to 80 mL of boiling and vigorously stirring water on a hot-plate, magnetic stirrer. Turn the heat off and continue stirring until the solution/slurry is homogenous (approx. 20 min). Add 5 mL of sodium acetate buffer (2 M, pH 4.5) and cool the solution to room temperature. Adjust the pH to 4.5 and the volume to 100 mL. Store this solution at 4 °C between use. Under these conditions, and if the solution is not accidentally contaminated with endocellulase, the substrate is stable for at least 12 months. Before use, mix the substrate solution vigorously by shaking the storage bottle. As the substrate solution is viscous, it should preferably be dispensed with a positive displacement dispenser (e.g., Eppendorf Multipette® with a 5.0 mL Combitip).

11. Precipitation solution: Dissolve 40 g of sodium acetate trihydrate and 4 g of zinc acetate in 150 mL of demineralized water. Adjust the pH to 5.0 with 5 M HCl and the volume to 200 mL with demineralized water. Add 200 mL of this solution to 800 mL of industrial methylated spirits (IMS, 95%) or ethanol (95%), mix well and store at room temperature in a well-sealed bottle.

12. *Aspergillus niger* cellulase (Megazyme catalog number E-CELAN).

---

3   Methods

3.1 Selection of the Time Point for Collection of Samples for Proteomics (See Notes 1 and 2)

1. Cultivate selected strain(s) of filamentous fungi on petri dishes containing agar medium with basic Czapek and either 2% Avicel or 2% carboxymethyl cellulose or 2% wheat bran or 2% glucose for 7 days at optimal temperature for each strain.

2. Cut out two agar plugs full of mycelium from each plate every day.

3. Inoculate agar plugs into 1 mL water in Eppendorf tubes.

4. Perform AZCL assay with dyed insoluble AZCL-HE-(hydroxyethyl)-Cellulose (Megazyme, product code # I-AZCEL).

3.2 AZCL Assay

1. Prepare AZCL-HE-Cellulose plates [4].

2. Pipet 15 µL of crude culture extracts into small round wells on AZCL plates.

3. Assess the enzyme activity by measuring the radius of the zone of released azurine dye.

4. Choose the time point which resulted in the highest enzyme activity for production and preparation of the samples for proteomics.

3.3 Production of Enzyme Extracts from Selected Fungi (See Note 3)

1. Cultivate selected strain(s) of filamentous fungi on PDA plates for 7 days at optimal temperature for each strain.

2. Harvest spores from the plates by addition of sterile Milli-Q water with 0.1% Tween-80.

3. Filter the spore suspension through Miracloth and count spores using hemocytometer.

4. Inoculate 1×10^6 spores per mL in 10 mL of potato dextrose broth in 50 mL Falcon tubes (*see* **Note 1**).

5. Incubate at 25 °C, 180 rpm for 2 days.

6. Collect the fungal biomass on Miracloth and wash it with sterile Milli-Q water in order to remove PDB carryover.

7. Distribute 20 mL of media from point 2.1.5–2.1.8 into 100 mL Erlenmeyer flasks.

8. Add the collected fungal biomass to the flasks.

9. Incubate the flasks with shaking at 25 °C, 180 rpm for the chosen amount of days, e.g., 5 days.

10. Collect the supernatant (=enzyme extract) by filtering it through Miracloth.

11. Freeze the enzyme extract in −20 °C (*see* **Note 3**).

3.3.1 Beta-Glucosidase Enzyme Assay with pNPG.

1. Mix 100 μL substrate (50 mM pNPG) with 10 μL enzyme extract in 1.5 mL microcentrifuge tube.

2. Incubate the tubes for 10 min at 50 °C on thermoshaker (no shaking).

3. Withdraw 30 μL, transfer to the microtiter plate that already contain 50 μL stop solution (1 M Na_2CO_3).

4. Prepare pNP standard curve by diluting the stock (10 mM pNP) to obtain 1 mM pNP (approx. 2 mL) and 0.1 mM pNP (approx. 1 mL).

5. Prepare pNP standards according to the Table 1.

6. Add 50 μL of the pNP solutions with different concentrations and transfer it to the microtiter plate already containing 50 μL stop solution (1 M Na_2CO_3).

7. Measure absorbance at 405 nm.

8. Calculate enzyme activity based on a pNP standard curve.

Table 1
Preparation of pNP standards

Final concentration (mM)	pNP (μL)	50 mM sodium acetate pH 5.0 (μL)
1.2	12 (10 mM)	88
1.1	11 (10 mM)	89
1.0	10 (10 mM)	90
0.9	9 (10 mM)	91
0.8	80 (1 mM)	20
0.75	60 (1 mM)	20
0.6	60 (1 mM)	40
0.5	40 (1 mM)	40
0.4	40 (1 mM)	60
0.25	20 (1 mM)	60
0.2	20 (1 mM)	80
0.1	20 (1 mM)	180
0.075	60 (0.1 mM)	20
0.05	40 (0.1 mM)	40
0.025	20 (0.1 mM)	60

3.3.2 Cellulase Enzyme Assay with AZO-CM-Cellulose

1. Mix 40 μL substrate (AZO CM-Cellulose 4 M) with 40 μL enzyme extract in a 1.5 mL tube.

2. Incubate for 10 min at 50 °C on a thermoshaker (no shaking).

3. Stop the reaction by adding 200 μL precipitation solution.

4. Vortex and centrifuge for 10 min at 1000 rpm (4000 × g).

5. Transfer 200 μL of supernatant to a microtiter plate.

6. Measure absorbance at 595 nm.

7. Calculate activity based on cellulose standard curve prepared with an *Aspergillus niger* cellulase.

3.4 Measurement of Protein Concentration (BCA Assay) (Thermo Scientific)

1. Perform BCA assay before sample cleanup/digestion (*see* **Note 4**).

2. Prepare protein standards by using dilutions of Bovine Serum Albumin (BSA) standard according to the Table 2 (*see* **Note 5**). Preferably, use the same diluent as the sample(s).

3. Pipette 25 μL of each standard or unknown sample replicate into a microplate well (working range = 20–2000 μg/mL).

Table 2
Preparation of diluted albumin (BSA) standards (Working range = 20–2000 μg/mL)

Vial	Volume of diluent (μL)	Volume and source of BSA (μL)	Final BSA concentration (μg/mL)
A	0	300 of stock	2000
B	125	375 of stock	1500
C	325	325 of stock	1000
D	175	175 of vial B dilution	750
E	325	325 of vial C dilution	500
F	325	325 of vial E dilution	250
G	325	325 of vial F dilution	125
H	400	100 of vial G dilution	25
I	400	0	0 = Blank

4. Add 200 μL of working solution to each well and mix plate thoroughly on a Thermoshaker for 30 s.

5. Cover plate and incubate at 37 °C for 30 min.

6. Cool plate to room temperature. Measure the absorbance at or near 562 nm.

7. Subtract the average 562 nm absorbance measurement of the Blank standard replicates from the 562 nm measurements of all other individual standard and unknown sample replicates.

8. Prepare a standard curve by plotting the average Blank-corrected 562 nm measurement for each BSA standard vs. its concentration in μg/mL. Use the standard curve to determine the protein concentration of each unknown sample.

3.5 Modified FASP Cleanup/Digestion Procedure for Proteins.

1. Keep samples on ice until first incubation at 37 °C, then manipulations can be performed at room temperature.

2. Transfer required amount of sample to appropriately labeled 1.5-mL sterile microcentrifuge tube.

3. Add urea pellets to obtain 8 M concentration in each sample.

4. Incubate sample at 800 rpm, 37 °C for 30 min in thermomixer to denature proteins.

5. Split sample(s) across an appropriate number of ultracentrifuge tubes (30 K MWCO filter) appropriate for speeds of 100 K rpm ($4000\,000 \times g$). Weigh tubes containing sample(s) and ensure there is no more than 10 mg difference between matched pairs.

6. Centrifuge at 100,000 rpm for 10 min at 4 °C.

7. Transfer supernatant from ultracentrifuge tube(s) to new 1.5-mL tubes. Quickly freeze in liquid N_2, and store overnight at $-80\,^{\circ}C$.

8. Thaw samples, transfer up to 500 μL to 30 K MWCO filters. If there is more than 500 μL, retain remainder in original tube. If there is less than 500 μL, add sufficient urea buffer to make up the volume.

9. Centrifuge at max speed for 30–45 min to obtain the minimum volume (using 0.5 mL 30 K MWCO filters, spin at $14,000 \times g$).

10. If there is remaining sample, add it to the appropriate centrifugal filter, bringing final volume to 500 μL with urea buffer. Otherwise, perform two additional rinses with urea buffer. Centrifuge at $14,000 \times g$ for the time noted above to obtain minimum volume.

11. Wash 3× with 50 mM NH_4HCO_3 (buffer), centrifuge as above, except extend the last wash centrifugation to 45 min.

12. Transfer 30 K MWCO filter to a new Amicon centrifugal filter tube. Add sufficient Buffer to cover filter entirely. Add 1 μg/μL trypsin to sample in filter so that a 1:50 ratio of trypsin:protein (w:w) is obtained based upon mass of protein applied to filter initially.

13. Incubate at $37\,^{\circ}C$ for 3 h in thermomixer (800 rpm).

14. Following incubation, centrifuge the sample at $14,000 \times g$ for 30 min, or until the sample is at dead volume of filter. Add an additional 100 μL of 50 mM NH_4HCO_3 buffer, vortex to wash down the filter surfaces, then centrifuge again as before. Perform BCA protein assay on the peptides in the same way as the sample when it is in its protein state (*see* **Note 6**).

15. Dilute the samples to 0.1 μg/μL in water prior to submitting to the MS for analysis.

3.6 Mass-Spectrometry Analysis

This can be carried out using different systems. We have used an LC system in-line with the mass spectrometer, which was custom built using two Agilent 1200 nanoflow pumps and one Agilent 1200 cap pump (Agilent Technologies, Santa Clara, CA), various Valco valves (Valco Instruments Co., Houston, TX), and a PAL autosampler (Leap Technologies, Carrboro, NC). Custom software provided full automation that allowed for parallel event coordination and through the use of two trapping and analytical columns (Fig. 2). Reversed-phase C18 columns were prepared in-house by slurry packing 3 μm Jupiter C18 (Phenomenex, Torrence, CA) into 40 cm × 360 μm o.d. × 75 μm i.d fused silica (Polymicro Technologies Inc., Phoenix, AZ) using a 3-mm sol-gel frit for media retention [5]. Trapping columns were prepared similarly by slurry

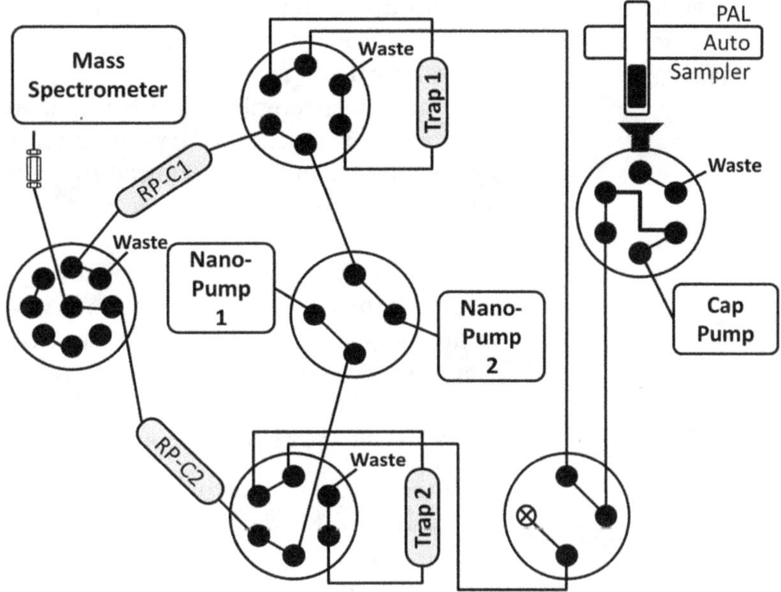

Fig. 2 Custom-built two-column LC system configuration with trapping

packing 5-μm Jupiter C18 into a 4-cm length of 150 μm i.d. fused silica and fritted on both ends. Mobile phases consisted of 0.1% formic acid in water (A) and 0.1% formic acid acetonitrile (B) operated at 300 nL/min with a gradient profile as follows (min: % B): 0:5, 2:8, 20:12, 75:35, 97:60, 100:85. Sample injections (5 μL) were trapped and washed on the trapping columns at 3 μL/min for 20 min prior to alignment with analytical columns. Data acquisition lagged the gradient start and end times by 15 min to account for column dead volume that allowed for the tightest overlap possible in two-column operation. Two-column operation also allowed for columns to be "washed" (shortened gradients) and regenerated offline without any cost to duty cycle.

MS analysis was performed using a Velos Orbitrap mass spectrometer (Thermo Scientific, San Jose, CA) outfitted with a custom electrospray ionization (ESI) interface. Electrospray emitters were custom made by chemically etching 150 μm o.d. × 20 μm i.d. fused silica [6]. The heated capillary temperature and spray voltage were 325 °C and 2.2 kV, respectively. Data was acquired for 100 min after a 15 min delay from when the gradient started. Orbitrap spectra (AGC 1×10^6) were collected from 400 to 2000 m/z at a resolution of 60 k followed by data dependent ion trap MS/MS (collision energy 35%, AGC 1×10^4) of the ten most abundant ions. A dynamic exclusion time of 45 s was used to discriminate against previously analyzed ions using a −0.55 to 1.55 Da mass window.

3.7 Identification of Cellulases Using Bioinformatics

Bottom-up protein identification uses an annotated genome to search and identify the MS2 spectra as peptide sequences. For this study, fungal and plant biomass genomes (translated to amino acid sequences) were combined with a standard set of common contaminant proteins (including trypsin autocleavage fragments, keratin, and bovine serum albumin, used for instrument calibration). MS2 spectra were searched against the protein sequences using the MSGFDB+ search engine and the resulting MSGF score was used to filter results to a 0.4% false discovery rate (FDR) of peptide to spectrum matches. Spectra counts per peptide sequence were summed to give protein abundances that were used for semiquantitative comparisons across conditions. MeV (http://mev.tm4.org/#/welcome) was used for Kmeans clustering of protein abundances to find trends across conditions.

1. Identify the subcellular localization of proteins by using the WolF PSORT database [7]

2. Identify the presence of signal peptides by using SignalP 4.1 Server [8]. Nonclassical secretory pathway using SecretomeP 2.0 Server can be used for prediction of extracellular proteins without signal peptides [9].

3. Identify protein families using Pfam 27.0 database [10], InterProScan 4 [11], and GenBank NCBI database [12] protein blastp search [13] against nonredundant protein sequences database.

---

4   Notes

1. The samples should be prepared in at least three biological replicates in order to obtain reliable results.

2. The time point of sample collection should be selected very precisely in order to avoid for example starvation of the fungi as the may result in cell lysis and excretion of intracellular proteins.

3. Carry out enzyme assays in order to select the optimal time point for sample collection. This could be supported by gene expression studies

4. Measurement of fungal protein content in lignocellulosic biomass samples such as wheat bran is difficult due to the presence of plant proteins.

5. Each 1 mL ampule of 2 mg/mL BSA standard is sufficient to prepare a set of diluted standards for either working range suggested in Table 1. There will be sufficient volume for three replications of each diluted standard.

6. Carry out enzyme assays on the concentrated samples for sample quality assessment.

7. Times for centrifugal filtration given in Subheading 3.5 are approximate. Adjust as appropriate for your sample type and filter size.

References

1. Punt PJ, van Biezen N, Conesa A et al (2002) Filamentous fungi as cell factories for heterologous protein production. Trends Biotechnol 20:200–206

2. Bouws H, Wattenberg A, Zorn H (2008) Fungal secretomes—nature's toolbox for white biotechnology. Appl Microbiol Biotechnol 80:381–388. https://doi.org/10.1007/s00253-008-1572-5

3. Samson A, Hoekstra ES, Frisvad JC (2004) Introduction to food-and airborne fungi. Centraalbureau voor Schimmelcultures (CBS), Utrecht

4. Pedersen M, Hollensted M, Lange L, Andersen B (2009) Screening for cellulose and hemicellulose degrading enzymes from the fungal genus *Ulocladium*. Int Biodeter Biodegrad 63:484–489

5. Maiolica A, Borsotti D, Rappsilber J (2005) Self-made frits for nanoscale columns in proteomics. Proteomics 5:3847–3850

6. Kelly RT, Page JS, Luo Q et al (2006) Chemically etched open tubular and monolithic emitters for nanoelectrospray ionization mass spectrometry. Anal Chem 78:7796–7801

7. Horton P, Park K, Obayashi T et al (2007) WoLF PSORT: protein localization predictor. Nucleic Acids Res 35:W585–W587

8. Petersen TN, Brunak S, von Heijne G et al (2011) SignalP 4.0: discriminating signal peptides from transmembrane regions. Nat Methods 8:785–786

9. Bendtsen JD, Jensen LJ, Blom N et al (2004) Feature-based prediction of non-classical and leaderless protein secretion. Protein Eng Des Sel 17:349–356

10. Punta M, Coggill PC, Eberhardt RY et al (2012) The Pfam protein families database. Nucleic Acids Res 40(D1):D290–D301

11. Quevillon E, Silventoinen V, Pillai S et al (2005) InterProScan: protein domains identifier. Nucleic Acids Res 33(suppl 2): W116–W120

12. Benson DA, Cavanaugh M, Clark K et al (2013) GenBank. Nucleic Acids Res 41: D36–D42

13. Altschul SF, Gish W, Miller W et al (1990) Basic local alignment search tool. J Mol Biol 215:403–410

Chapter 9

Identification of Key Components for the Optimization of Cellulase Mixtures Using a Proteomic Strategy

Jingyao Qu, Jing Zhu, Guodong Liu, and Yinbo Qu

Abstract

Efficient degradation of complex lignocellulosic materials requires the synergistic action of different types of enzymes. Characterizing the compositions of lignocellulolytic enzyme mixtures could provide comprehensive understandings about the enzymatic degradation of lignocelluloses. In this chapter, we present a proteomic strategy for the analysis of enzyme mixtures produced by lignocellulolytic fungi. The described method is easy to carry out and is suitable to determine the composition of lignocellulolytic enzyme mixtures in a semiquantitative manner. Comparison of the compositions of enzyme mixtures with different degrading efficiencies allows for the identification of candidate targets for the optimization of lignocellulolytic enzyme mixtures.

Key words Cellulase, Fungi, Secretome, Proteomics, Tandem mass spectrometry

1 Introduction

Industrial cellulase preparations are mainly produced by filamentous fungi [1]. The cellulase mixtures secreted by fungi usually contain dozens or even more than one hundred proteins involved in the degradation of lignocellulosic materials. These proteins include cellulases and hemicellulases, which have different substrate specificities and act synergistically with each other [2]. Optimization of the composition of native cellulase mixtures has been successfully employed to improve their efficiency in hydrolysis, and is still to be performed to reduce the cost of enzymes for bioconversion of lignocellulosic materials to fuels and chemicals [3, 4].

Proteomic analysis is an efficient tool in the optimization of complex cellulase mixtures. On one hand, the components in the "starting" cellulase mixture need to be profiled to provide a comprehensive understanding about its composition. On the other hand, comparison of the compositions of cellulase mixtures with different hydrolysis efficiencies could identify potential determining components for the degradation, thus providing candidate targets

Mette Lübeck (ed.), *Cellulases: Methods and Protocols*, Methods in Molecular Biology, vol. 1796,
https://doi.org/10.1007/978-1-4939-7877-9_9, © Springer Science+Business Media, LLC, part of Springer Nature 2018

for the optimization of cellulase preparations. For example, Lehmann et al. studied the protein compositions of a series of *Trichoderma reesei* secretomes, and through a regression analysis of the contents of individual proteins and cellulose conversions, they identified proteins potentially important in the hydrolysis of pretreated corn stover [5]. Another example is that proteomic analysis of the secretome of *Ustilago maydis* revealed that it was distinctively rich in arabinoxylan-degrading enzymes and oxidoreductases, which might account for its enhancing effect on the hydrolysis by *T. reesei* enzymes [6].

So far, various proteomic technologies including two-dimensional gel electrophoresis [7], stable isotope labeling with amino acids in cell culture (SILAC) [8], and isobaric tag for relative and absolute quantification (iTRAQ) [9] have been used for the (semi)quantitative analysis of the composition of cellulase mixtures from different fungal species. Here, we present a label-free liquid chromatography-tandem mass spectrometry (LC-MS/MS) strategy for proteomic analysis of fungal cellulase mixtures. The method is sensitive, relatively less costly, and easy to implement. Potentially key components in lignocellulose hydrolysis inferred by proteomic analysis could be subsequently expressed and used for supplementation or partial replacement of existing cellulase mixtures to validate their functions. The related methods are described in Chapters 8, 10 and 20 in this book.

2 Materials

All the reagents are dissolved in ultrapure water, unless specifically clarified.

2.1 Collection and Concentration of Proteins

1. Refrigerated benchtop centrifuge with a fixed-angle rotor. We use Eppendorf Centrifuge 5804 R for 50 mL-tubes and Eppendorf Centrifuge 5427 R for 1.5 mL-tubes.

2. Filter syringes or vacuum filter units containing polyethersulfone membrane (pore size 0.22 μm).

3. Protein concentration determining reagent/kit. We use Bradford reagent [10] (Sangon Biotech, China).

4. Bovine serum albumin.

5. Ultrapure water.

6. Spectrophotometer. We use SpectraMax Plus 384 Microplate Reader (Molecular Devices, USA).

7. Centrifugal ultrafiltration concentrator with a molecular cutoff of 3 kDa. We use Macrosep® Advance centrifugal device (Pall Corporation, USA).

8. Precipitation solution: 10% (w/v) trichloroacetic acid (TCA) and 0.1% (w/v) dithiothreitol (DTT) in cold acetone. Stored at −20 °C until use.

9. Vortex mixer.

10. Vacuum evaporator centrifuge (e.g., Eppendorf Vacufuge Concentrator 5301).

2.2 Sample Preparation for LC-MS/MS

1. Thermostatic water bath.

2. Denaturation buffer: 0.5 M Tris–HCl, 2.75 mM EDTA, 6 M guanidine hydrochloride (*see* **Note 1**), pH 8 (adjusted with 2 M HCl).

3. 100 mM DTT. Stored at −20 °C until use.

4. 100 mM iodoacetamide. Prepare immediately before use and keep protected from light.

5. Microcon 10 kDa Centrifugal Filter Unit with Ultracel-10 membrane (Millipore Corporation, USA).

6. 25 mM NH_4HCO_3.

7. Trypsin solution: 0.5 μg/μL of proteomics grade trypsin (Sigma-Aldrich, USA) in 50 mM acetic acid solution. Make 5 μL aliquots and store at −20 °C for up to 1 month.

8. ZipTip C18 column (Millipore Corporation, USA).

9. 50% (v/v) acetonitrile (ACN) solution.

10. 0.1% (v/v) trifluoroacetic acid (TFA) solution.

11. Peptide elution buffer: 50% (v/v) ACN and 0.1% (v/v) TFA.

12. Freeze-dryer or vacuum evaporator centrifuge fitting 1.5 mL-tubes. We use Heto Maxi Dry Lyo Freeze-dryer (Gemini BV, Netherlands).

2.3 LC-MS/MS and Data Processing

1. Injection vials with 0.1 mL glass micro-inserts (31 × 5 mm).

2. Prominence nano LC system (Shimadzu, Tokyo, Japan) equipped with a custom-made trapping column (150 μm × 2 cm) packed with Reprosil-Pur 120 C18-AQ (particle size 5 μm; Dr Maish, Germany), and an analytical column (75 μm × 15 cm) packed with Reprosil-Pur 120 C18-AQ (particle size 3 μm).

3. Solvent A: 2.0% (v/v) ACN and 0.1% (v/v) formic acid.

4. Solvent B: 98.0% (v/v) ACN and 0.1% (v/v) formic acid.

5. Linear Trap Quadrupole (LTQ)-Orbitrap Velos Pro Electron Transfer Dissociation (ETD) mass spectrometer (Thermo Scientific, MA, Germany) connected to the LC system.

6. Xcalibur 2.2.0 software (Thermo Fisher Scientific, USA).

7. Proteome Discoverer software 1.4 (Thermo Scientific) with the SEQUEST search engine.

8. Protein sequence database (protein sequences of *Penicillium oxalicum* in this study, *see* **Note 2**).

3   Methods

We use the analysis of enzymes produced by *P. oxalicum* as an example for describing the methods. The same operation procedures can be used for the profiling of cellulase mixtures of other sources.

3.1   Collection and Concentration of Cellulase Mixtures

1. Centrifuge the fermentation broth of *P. oxalicum* cultivated in cellulase-inducing medium (*see* **Note 3**) at 12,000 × *g* for 10 min at 4 °C.

2. Filter the supernatant over a 0.22 μm polyethersulfone membrane (*see* **Note 4**) to further remove mycelia and other solids.

3. Determine the concentration of proteins in the sample using the Bradford method (*see* **Note 5**) according to the manufacturer's instructions. Use bovine serum albumin as the standard.

4. Optionally, concentrate the sample using a centrifugal ultrafiltration concentrator with a 3 kDa molecular mass cutoff membrane to reduce the volume of protein precipitation in the next steps (*see* **Note 6**).

5. Divide the sample to three equal parts as technical replicates.

6. Add 5 volumes of precipitation solution (*see* **Note 7**). Vortex well and then incubate at −20 °C overnight (or for at least 2 h).

7. Centrifuge at 12,000 × *g* for 15 min at 4 °C. Discard supernatant.

8. Add maximum volume of cold acetone and vortex to resuspend the pellet. Incubate at −20 °C for 20–30 min. Centrifuge at 12,000 × *g* for 15 min at 4 °C and discard supernatant.

9. Repeat **step 8** twice. Check the pH of supernatant to confirm the removal of TCA. Do more washes if necessary.

10. Dry the pellet with vacuum evaporator centrifuge at room temperature to remove any residual acetone. Add appropriate volume of ultrapure water and vortex to dissolve the pelleted proteins to reach an estimated protein concentration of about 10 mg/mL. Use a water-bath sonicator to improve protein solubilization if necessary.

11. Determine the protein concentration of obtained protein solution by appropriate dilution.

3.2 Sample Preparation for LC-MS/MS

1. To 10 µL of protein solution containing 50–100 µg protein, add 50 µL of denaturation buffer. Vortex.

2. Add 30 µL of 100 mM DTT as a reducing agent, vortex. Incubate at 37 °C for 2 h.

3. Add 50 µL of fresh-made 100 mM iodoacetamide. Incubate in darkness at room temperature for 30 min for alkylation.

4. Add 500 µL of ultrapure water to the Microcon-10 kDa centrifugal filter unit, and then centrifuge at $14,000 \times g$ for 10 min at 4 °C to wash the device.

5. Add 360 µL of 25 mM NH_4HCO_3 to the protein sample, and then centrifuge with the Microcon-10 kDa centrifugal filter unit at $14,000 \times g$ for 10–15 min at 4 °C to concentrate proteins.

6. Repeat **step 5** three times more to reduce salt concentration in the sample. Keep the final volume of sample at 100–200 µL by reconstituting the last concentrate with 25 mM NH_4HCO_3.

7. Invert the filter device and centrifuge at $1000 \times g$ for 3 min at 4 °C to transfer the desalted sample to a new tube.

8. Add trypsin at a ratio of 1:20 (w/w) of protein. Incubate at 37 °C overnight. Then transfer the tube to ice.

9. Wash the ZipTip C18 column with 200 µL of 50% ACN solution three times. Then wash the column with 200 µL of 0.1% TFA solution three times.

10. Pass 10 µL of peptide sample through the ZipTip column by pipetting in and out 15 times to load the peptides to the resin.

11. Wash the ZipTip column with 200 µL of 0.1% TFA solution seven times.

12. Elute the peptides from the ZipTip column with 10 µL of peptide elution buffer by pipetting in and out 15 times.

13. Dry the peptide mixture in freeze-dryer or vacuum evaporator centrifuge (at room temperature).

14. Dissolve peptides in 10 µL of ultrapure water and transfer to a micro-insert vial.

3.3 LC-MS/MS

1. Put the vials containing peptide samples into the autosampler on LC system. Load 5 µL of sample onto the trap column with ultrapure water at a flow rate of 5 µL/min for 3 min.

2. Elute peptides with 2% (v/v) solvent B (0.0 to 5.0 min), then a linear gradient from 2% to 15% solvent B (5.0 to 25.0 min), 15% to 40% solvent B (25.0 to 55.0 min), 40% to 98% solvent B (55.0 to 60.0 min), 98% solvent B (60.0 to 70.0 min), 98% to 2% solvent B (70.0 to 75.0 min), and 2% solvent B (75.0 to 90.0 min) at a flow rate of 300 nL/min.

3. Analyze the eluted peptides on mass spectrometer controlled with Xcalibur 2.2.0 software. Use an electrospray voltage of 2.0 kV and transfer capillary temperature of 275 °C. Detect full-scan MS spectra over the range of 400–1800 *m/z* in the Orbitrap with a resolution of 60,000 at 400 *m/z*.

4. Run the MS/MS analyzer in data-dependent acquisition mode. Select the 10 most intense precursor ions (minimum threshold of 5000 counts) in the linear ion trap for MS/MS fragmentation analysis at a normalized collision energy of 35%. Use dynamic exclusion within 60 s to avoid repetitively selecting peptides.

3.4 Protein Identification and Semiquantitation

1. Search the data from LC-MS/MS against the in-house *P. oxalicum* protein sequence database using Proteome Discoverer software with the SEQUEST search engine (*see* **Note 8**) and the following parameters: Enzyme Name: Trypsin (Full); Max. Missed Cleavage Sites: 2; Min. Peptide Length: 6; Max. Peptide Length: 144; Max. Delta Cn: 0.05; Precursor Mass Tolerance: 10 ppm; Fragment Mass Tolerance: 0.8 Da; Dynamic Modifications: methionine oxidation, N-terminal pyroglutamination; Static Modifications: cysteine carbamidomethylation.

2. Use the Precursor Ions Area Detector node in Proteome Discoverer to calculate the peak area of identified proteins based on the extracted ion chromatograms [11]. Estimate the relative abundance of individual proteins in the sample according to the following formula: Protein content (mol%) = (Protein peak area/total peak area of all identified proteins) × 100 (*see* **Note 9**).

3. Compare the compositions of cellulase mixtures of different saccharification efficiencies. Speculate proteins with potentially key functions in the degradation of lignocellulosic materials by correlation analysis of protein contents and the saccharification efficiency (*see* **Note 10**).

4 Notes

1. Alternatively, use 8 M urea as the denaturation solution.

2. Remove the peptide signal sequence of secreted proteins as predicted by SignalP server [12].

3. The sample could be fermentation broth, extract of solid-state culture, commercial cellulase solution, or the proteins taken from the saccharification system of lignocellulosic materials, etc. We suggest a starting amount of proteins ≥1 mg.

4. Avoid the use of cellulose membranes as they can be hydrolyzed by cellulases in the samples.

5. Other methods (e.g., Lowry assay) for determining protein concentrations can also be used. Bicinchoninic acid-based method is not suitable for samples containing reducing sugars, which is often the case in cellulase-producing cultures [13].

6. For protein samples containing high concentrations of salt or detergent (e.g., commercial cellulase preparations), use the centrifugal ultrafiltration concentrator for desalination.

7. For concentrated samples with protein concentrations $\geq$5 mg/mL, skip the TCA–acetone precipitation steps and proceed with operations in Subheading 3.2.

8. Other software (e.g., Mascot) for the identification and quantitation of proteins using mass spectrometry data could also be used.

9. Relative protein abundances could also be estimated by the spectral counting method, where the number of observed peptides per protein normalized by the theoretical number of peptides was used for calculation [14]. Although the calculation methods cannot provide accurate protein contents, we did confirm cellobiohydrolase I as the most abundant protein in the secretome of both *P. oxalicum* and *T. reesei*.

10. One may also predict the lacking of component(s) in a single cellulase mixture based on the knowledge about the composition of substrates and substrate specificities of enzymes. It should be noted that the enzymatic properties of individual proteins, which cannot be revealed by proteomics analysis, should be also critically considered during the optimization of cellulase mixtures.

Acknowledgments

We thank Dr Xiaolong Han for helpful comments on the protocol. The work is supported by the National Natural ScienceFoundation of China (No. 31700053), the Project funded by China Postdoctoral Science Foundation (No. 2017M612260) and the Fundamental Research Funds of Shandong University (No. 2016GN022).

References

1. Gusakov AV (2011) Alternatives to *Trichoderma reesei* in biofuel production. Trends Biotechnol 29:419–425. https://doi.org/10.1016/j.tibtech.2011.04.004

2. Van Dyk JS, Pletschke BI (2012) A review of lignocellulose bioconversion using enzymatic hydrolysis and synergistic cooperation between enzymes–factors affecting enzymes, conversion and synergy. Biotechnol Adv 30:1458–1480. https://doi.org/10.1016/j.biotechadv.2012.03.002

3. Liu G, Qin Y, Li Z, Qu Y (2013) Development of highly efficient, low-cost lignocellulolytic enzyme systems in the post-genomic era.

Biotechnol Adv 31:962–975. https://doi.org/10.1016/j.biotechadv.2013.03.001

4. Harris PV, Xu F, Kreel NE et al (2014) New enzyme insights drive advances in commercial ethanol production. Curr Opin Chem Biol 19:162–170. https://doi.org/10.1016/j.cbpa.2014.02.015

5. Lehmann L, Ronnest NP, Jorgensen CI et al (2016) Linking hydrolysis performance to *Trichoderma reesei* cellulolytic enzyme profile. Biotechnol Bioeng 113:1001–1010. https://doi.org/10.1002/bit.25871

6. Couturier M, Navarro D, Olive C et al (2012) Post-genomic analyses of fungal lignocellulosic biomass degradation reveal the unexpected potential of the plant pathogen *Ustilago maydis*. BMC Genomics 13:57. https://doi.org/10.1186/1471-2164-13-57

7. Herpoel-Gimbert I, Margeot A, Dolla A et al (2008) Comparative secretome analyses of two *Trichoderma reesei* RUT-C30 and CL847 hypersecretory strains. Biotechnol Biofuels 1:18. https://doi.org/10.1186/1754-6834-1-18

8. Phillips CM, Iavarone AT, Marletta MA (2011) Quantitative proteomic approach for cellulose degradation by *Neurospora crassa*. J Proteome Res 10:4177–4185. https://doi.org/10.1021/pr200329b

9. Adav SS, Ravindran A, Sze SK (2012) Quantitative proteomic analysis of lignocellulolytic enzymes by *Phanerochaete chrysosporium* on different lignocellulosic biomass. J Proteome 75:1493–1504. https://doi.org/10.1016/j.jprot.2011.11.020

10. Bradford MM (1976) A rapid and sensitive method for the quantitation of microgram quantities of protein utilizing the principle of protein-dye binding. Anal Biochem 72:248–254

11. Latosinska A, Vougas K, Makridakis M et al (2015) Comparative analysis of label-free and 8-Plex iTRAQ approach for quantitative tissue proteomic analysis. PLoS One 10(9): e0137048. https://doi.org/10.1371/journal.pone.0137048

12. Petersen TN, Brunak S, von Heijne G, Nielsen H (2011) SignalP 4.0: discriminating signal peptides from transmembrane regions. Nat Methods 8:785–786. https://doi.org/10.1038/nmeth.1701

13. Chundawat SP, Lipton MS, Purvine SO et al (2011) Proteomics-based compositional analysis of complex cellulase-hemicellulase mixtures. J Proteome Res 10:4365–4372. https://doi.org/10.1021/pr101234z

14. Arike L, Peil L (2014) Spectral counting label-free proteomics. In: Martins-de-Souza D (ed) Shotgun proteomics. Methods Mol Biol, vol 1156. Humana Press, New York, pp 213–222. https://doi.org/10.1007/978-1-4939-0685-7_14

Chapter 10

Cloning and Expression of Heterologous Cellulases and Enzymes in *Aspergillus niger*

Morgann C. Reilly, Saori Amaike Campen, Blake A. Simmons, Scott E. Baker, John M. Gladden, and Jon K. Magnuson

Abstract

Cellulases and other enzymes are needed for saccharification of plant biomass in the biorefinery industry. Expression, characterization, and eventual large-scale production of known and novel cellulases requires the ability to express and secrete heterologous enzymes in relevant protein production platforms like *Aspergillus niger*. A method for cloning and expression of genes for these desirable enzymes in *A. niger* is presented in this Chapter.

Key words *Aspergillus niger*, Heterologous protein expression, Protoplast transformation, Cellulases, Enzymes

1 Introduction

The biorefinery industry, whether utilizing starch, sucrose, or more recently lignocellulose-based feedstocks, requires enzymes for the release of monomeric sugars. Examples of these enzymes include glucoamylase and α-amylase for starch saccharification, and cellulase and hemicellulase cocktails to release hexoses and pentoses from lignocellulose. Historically, particular Ascomycete fungi have been widely used commercially for the production of native enzymes for such applications, with *Aspergillus* spp. and *Trichoderma reesei* the most commonly employed. These systems have reportedly been used to produce approximately 100 g/L of native enzymes (actual values are held as trade secrets). More recently, heterologous protein expression has become a desirable capability in order to supplement the cocktails of natively produced enzymes with enzymatic activities to enhance performance of cellulolytic cocktails [1] or to produce unusual enzymes with desirable thermodynamic stability or kinetic characteristics sourced from other organisms. However, expression of heterologous genes and

Mette Lübeck (ed.), *Cellulases: Methods and Protocols*, Methods in Molecular Biology, vol. 1796,
https://doi.org/10.1007/978-1-4939-7877-9_10, © Springer Science+Business Media, LLC, part of Springer Nature 2018

secretion of the resulting enzymes in an active conformation can be challenging. Herein, we describe a system that has been successfully used to express and secrete active bacterial and fungal enzymes in the heterologous fungal system *Aspergillus niger* ATCC11414 [2]. The sequencing of fungal genomes has greatly enabled the use of homologous integration of expression constructs in various fungi of interest including industrial strains [3]. The ATCC11414 strain of *A. niger* that we utilize is derived from ATCC1015 [4] and resequencing shows there are relatively few differences between the two strains, primarily single nucleotide polymorphisms. The expression construct for the heterologous gene of interest (Fig. 1a) is targeted to the *glaA* locus (glucoamylase) of *A. niger*, utilizing the *glaA* promoter to drive expression while also eliminating the native *glaA* gene that would otherwise compete for metabolic flux to secreted proteins. The heterologous gene sequence is fused in-frame with the secretion signal of the GlaA protein at the N-terminal and followed by a FLAG tag at the C-terminal for localization studies or purification. The terminator sequence from the *trpC* gene is commonly employed, although the choice of terminator is less critical than the promoter. Finally, the hygromycin phosphotransferase (*hph*) gene confers resistance to the antifungal drug hygromycin, to which *A. niger* is quite sensitive with the correct media.

2 Materials

Prepare all solutions using ultrapure (e.g., filtered deionized 18 MΩ) water and store at room temperature, unless otherwise indicated.

2.1 Construct Design and Preparation

1. Primer Pair A (see Fig. 1a).
2. High-fidelity DNA polymerase.
3. Agarose gel, DNA stain, DNA size ladder.
4. QIAquick PCR Purification Kit (from Qiagen).
5. Spectrophotometer (e.g., NanoDrop™, from Thermo Scientific).
6. Centrifugal evaporator.

2.2 Protoplast Preparation and Transformation

1. PD Agar (PDA) slants: 0.4% potato extract; 2% glucose; 1.8% agar; autoclave to sterilize.
2. 0.02% Tween-20; filter-sterilized.
3. Hemocytometer.
4. YPD Medium: 1% yeast extract; 2% peptone; 2% glucose; autoclave to sterilize.
5. Büchner funnel.

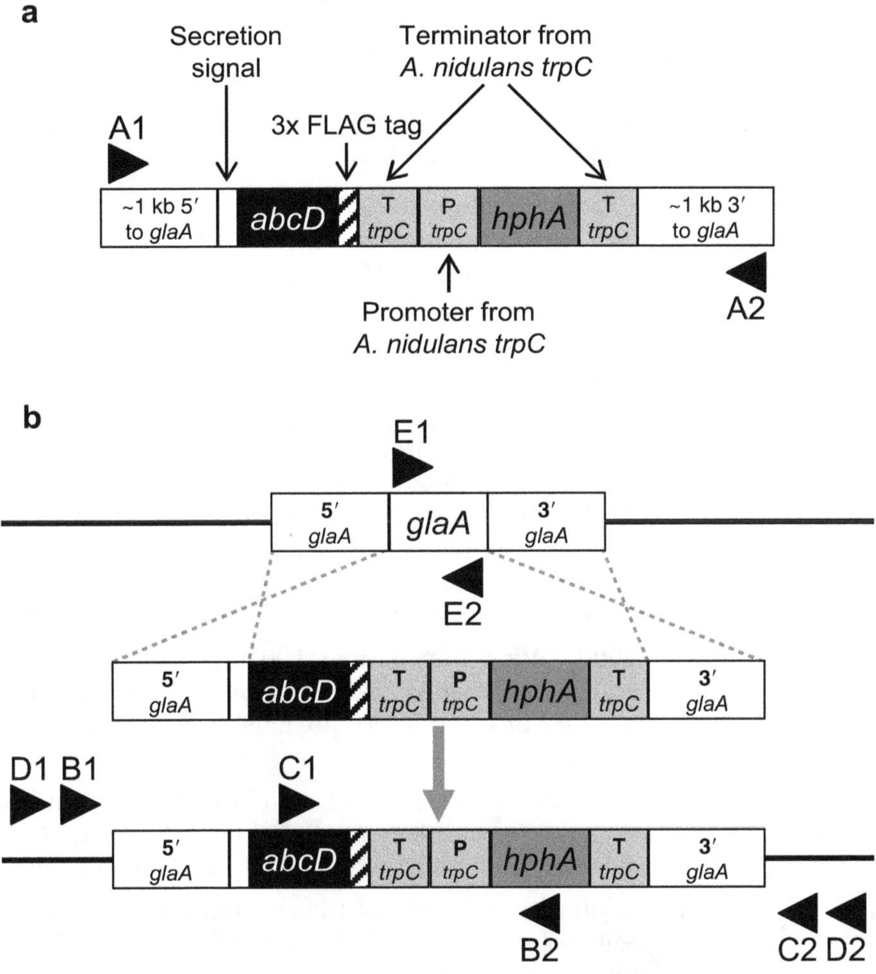

Fig. 1 Generation of an *abcD* expression construct. The expression construct for a given locus, *abcD*, is represented in Panel (**a**). The "5′ to *glaA*" fragment includes the *glaA* promoter while the "3′ to *glaA*" fragment includes the *glaA* terminator. The "Secretion signal" is 78 bp of sequence encoding the first 22 aa of GlaA. An affinity tag ("3× FLAG tag") is indicated at the end of the *abcD* sequence just prior to the stop codon, followed by the *A. nidulans trpC* terminator. The P_{trpC}-*hphA*-T_{trpC} fragment represents a hygromycin resistance construct that can be used in selection. The forward (▶) and reverse (◀) primers used to amplify the expression construct are indicated. Panel (**b**) shows the homologous recombination that should result following transformation of A. niger with the *abcD* expression construct. The solid black line represents the chromosome on which *glaA* is located; the primers used to screen for insertion of the expression construct are indicated as in (**a**). Not drawn to scale

6. Sterilized squares of Miracloth (roughly 8 × 8 in.).

7. Sterile ddH$_2$O (autoclaved).

8. Protoplasting Buffer: 0.6 M ammonium sulfate; 50 mM maleic acid in ddH$_2$O; pH adjusted to 5.5; filter-sterilized.

9. VinoTaste® Pro (from Novozymes).

10. Wide-bore, filtered p1000 pipette tips.

11. 1 M Tris; pH adjusted to 8.0.

12. ST Solution: 1 M sorbitol in 50 mM Tris (pH 8.0), filter-sterilized.

13. STC Solution: 1 M sorbitol and 50 mM $CaCl_2$ in 50 mM Tris (pH 8.0), filter-sterilized.

14. PEG Solution: 40% PEG 4000 in STC Solution; autoclaved.

15. 20× Nitrate Salts Solution: 120 g sodium nitrate; 10.4 g potassium chloride; 10.4 g magnesium sulfate heptahydrate; 30.4 g potassium dihydrogen phosphate; bring total volume to 1 L; filter-sterilize and store at 4 °C.

16. 1000× Trace Elements Solution: 562.5 mg zinc sulfate heptahydrate, 2.75 g boric acid, 1.25 g manganese chloride tetrahydrate, 1.25 g iron sulfate heptahydrate, 425 mg cobalt chloride hexahydrate, 400 mg copper sulfate pentahydrate, 21.25 mg ammonium molybdate dihydrate, and 12.5 g tetrasodium ethylenediaminetetraacetic acid; adjust pH to 6.7 with potassium hydroxide pellets; bring total volume to 250 mL ddH_2O; filter-sterilize, cover with foil, and store at 4 °C.

17. 1000× Vitamin Solution: 100 mg each of biotin, nicotinic acid, *p*-aminobenzoic acid, pyridoxine, riboflavin, and thiamine in a total of 100 mL ddH_2O; filter-sterilize, cover with foil, and store at 4 °C.

18. Minimal media + sorbitol: 1% glucose; 1× Nitrate Salts Solution; 1× Trace Elements Solution; 1 M sorbitol; 1.8% agar; autoclave to sterilize; add 1 mL/L of 1000× Vitamin Solution.

19. Minimal agar + sorbitol + hygromycin: 1% glucose; 1× Nitrate Salts Solution; 1× Trace Elements Solution; 1× Vitamin Solution; autoclave to sterilize, cool, then add hygromycin-B to 100 mg/mL.

2.3 Screening of Colonies

1. PDA slants: see Subheading 2.1.

2. YPD Medium: see Subheading 2.1.

3. Microcentrifuge tube containing a nylon filter insert (0.45 μm pore size; e.g., from VWR).

4. High-fidelity DNA polymerase.

5. Agarose gel, DNA stain, DNA ladder.

2.4 Growth for Expression

1. PDA slants: see Subheading 2.1.

2. 0.02% Tween-20: see Subheading 2.1.

3. 50% maltose; filter-sterilized.

4. CSL Medium: 100 g corn steep liquor (50% (w/v) solids; from Sigma), 50 g fructose, 10 g glucose, 1 g sodium phosphate, 0.5 g magnesium sulfate, and 0.05 mL Antifoam 204; dissolve

in ddH$_2$0; adjust pH to 5.8 using NaOH; dilute to 760 mL with ddH$_2$0, autoclave. After autoclaving, add 240 mL of sterile 50% (w/v) maltose solution.

5. HMM Medium: 120 g maltose, 70 g sodium citrate (tribasic dihydrate), 15 g ammonium sulfate, 1 g sodium phosphate, 1 g magnesium sulfate (anhydrous), and 3 g SC Complete Media (Sunrise Science; San Diego, CA USA) were added to ~800 mL ddH$_2$0, the pH adjusted to 6.2 using HCl, diluted to 1 L with ddH$_2$0, and filter-sterilized.

6. Büchner funnel.

7. Squares of Miracloth (roughly 8 × 8 in.).

8. Preweighed 15 mL conical tubes.

2.5 Assay Culture Supernatant

pNPG, Bradford, SDS-PAGE, immunoblot, or other procedures described elsewhere in this book.

3 Methods

Carry out all procedures at room temperature unless otherwise specified. When handling *A. niger*, always use filtered pipette tips (*see* **Note 1**).

3.1 Construct Design and Preparation

1. Generate a plasmid as in Fig. 1a for heterologous protein production (*see* **Note 2**).

2. Amplify the expression construct from the plasmid template using the PRIMER PAIR AA (see Fig. 1b) using a high-fidelity DNA polymerase according to the manufacturer's instructions (*see* **Note 3**).

3. Run an aliquot of the reaction on an agarose gel to verify the purity of the PCR product.

4. Purify the product from the reaction using the QIAquick PCR Purification Kit (from Qiagen) according to the manufacturer's protocol.

5. Determine the concentration of the sample using a spectrophotometer. Aim for a concentration of ~250–500 ng/μL.

   (a) If necessary, amplify the deletion construct using multiple PCR reactions. Combine the purified reaction products into a single 1.5 mL microcentrifuge tube and concentrate the DNA using a centrifugal evaporator (i.e., a Savant SpeedVac Concentrator (from ThermoScientific)) to reduce the volume.

3.2 Protoplast Preparation and Transformation (All Steps Performed in a Biosafety Cabinet)

3.2.1 Day 1: Late in the Afternoon

1. Harvest spores from PDA slant tubes of the *A. niger* strain using a 0.05% Tween-20 solution and counted with a hemocytometer (*see* **Note 4**).

   (a) Pipette 1 mL 0.05% Tween-20 down the inside edge of the PDA slant tube (*see* **Note 5**).

   (b) Recap the tube and then gently vortex to resuspend the spores (*see* **Note 6**).

   (c) Transfer the resuspended spores to a sterile 15 mL conical tube.

   (d) Determine the concentration of the spores using a hemocytometer (*see* **Note 7**).

2. Inoculate the *A. niger* at a final concentration of 5×10^5 spores/mL into 100 mL of YPD Medium in a 250 mL glass culture flask with cap.

   (a) Incubate the culture overnight (~16 h) in a shaker set at 150 rpm at 30 °C.

3.2.2 Day 2: First Thing in the Morning

1. Prepare fresh 40 mL of 30 mg/mL VinoTaste Pro (from Novozymes) in Protoplasting Buffer (*see* **Note 8**).

2. Filter the overnight YPD culture of the *A. niger* strain through a single layer of sterilized Miracloth (from Millipore) using a Buchner funnel and gravity filtration system (into a sterile 500 mL glass bottle).

3. Rinse the mycelia retained on the Miracloth three times with 100 mL sterile ddH$_2$O; discard the filtrate (flow-through).

4. Transfer the washed mycelia to a sterile 250 mL glass culture flask containing 40 mL of the filter-sterilized solution of the VinoTaste Pro/Protoplasting Buffer solution.

5. Incubate the mycelia cell wall digestion mixture in a shaker set at 70 rpm at 30 °C for 4–6 h.

   (a) Every 1–2 h, examine a small aliquot of the mycelia suspension under a light microscope to detect the progress of protoplast formation (*see* **Note 9**).

3.2.3 Day 2: Mid-Afternoon

1. When a significant quantity of protoplasts have been generated (this is necessarily a qualitative observation; there is a balance between over-digestion and lysis of protoplasts vs. too much nondigested mycelia; both result in low yields of viable protoplasts), decant the digested mycelia suspension over a single layer of sterilized Miracloth in a 50 mL conical tube. Discard the mycelia retained on the Miracloth.

2. Centrifuge the filtrate containing the protoplasts at 800 rcf for 10 min. Remove the resulting supernatant by decanting and discard it.

3. Protoplast washing: resuspend the pellet in 1 mL ST Solution by pipetting up and down slowly using a wide-bore pipette tip to minimize shear forces on the delicate protoplasts. Add 25 mL ST Solution to the sample. Centrifuge the sample at 800 rcf for 10 min. Remove the resulting supernatant by decanting and discard it.

4. Repeat the protoplast washing two additional times.

5. Finally, resuspend the pellet in 0.5 mL STC Solution by pipetting up and down using a wide-bore pipette tip. Determine the concentration of the protoplasts by counting in a hemocytometer. Dilute with STC Solution if necessary to obtain a suitable concentration of protoplasts for counting. STC Solution is added to the sample to dilute to a concentration of 1.2×10^7 protoplasts/mL (*see* **Note 10**).

6. Add 0, 1, 3, and 5 µg of the purified expression construct (Subheading 3.1, **step 5**) to 4 × 100 µL aliquots of protoplasts in 15 mL screw top conical centrifuge tubes, gently mix by stirring, and then incubate on ice for 15 min (the no DNA sample serves as a control for the selection) (*see* **Note 11**).

7. Add 1 mL PEG Solution to the protoplasts + DNA samples and mix by gently pipetting up and down using a wide-bore pipette. Incubate the samples at room temperature for 15 min.

8. Add 10 mL Minimal Medium + Sorbitol to the protoplasts + DNA + PEG samples and incubate in a shaker set at 80 rpm at 30 °C for 1 h.

9. Centrifuge the flow-through at 800 rcf for 15 min. Remove the resulting supernatant by decanting and discard it.

10. Resuspend the protoplasts in 12 mL Minimal Agar + Sorbitol + Hyg and transfer to sterile Petri dish. Once the agar containing the protoplasts solidifies, add another 12 mL Minimal Agar + Sorbitol + Hyg to each Petri dish to form an overlay. Incubate the plates at 30 °C for 3–5 days.

3.3 Screening of Colonies

1. Once colonies appear on the plates (3–5 days after transformation), use a sterile needle to pick individual colonies onto fresh PDA slants, working next to a Bunsen burner (*see* **Note 12**). The needle can be reused if heat-sterilized then cooled between picking colonies. Incubate the slants at 30 °C for at least 2 days to allow for hyphal growth and ascospore formation.

2. Transfer conidia from the PDA slants of the strains to 3 mL of YPD Medium aliquoted in 14 mL Falcon polyethylene snap-cap test tubes using a sterile wooden stick. Also include a culture of the parent *A. niger* strain to be used as a control. The tubes are capped and then incubated upright in a shaker set at 200 rpm at 30 °C for 2–3 days.

3. Harvest biomass from the YPD cultures by transferring culture to a microcentrifuge tube containing a nylon filter insert (0.45 μm pore size; e.g., from VWR). Centrifuge the sample at 10,000 rcf for 10 min. Freeze the retained biomass then store at −20 °C until DNA can be prepared from it (*see* **Note 13**).

4. Prepare DNA from the biomass samples using the ZR Fungal/ Bacterial DNA MiniPrep (Zymo Research) according to the manufacturer's protocol (any effective fungal gDNA isolation technique can be used).

5. To screen the gDNA to determine if the isolates contain the intended DNA insertion, first amplify the 5′ side using Primer Pair B and the 3′ side using Primer Pair C (see Fig. 1b) using an inexpensive DNA polymerase according to the manufacturer's instructions. Run an aliquot of the reaction on an agarose gel to verify the presence and confirm the expected size of the PCR products.

6. For those samples that yielded positive results from both the 5′ and 3′ side PCR reactions, amplify across the entire construct using Primer Pair D, again using an inexpensive DNA polymerase according to the manufacturer's instructions. Run an aliquot of the reactions on an agarose gel to verify the presence and size of the Primer Pair D PCR product (see **Note 14**).

7. It is possible to proceed with the following phenotyping studies at this stage, but in order to confirm the homogeneity of any positive strains, it is highly recommended to isolate a single colony of that strain. Use a sterile wooden stick to transfer a small amount of spores to a 1.5 mL centrifuge tube containing 1 mL of 0.02% Tween-20. Determine the concentration of the spores using a hemocytometer. Prepare dilutions as necessary in order to plate ~50 colonies per plate onto Minimal Agar + Sorbitol + Hyg. Repeat picking single colonies from each strain, preparing gDNA, and PCR-screening as above.

3.4 Growth for Expression

3.4.1 Day 1: Inoculate Starter Culture

1. Harvest spores from PDA slant tubes of the *A. niger* strains as in Subheading 3.2. In addition to whatever strain is being tested for its production of heterologous enzyme, also include a negative control (the parent strain) and a positive control (a strain known to produce the desired enzyme) if possible.

2. For each *A. niger* strain, inoculate 1×10^6 spores/mL into 50 mL of CSL Medium in a 250 mL glass culture flask with cap. Incubate the cultures for 48 h at 30 °C in a shaker set at 200 rpm.

3.4.2 Day 3: Inoculate Expression Culture

1. For each strain, at 48 h post-inoculation, transfer 5 mL of the CSL Medium culture into 50 mL of HMM Medium in a 250 mL glass culture flask with cap; repeat this twice more for

a total of three HMM Medium culture flasks per strain (*see* **Note 15**). Incubate the cultures for 120 h at 30 °C in a shaker set at 200 rpm.

3.4.3 Day 7: Harvest Culture Supernatant and Biomass

1. For each of the HMM Medium cultures, at 120 h post-inoculation, swirl the flask to wash down any biomass that has accumulated on the side of the flask along the air–liquid interface of the culture. In order to harvest an aliquot of the culture supernatant, decant the culture into a 50 mL conical tube and centrifuge at 3,000 rcf for 10 min.

2. Transfer 750 μL of the resulting supernatant to a microcentrifuge tube containing a nylon filter insert (0.45 μm pore size; from VWR).

3. Centrifuge this sample for 10 min at 10,000 rcf, allowing the liquid to pass through the filter while leaving any residual fungal biomass behind. Discard the filter and any associated biomass and store the culture supernatant sample at 4 °C until it can be assayed for enzyme activity, protein concentration, or SDS-page (e.g., using techniques described elsewhere in this book).

4. In order to harvest the fungal biomass from the culture, first vortex the contents of the 50 mL conical tube to resuspend the biomass and then decant the suspension over a single layer of nonsterile Miracloth (from Millipore) in a Büchner funnel attached to a vacuum system. Rinse any remaining biomass from the conical tube and its lid into the funnel using a wash bottle containing ddH$_2$O. Also rinse any remaining biomass from the 250 mL culture flask in the same manner. Use the wash bottle to rinse the mycelia retained on the Miracloth three times with a total of ~100 mL water to remove any residual HMM Medium; discard the flow-through.

5. Remove the Miracloth from the funnel and fold it around the mycelia to form a small packet. Press the packet between several layers of paper towels until no more moisture is released. Transfer the semidry patty of mycelia to a preweighed 15 mL conical tube and store at −80 °C for at least 12 h; the samples can then be lyophilized using a freeze-dryer.

6. Once lyophilized, weigh the 15 mL conical tube plus biomass and subtract the weight of the tube to calculate the weight of the fungal biomass.

4 Notes

1. The spores of filamentous fungi are easily spread. Filtered pipette tips help to prevent contamination of the pipette and therefore subsequent samples. Slant tubes of *A. niger* should

only be opened at the bench next to the flame of a Bunsen burner or within the confines of a biosafety cabinet.

2. In order to express a heterologous protein from the *glaA* locus, a construct like that shown in Fig. 1a should be assembled. The heterologous expression construct should be flanked by approximately 1 kb of sequence identical to the target locus to facilitate homologous recombination at the target site.

3. Linear DNA fragments without extraneous sequence are preferred for transformation.

4. Use of Δ*kusA* (aka, Δ*ku70*) strains greatly increases the rate of homologous recombination in filamentous fungi by disrupting the nonhomologous end-joining (NHEJ) system.

5. The spores of *A. niger* are hydrophobic and are not easily resuspended in water alone. The addition of low concentrations of Tween-20, a nonionic surfactant, serves as a wetting and dispersing agent.

6. Set the vortex to a speed that allows the Tween-20 solution to cover the entirety of the agar slant but does not splash the suspension up into the cap of the tube.

7. It will likely be necessary to dilute an aliquot of the spore suspension 1:100 in 0.05% Tween-20 in order to obtain an accurate count in the fields of the hemocytometer.

8. The VinoTaste Pro enzymes should be freshly dissolved in the Protoplasting Buffer and sterile-filtered through a 0.2 μm syringe filter.

9. Viable protoplasts are spherical but not completely transparent, so transparent spheres indicate undesirable loss of integrity of the plasma membrane and consequent release of cytosol and organelles. Complete conversion of mycelia to protoplasts is not possible.

10. In general, fresh preparation of protoplasts will lead to highest viability and subsequent transformation success. However, protoplasts can be successfully stored frozen in PEG/DMSO for later use. PEG Solution (40% PEG 4000 in STC Solution) is added to the protoplast sample to a final volume of 20%; next, DMSO is added to the protoplasts + PEG sample to a final volume of 7%, store protoplasts preserved in DMSO at −20 °C.

11. For the volume of DNA solution added, do not exceed 10% of the total volume of the protoplast suspension.

12. Colonies toward the edge of the agar plate may be false positives. First pick colonies and transfer to selective medium, then to PDA.

13. Freezing the fungal cell mass helps to break down the cells.

14. Primer Pair E designed to amplify the targeted locus can be used to verify its absence. Caution, negative control. The presence of an amplified band could indicate nonhomologous integration external to the target locus.

15. The replicates of the cultures are useful to account for the natural variation often seen with fungal cultures. Having biological triplicates for each strain will allow for greater confidence in the observed results.

Acknowledgment

This work was in part from the DOE Joint BioEnergy Institute (http://www.jbei.org) supported by the US Department of Energy, Office of Science, Office of Biological and Environmental Research, through Contract DE-AC02-05CH11231 between Lawrence Berkeley National Laboratory and the US Department of Energy. Many of the techniques were also initially developed with support from the US Department of Energy under Contract No. DE-AC05-76L01830 at the Pacific Northwest National Laboratory, particularly BETO sponsored Fungal Biotechnology research at PNNL.

References

1. Harris PV, Welner D, McFarland KC et al (2010) Stimulation of lignocellulosic biomass hydrolysis by proteins of glycoside hydrolase family 61: structure and function of a large, enigmatic family. Biochemistry 49 (15):3305–3316

2. Reilly MC, Kim J, Lynn J et al (2018) Forward genetics screen coupled with whole-genome resequencing identifies novel gene targets for improving heterologous enzyme production in *Aspergillus niger*. Appl Microbiol Biotechnol 102(4):1797–1807. In revision

3. Pel HJ, de Winde JH, Archer DB et al (2007) Genome sequencing and analysis of the versatile cell factory *Aspergillus niger* CBS 513.88. Nat Biotechnol 25(2):221–231

4. Andersen MR, Salazar MP, Schaap PJ et al (2011) Comparative genomics of citric-acid-producing *Aspergillus niger* ATCC 1015 versus enzyme-producing CBS 513.88. Genome Res 21(6):885–897

Chapter 11

Advanced Cloning Tools for Construction of Designer Cellulosomes

Amaranta Kahn, Edward A. Bayer, and Sarah Moraïs

Abstract

Cellulose deconstruction is achieved in nature through two main enzymatic paradigms, i.e., free enzymes and enzymatic complexes (called cellulosomes). Gaining insights into the mechanism of action and synergy among the different cellulases is of high interest, notably in the field of renewable energy, and specifically, for the conversion of cellulosic biomass to soluble sugars, en route to biofuels. In this context, designer cellulosomes are artificially assembled, chimaeric protein complexes that are used as a tool to comparatively study cellulose degradation by different enzymatic paradigms, and could also serve to improve cellulose deconstruction. Various molecular biology techniques are employed in order to design and engineer the various components of designer cellulosomes. In this chapter, we describe the cloning processes through which the appropriate modules are selected and assembled at the molecular level.

Key words Enzymatic complex, Scaffoldin, Dockerin, Cohesin, CPEC, Chimaeric enzyme, Cloning

1 Introduction

Today, the production of second generation biofuels from biomass wastes as an alternative energy source remains an industrial challenge [1, 2]. The limiting step of second generation biofuel production is the deconstruction of the biomass substrate to simple valuable pentoses and hexoses [1, 3]. This step is achieved by various types of enzymes, such as cellulases and xylanases, that can act either as free single enzymes [4] (i.e., secreted or attached directly to the cell surface) or combined together in an enzymatic complex called the cellulosome [5, 6]. Cellulosomes were reported to be very efficient systems for cellulose deconstruction [7]. Cellulosomes are very large enzymatic complexes that possess nonenzymatic, multimodular proteins called scaffoldins, to which the enzymes are strongly attached. This strong interaction is mediated through interacting modular components, namely dockerins (a module harbored by the individual enzymes) and cohesins (modules within the enzyme-integrating scaffoldin subunit). The

Mette Lübeck (ed.), *Cellulases: Methods and Protocols*, Methods in Molecular Biology, vol. 1796,
https://doi.org/10.1007/978-1-4939-7877-9_11, © Springer Science+Business Media, LLC, part of Springer Nature 2018

cohesin–dockerin interaction is species-specific and two or more different specificities can be found in the same bacterial species [8–10]. The architecture of the cellulosome, that brings the enzymes into close proximity, provides a high synergistic effect among the enzymes. The scaffoldins can contain an additional module, i.e., the carbohydrate binding module (CBM), that targets the entire complex to the substrate, thereby bringing the enzymes in close proximity to their substrate. The complex can also be attached to the cell surface of the bacterium through another type of dockerin interaction with an anchoring protein or additional scaffoldin, and this would result in minimal diffusion loss of enzymes and hydrolytic products.

In natural cellulosomes, the specific position and number of enzymes in the complex cannot be controlled, since the dockerins and cohesins have the same specificity. Thus, the concept of designer cellulosome has been proposed [11–17], both to control the specific placement and copy number of the enzymes within the complex and to engineer cellulosomes that could be even more efficient than the native ones. For the purpose of constructing designer cellulosomes, chimaeric scaffoldins are designed to contain various cohesins, originating from different bacterial species. Accordingly, the enzymes are provided with dockerins of matching binding specificities. (An example of designer cellulosome is given in Fig. 1.) A CBM module is also included in one or more of the chimaeric scaffoldins for substrate targeting of the complex [17–19]. Designer cellulosomes allow the study of the arrangement of each glycoside hydrolases (GHs) and other carbohydrate-active enzymes within the cellulosome. In addition, the incorporation of different family types and specificities of enzymes into a single complex to maximize enzyme synergy can be examined experimentally [15, 20]. Any enzyme can thus be integrated precisely into designer cellulosomes. Free enzymes can be converted into the cellulosomal mode by addition of a dockerin module or, alternatively, the native dockerin of cellulosomal enzymes can be exchanged with a dockerin of divergent specificity. To do so, various advanced cloning strategies must be used. Here we describe in details, the cloning by restriction enzymes for expression of chimaeric enzymes that allow an easy interchange of dockerin modules. We also describe the CPEC method (circular polymerase extension cloning) for fast and effective construction of chimaeric scaffoldins.

In addition, we provide the repertoire of cohesins and dockerins sequences currently in use for construction of designer cellulosomes.

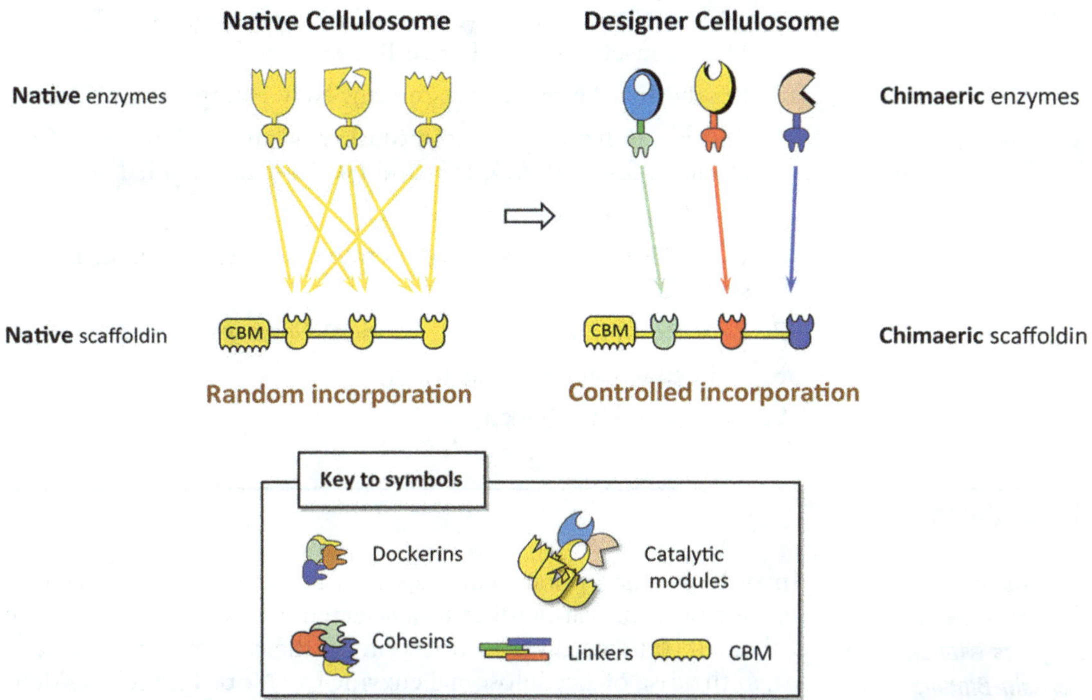

Fig. 1 Simplified schematic view of the molecular architecture of the two wild-type free enzymes (here, GH5 and GH9 from *C. bescii*) converted into a divalent (2-cohesin and two chimaeric enzymes) designer cellulosome. In a given bacterial species, the native cohesins exhibit similar specificities for their dockerins. Here, two color-coded cohesin–dockerin pairs with divergent specificity (originating from two different bacterial species) were used to fabricate the designer cellulosomes. A key to the symbols is given, and numbers in the pictograms refer to GH or CBM families. In this example, the two CBM2 of the wild-type enzymes were replaced by the selected dockerin modules. The GH9 cloning is conducted as explained in Subheading 3.1 for GH5

2   Materials

2.1   Media

Luria–Bertani (LB) medium (for 1 L):

1. Weigh 25 g of LB broth powder (Sigma-Aldrich, St. Louis, Missouri, USA) in a 1-L glass bottle.

2. Bring to 1 L with double-distilled water.

3. Autoclave (121 °C, 20 min) on liquid cycle.

2.2   Materials and Kits

This section lists the materials to purchase. Although specific products and companies are herein suggested, alternatives can be purchased and used according to each manufacturer's protocol.

1. FastDigest enzymes (ThermoFisher Scientific according to designed primers and the methylation-specific restriction endonucleases *Dpn*I. Green buffer provided.

2. T4 DNA ligase (New England Biolabs, Ipswich, Massachusetts, USA). Ligase Buffer provided.

3. ReddyMix (ABgene, Portsmouth, New Hampshire, USA).

4. Phusion polymerase (ThermoFisher Scientific, Lafayette, Colorado, USA). DMSO, HF, and GC buffers provided.

5. dNTPs at 10 mM (2.5 mM each).

6. KOD Hot Start DNA Polymerase (Novagen, Merck, Darmstadt, Germany).

7. PCR cleaning kit (Qiagen, Hidden, Germany).

8. Gel extraction kit (Qiagen).

9. Mini-prep kit (Qiagen).

3   Methods

3.1   Construction of Dockerin-Grafted Enzymes with Specific Cohesin-Binding Ability

In order to incorporate enzymes into designer cellulosomes in a specific position via binding to a selected cohesin, they should be either supplemented with a dockerin (in the case of a free enzyme) or, in the case of a cellulosomal enzyme (*see* **Note 1**), their resident dockerin should be replaced with an alternative dockerin of divergent specificity. To do so, two-step restriction cloning can be employed (Fig. 2). The catalytic module (with or without an adjacent CBM, *see* **Note 2**) is first cloned in a plasmid with an additional restriction enzyme site on its 3′ end (*see* **Note 3**), in order to enable further addition of a selected divergent dockerin by simple insertion of the selected "dockerin cassette." The process can be repeated using various dockerins in order to obtain a multiplicity of chimaeras. Here, we detail an example of a *Caldicellulosiruptor bescii* enzyme, Cel5D, to which a dockerin from *Clostridium thermocellum* has been appended (Table 1).

3.1.1   Cloning of the Catalytic Module

1. Use Clustal Omega tool (http://www.ebi.ac.uk/Tools/msa/) [21] to align the selected enzyme sequence with various other enzyme sequences from the same GH family originating from different organisms. Use the sequence alignment to define the exact positions of the different modules (catalytic site, CBM, dockerin) and linkers. Here, in *C. bescii* Cel5D, the inherent CBM28, which has cellulose-binding ability, is replaced by the dockerin (*see* **Note 4**), and the original linker between the GH5 and CBM28 is retained.

2. Define the signal peptide region using the program SignalP [22], and remove it from the sequence.

3. Check that restriction enzyme sequences do not appear in the sequence of the selected enzyme (*Nco*I and *Xho*I, are the standard choices in the pET28a plasmid used here). An additional

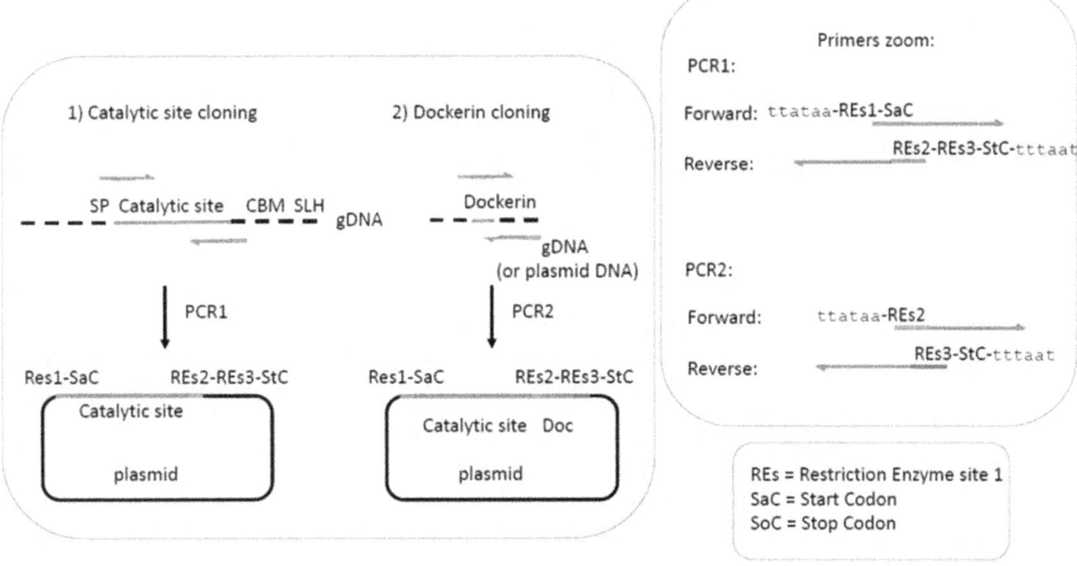

Fig. 2 Simplified schematic view of the two-step cloning for the fusion of an enzyme to a specific dockerin module. *See* text Subheading 3.1 for details

Table 1
Chimaeric cellulosomal Cel5D enzyme

Module	Sequence
Cel5D (*Caldicellulosiruptor bescii*)	MAQSILYEKEKYPHLLGNQVVKKPSVAGRLQIIEKDGKKYLAD QKGEIIQLRGMSTHGLQWYGDIINKNAFKALSKDWECNVIRLAM YVGEGGYASNPSIKEKVIEGIKLAIENDMYVIVDWH VLNPGDPNAEIYKGAKDFFKEIATSFPNDYHIIYELCNEPNPNEPG VENSLDGWKKVKAYAQPIIKMLRSLGNQNIIIVGSPNWSQRPDFAI QDPINDKNVMYSVHFYSGTHKVDGYVFENMKNAFENGVPIFVSE WGTSLASGDGGPYLDEADKWLEYLNSNYISWVNWSLSNKNE TSAAFVPYINGMHDATPLDPGDDKVWDIEELSISGEYVRARIKGIA YQPIKRDNKIK
Restriction enzyme and linker	**AVP**TYKVPGTPSTKL
Doc T (*Clostridium thermocellum*)	YGDVNDDGKVNSTDAVALKRYVLRSGISINTDNADLNEDGRVN STDLGILKRYILKEIDTLPYKN

restriction site that does not appear either in the plasmid or insert sequence should be included after the sequence of the catalytic module to allow subcloning (in the example we describe here, *Kpn*I is used).

4. Design reverse and forward primers of about *20 bases identical to the catalytic module*. The lower primer is converted into reverse complement. Restriction sites are added to the 5′ end

of each primer, as well as 5–6 additional random bases to allow restriction. The addition of the restriction enzyme site may shift the reading frame, in which case *an appropriate number of bases* must be added to remain in the proper reading frame:

Forward primer:

5′ aatgc **CCA TGG ca** *cac cat cac cat cac cat* **cag agc ata ctg tat gaa aag g** 3′, Restriction enzyme: *Nco*I Tm = 53 °C (*see* **Note 5**).

Reverse primer:

5′ ataga **CTC GAG GGT ACC gc ttt tat ttt gtt atc tct ctt a** 3′, Restriction enzyme: *Xho*I and *Kpn*I Tm = 53 °C (*see* **Note 5**).

For subsequent protein purification, a *His-tag* can be added in the upper primer after the restriction site. Alternatively, the stop codon can be omitted in the lower primer, and the His-tag present in the plasmid can be used instead (*see* **Note 6**).

5. Prepare a PCR mix following the Phusion manufacturer's instructions (ThermoFisher Scientific, Lafayette, Colorado, USA) in a PCR tube on ice, and run the PCR accordingly (*see* **Note 7**):

- Phusion buffer 5×: 10 μL (*see* **Note 8**).
- dNTPs 10 mM (2.5 mM each): 5 μL.
- Forward primer: 0.5 μM.
- Reverse primer: 0.5 μM.
- Template DNA 1–10 ng for plasmid, 50–250 ng for genomic DNA.
- Dimethyl sulfoxide (DMSO) 5%: 2.5 μL.
- Phusion 1 μL (in the end, and on ice).
- Double-distilled water up to 50 μL.

Program:

- Initial DNA denaturation: 98 °C, 30 s.
- DNA Denaturation: 98 °C, 10 s.
- Primer annealing: 52 < 60 °C, according to the melting temperature of the primers, 30 s.
- Elongation: 72 °C, 30 s/1 kb.
- Final elongation: 72 °C, 5 min.
- Repeat from **steps 2–4**: 29 cycles.
- Pause 8 °C.

6. Electrophoresis on an agarose gel: 3 μL of PCR product with 2 μL of 6× sample buffer and 7 μL of double-distilled water. Run for 20–30 min at 130 V. Usually 1% agarose gel is used,

but the percentage can be increased or reduced according to the size of the amplified DNA fragment.

7. If the PCR product is at correct size and in a single band, clean it with a PCR cleaning kit (Qiagen, Hidden, Germany). Elute with 20 μL of elution buffer or water, and measure the concentration with 1 μL on a NanoDrop spectrophotometer (ThermoFisher Scientific).

8. Digest the PCR product as well as the plasmid containing the vector with the appropriate FastDigest enzymes (Thermo-Fisher Scientific) as follows (*see* **Note 6**):

   Final reaction volume: 30 μL.

   - 3 μL Green buffer 10×.

   - 1 μL per restriction enzyme.

   - 200 ng of insert PCR or 1000 ng of plasmid.

   - Double-distilled water up to 30 μL.

   The preparation of the two mixes (one for vector and one for insert) is performed on ice, into clean Eppendorf tubes.
   Place the Eppendorf tubes in a 37 °C incubator for 30 min.

9. Run the restriction products on a 0.8% agarose gel, and allow the bands to migrate for 20–30 min at 130 V. Cut the band corresponding to the desired product size (i.e., at the expected size) in an Eppendorf tube, extract and clean it with a gel extraction kit (Qiagen). Measure the DNA concentration with 1 μL of DNA on a NanoDrop spectrophotometer (ThermoFisher Scientific).

10. Ligate the restricted insert with the restricted vector with a T4 DNA ligase (New England Biolabs, Ipswich, Massachusetts, USA) as described below:

    Final reaction volume, on ice: 20 μL.

    - 37.5 ng of cut clean insert (for 1 kb standard size).

    - 40 ng of cut clean vector (for a 4 kb standard size).

    - 2 μL ligase buffer.

    - 1 μL ligase.

    - Double-distilled water up to 20 μL.

    Store the tube at 4 °C overnight for the ligation reaction to proceed. Perform the same reaction with only the restricted plasmid vector as a control.

11. Transform 10 μL of both the vector alone and the vector plus insert ligation product in 100 μL of *E. coli* competent cells (such as DH5α, XL1, and TG1), by heat shock as follows:

    - 20 min on ice.

    - Heat-shock for 2 min at 42 °C.

    - Place the tubes back on ice for 1 min.

- Add 1 mL LB in each tube.

- Incubate the tubes for 1 h at 37 °C, using a shaking incubator (recovery phase).

- Centrifuge the tubes for 2 min at 13,000 rpm ($17,900 \times g$).

- Remove 900 μL of the supernatant fluids, resuspend the pellet, and plate it on LB-petri dishes, supplemented with the desired antibiotic (toward which the vector contains the selective resistant gene). Allow the cells to grow overnight at 37 °C.

12. Perform colony PCR with up to 18 colonies:

    - Prepare an agarose gel with two lines of ten wells (including two markers).

    - Create a mix for 18 tubes with 100 μL of ReddyMix (ABgene, Portsmouth, New Hampshire, USA, *see* **Note 6**): 0.5 μL of forward primer, 0.5 μL of reverse primer, and 99 μL of double-distilled water.

    - Aliquot 10 μL of the mix in 18 PCR tubes.

    - Select a single colony from the plate (circle and number it for identification). Apply the colony to the PCR tube and dispense into the corresponding Eppendorf tube.

    - Run the PCR as follows:

        - Initial DNA denaturation: 95 °C 2 min.

        - DNA denaturation: 95 °C 25 s.

        - Primer annealing: $48 < 63$ °C according to the melting temperature of the primer, 35 s.

        - Elongation 72 °C 1 min/1 kb.

        - Final elongation: 72 °C, 5 min.

        - Repeat from **steps 2–4**: 29 cycles.

        - Pause 8 °C.

    - Run the PCR reaction products in 1.5% agarose gel.

    - Select the transformed colonies according to their size. Grow them in 10 mL LB supplemented with 10 μL of antibiotic, overnight at 37 °C in a shaking incubator.

    - Extract and purify the plasmid using a mini-prep kit (Qiagen). Measure the concentration using the NanoDrop instrument, and send the plasmid for sequencing to select plasmids without mutations.

3.1.2 Cloning of the Dockerin Module

Several dockerins from different bacterial species have been used to convert enzymes to the cellulosomal mode. We list in Table 2 all dockerins that have been functional upon incorporation into chimaeric enzymes to form designer cellulosomes (*see* **Note 7**).

Table 2
Cohesin–dockerin pairs used in designer cellulosomes

Organism	Origin of the protein (cohesin/dockerin)	Cohesin sequence	Dockerin sequence
Acetivibrio cellulolyticus ATCC 33288	ScaC (cohesin 3)/ScaB	LQVDIGSTSGKAGSVVSVPITFTNVPKSGIYALSFRTNFDPQK VTVASIDAGSLIENASDFTTYYNNENGFASMTFEAPVDRARIID SDGVFATINFKVSDSAKVGELYNITTNSAYTSFYYSGTDEIKN VVYNDGKIEVIAS	TATTTPTTTPTTTPTPKFIYGD VDGNGSVRINDAVLIRD YVLGKINEFPYEYGMLAADVDGNG SIKINDAVLVRDYVLGKIFLFP VEEKEE
Bacteroides cellulosolvens DSM 2933	ScaB (cohesin 3)/ScaA	GKSSPGNKMKIQIGDVKANQGDTVIVPITFNEVPVMGVNNCNF TLAYDKNIMEFISADAGDIVTLPMANYSYNMPSDGLVKFL YNDQAQGAMSIKEDGTFANVKFKIKQSAAFGKYSVGIKAIGSI SALSNSKLIPIESIFKDGSITVT	PKGTATVLYGDVDNDGNVDSDDYA YM RQWLIGMIADFPGGDIGLANAD VDGDGNVDSDDYAYMRQWLIGMI SEFPAEQKA
Clostridium cellulolyticum	CipC (cohesin 1)/Cel5A	GDSLKVTVGTANGKPGDTVTVPVTFADVAKMKNVGTCNFY LGYDASLLEVVSVDAGPIVKNAAVNFSSSASNGTISFLFLDNTI TDELITADGVFANIKFKLKSVTAKTTTPVTFKDGGAFGDGTM SKIASVTKTNGSVT I	PVIVYGDYNNDGNVDALDFAGLKK YIMAADHAYVKNLDVNLDNE VNAFDLAILKKYLLGMVSKLP
Ruminoccocus flavefaciens FD1	ScaB (cohesin 1)/ScaA	GTVEWLIPTVTAAPGQTVTMPVVVKSSSLAVAGAQFKIQAATG VRYSSKTDGDAYGSGIVYNNSKYAFGQGAGRGIVAADDSVVL TLAYTVPADCAEGTYDVKWSDAFVSDTDGQNITSKVTL TDGAIIV	GPGGPGEPGGPGDGT CAVNYTVVNDWGHGMQGAITVSN TGSSPINNWTLQFSFSGVNISNG WNGEWSQSGSQITVRAPAWNSTL QPGQSVELGFVADKTGNVSPPSQF TLNGATCS
Archeoglobus fulgidus	Orf 2375	PKTTIIAGSAEAPQGS DIQVPVKIENADKVGSINLILSYPNVLEVEDVLQGSLTQNSLFD YNVEGNQIKVGIADSNGISGDGSLFYVKFRVTGNEKAEQAEN VKGKLRGLGQQLSEITLRNSHALTLQGIEIYDIDGNSVKVA TINGTFRIVSQEEA	EEANKGDVNGDGEINSLDALLALQM SIGKVEPNPVADMDGDGKVLAKDA TEIMKMATDMMIRRTAEII SQNGLLGK

(continued)

Table 2
(continued)

Organism	Origin of the protein (cohesin/dockerin)	Cohesin sequence	Dockerin sequence
Clostridium thermocellum YS (type I)	CipA (cohesin 3)/Cel48S	IKIKVDTVNAKPGDTVNIPVRFSGIPSKGIANCDFVYSYDPN VLEIIEIKPGELIVDPNPDKSFDTAVYPDRKIIVFLFAEDSGTGA YAITKDGVFATIVAKVKSGAPNGLSVIKFVEVGGFANNDLVE QRTQFFDGGVNV	TYKVPGTPSTKLYGDVNDDGKVN STDAVALKRYVLRSGISIN TDNADLNEDGRVNSTDLGILK RYILKEIDTLPYKN
Clostridium thermocellum YS (type II)	OlpB/CipA	VALELDKTKVKVGDIITATIKIENMKNFAGYQLNIKYD PTMLEAIELETGSAIAKRTWPVTGGTVLQSDNYGKT TAVANDVGAGIINFAEAYSNLTKYRETGVAEETGIIGKIGFRV LKAGSTAIRFEDTTAMPGAIEGTYMFDWYGENIKGY SVVQPGEIVVEgeepgeepteepvptetsvd ptptvteepvpselpdsy	TNKPVIEGYKVSGYILPDFSFDA TVAPLVKAGFKVEIVGTELYA VTDANGYFEITGVPANASGYTLKI SRATYLDRVIANVVTGD TSVSTSQAPIMMWVGDIVKDN SINLLDVAEVIRCFNATKGSAN YVEELDINRNGAINMQDIMI VHKHFGATSSDYDAQ
Clostridium clariflavum DSM 19732	ScaC (cohesin 1)	GQLQIDIGRVSGEPGSIVTVPVTFSNVPATGIYALSLNLNFDSTKI SVVSVEPGSLVEDPDDFALFTNNEHGFTSMSFMAPADRSRI VDKDGVFAAIKFKISEANPVGEVYDISVNYSRTSFYSTG TEEIKNVLYNDGAILVGSN	SHKFIYGDVDGNESVRINDAVLVRD YVLGKIDEFPYEYGMLAAD VDGDGNIRINDSVLIRDFVLGKI SLFPVEEQ

1. Use the amino-acid sequences listed in Table 2 to clone a dockerin that has been previously used in designer cellulosomes studies. If you plan to select a dockerin that has never been used previously, perform a sequence alignment (Clustal Omega) with other dockerins, in order to determine its starting and ending amino acids. The natural linker of the dockerin should be included (*see* **Note 9**).

2. Design primers to contain the previously selected restriction sites (ensure that they are absent from the dockerin sequence) (*see* **Note 10**).

   5′ atttat **GGT ACCg** <u>aca tat aaa gta cct ggt act c</u> 3′, restriction enzyme: *Kpn*I Tm = 52 °C.
   5′ aattat **CTC GAG** <u>gtt ctt gta cgg caa tgt atc</u> 3′, restriction enzyme: *Xho*I Tm = 51 °C

3. Follow the rest of the above protocol in Subheading 3.1.1 from **step 3** to **step 12**.

3.2 Cloning of the Chimaeric Scaffoldins

Chimaeric scaffoldins are commonly used to target the enzymes to the cellulosic substrate. Hence, they would preferably possess a CBM that targets the complex to the substrate, as the targeting effect was shown many times to be essential for proper cellulose degradation [23, 24]. For this purpose, CBM3a from scaffoldin CipA of *C. thermocellum* has generally been used. However, an alternative CBM could be considered as well. A monovalent scaffoldin (bearing a single cohesin and a single CBM), in which the cohesin that match the dockerin of a chimaeric enzyme, should be constructed for each dockerin used in a designer cellulosome study. The monovalent scaffoldins will be used as controls for the targeting effect in the activity experiments. Designer cellulosomes should demonstrate both targeting and proximity effects, where the neighboring cohesins bring the enzymes together in space [14, 20]. Longer scaffoldin can be further designed [16], containing a desired number of cohesins (*see* **Note 11**). Each cohesin on the chimaeric scaffoldin should possess a different dockerin-binding specificity. Cohesins that are incorporated into a scaffoldin are usually type I cohesins (*see* **Note 12**).

Circular polymerase extension cloning (CPEC) method [25], reported to be effective for addition and deletion of protein modules inside plasmids, is used to clone the chimaeric scaffoldins. It is a restriction-free method that is performed using three different PCR reactions: The first amplifies the vector, the second amplifies the insert, and the third allows integration of the insert into the vector by sequence complementarity without the need of ligation (Fig. 3).

3.2.1 Cloning of a Monovalent Scaffoldin (One Cohesin)

1. The CBM3a from *C. thermocellum* CipA is preferably used, the sequence is available in Table 3. For the use of a different CBM, proceed to perform a sequence alignment with known CBMs

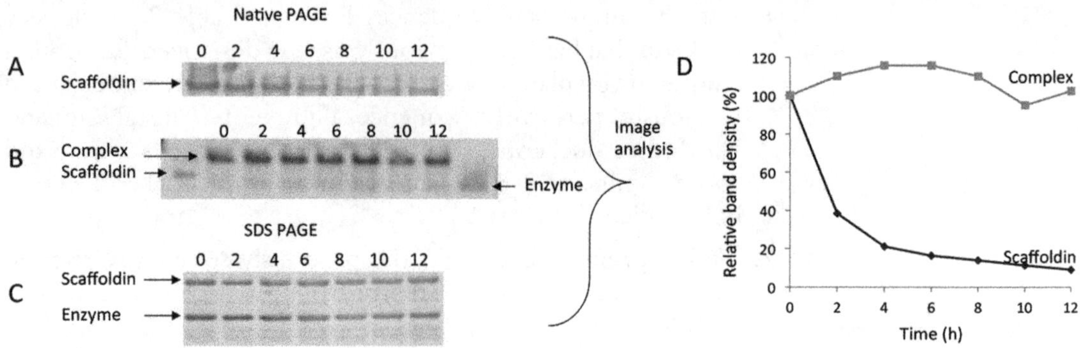

Fig. 3 Simplified schematic view of the three PCRs performed using the CEPEC (circular polymerase extension cloning) method. *See* text Subheading 3.2 for details

Table 3
Sequence of the chimaeric divalent scaffoldin G-CBM-T

Module/linker	Sequence
Cohesin G (*A. fulgidus*)	PPKTTIIAGSAEAPQGSDIQVPVKIENADKVGSINLILSYPNVLEVEDVL QGSLTQNSLFDYNVEGNQIKVGIADSNGISGDGSLFYVKF RVTGNEKAEQAENVKGKLRGLGQQLSEITLRNSHALTLQGIEI YDIDGNSVKVATINGTFRIVSQEEA
Linker	VPTNTPTNTP
CBM3a	ANTPVSGNLKVEFYNSNPSDTTNSINPQFKVTNTGSSAIDLSKLTL RYYYTVDGQKDQTFWCDHAAIIGSNGSYNGITSNVKGTFVKM SSSTNNADTYLEISFTGGTLEPGAHVQIQGRFAKNDWSNYTQSND YSFKSASQFVEWDQVTAYLNGVLVWGKEPGGS
Linker	VVPSTQPVTTPPATTKPPATTIPPSDDPNA
Cohesin T (*C. thermocellum*)	IKIKVDTVNAKPGDTVNIPVRFSGIPSKGIANCDFVYSYDPN VLEIIEIKPGELIVDPNPDKSFDTAVYPDRKIIVFLFAEDSGTGAYAI TKDGVFATIVAKVKSGAPNGLSVIKFVEVGGFANNDLVE QRTQFFDGGVNVGD

from the same family, in order to define the borders of the module.

2. Insert the gene for the CBM3a into a desired plasmid as described in Subheading 3.1.1 (*see* **Note 12**).

3. Design appropriate primers for the PCR amplification of:

The Vector

The forward primer is designed with the *20 last nucleotides of the cohesin* (insert) sequence (shown in orange on the Fig. 3 vector PCR, Vf) and the **20** first nucleotides (shown in blue in Fig. 3 vector PCR, Vf) of the CBM3a, contained in the plasmid (for addition of a cohesin at the N-terminus).

3′ **gattgtctctcaggaagaagc** CCATGGgacaaacacaccgac 5′.

The reverse primer is designed to contain the **20 nucleotides complementary to the first 20 nucleotides of the cohesin** (orange on the Fig. 3 vector PCR, Vr) followed by the 20 nucleotides complementary to the plasmid positioned before the site of insertion of the cohesin (blue on the Fig. 3 vector PCR, Vr).

3′**caatgatggtagttttcggaggcat** TATATCTCCTTCTTAAAGT-TAA 5′.

The PCR is conducted using KOD Hot Start DNA Polymerase (Novagen, Merck, Darmstadt, Germany) as follows:

- 5 μL of 10× KOD buffer.

- 5 μL of 10 mM dNTPs (2.5 mM each).

- 0.5 μM of forward primer.

- 0.5 μM of reverse primer.

- 10 ng of DNA.

- 5 μL 25 mM Mg^{2+}.

- 1.5 μL DMSO.

- 1 μL KOD polymerase.

- Double-distilled water up to 50 μL.

Divide the mix into two different PCR tubes, for two different annealing temperatures, and program the PCR machine as follows:

- Initial denaturation 95 °C, 5 min.

- Denaturation 95 °C, 30 s.

- Annealing 55 and 62 °C, 30 s.

- Elongation 68 °C, 4 min (different from the manufacturer's protocol, which recommends only 10–25 s/kb).

- Repeat from **steps 2** to **4**: 19 cycles.

- Pause 8 °C.

The Insert

The sequence of the cohesin (insert) is determined by sequence alignment of multiple type I cohesins (Clustal Omega). Table 2 provides a list of the cohesins that have been reported in the literature, and are complementary to the dockerins in the list. The cohesin sequences can be amplified either from genomic DNA or plasmids. Natural linkers of the cohesin should be included in the sequence. It is possible to reduce the length of the linker by removing some of the amino acids if the linker exceeds 20 amino acids [26, 27].

The insert's forward primer is designed with **20 nucleotides of the plasmid preceding the insertion of the dockerin** (blue on the Fig. 3 insert PCR, If) and the 20 first nucleotides of the dockerin (orange on the Fig. 3 insert PCR, If). This primer is the reverse complement of the reverse primer of the linker.

3′ **TTAACTTTAAGAAGGAGATATA** atgcctccgaaaactaccat-cattg 5′.

The insert's reverse primer is designed with **20 nucleotides that are complementary to the first 20 nucleotides of the CBM** (blue on the Fig. 3 insert PCR, Ir) and of 20 nucleotides complementary to the 20 last nucleotides of the dockerin (orange on the Fig. 3 insert PCR, Ir).

3′ **gtcggtgtgtttgtcCCATGG** gcttcttcctgagagacaatc 5′.

The PCR is conducted exactly as for the vector's program (Subheading 3.1) using the KOD polymerase (Novagen, Merck, Darmstadt, Germany).

Vector + Insert

The primers and DNA template are the vector PCR product and the insert PCR product, respectively. After completion of the first two PCRs, the two products can be run on a gel to assess their size and purity as in Subheading 3.1.1, **step 5** (*see* **Note 13**).

- Pipet 10 μL of the PCR reaction into an Eppendorf tube, supplemented with 1 μL of *Dpn*I FastDigest (ThermoFisher Scientific), and leave for 2 h at 37 °C.

- Clean the 11 μL solution as in Subheading 3.1.1, **step 6**, and measure the concentration with a NanoDrop spectrophotometer.

- The third PCR is conducted following the same program as for the two first PCRs (but only 10 cycles are performed). The DNA template and primers is the product of the two first PCRs: use an initial 1:1 vector–insert ratio.

- The intact PCR product is used to transform *E. coli* competent cells: follow Subheading 3.1.1, **steps 11** and **12** (*see* **Note 14**).

3.2.2 Cloning of a Multivalent Scaffoldin

Each additional cohesin can be added according to Subheading 3.2.1, **steps 3** and **4**. The arrangement of the modules has to be predetermined, according to the purpose of the study. There are no rules for defining the design of a scaffoldin, and several arrangements can be tested for the same components. Only one published study has defined preferable positions for three different families of glycoside hydrolase (i.e., GH9, GH5, and GH48) for *T. fusca* and *C. thermocellum* enzymes [13]. In Table 3 the amino-acid sequences of a divalent scaffoldin is given with the borders of the different modules.

4 Notes

1. The CBM can be retained or replaced by the dockerin module when converting a non-cellulosomal enzyme to the cellulosomal mode. The basis of the decision would be related to the

binding properties of the CBM. If it has different binding specificity than the CBM3a generally used in the chimaeric scaffoldin (i.e., strong binding to crystalline cellulose), the enzyme-bearing CBM should be retained [28].

2. If the enzyme will be used in vitro, an SLH domain, which serves to attach the parent enzyme to the cell surface via interaction with the peptidoglycan, should be removed in order to reach optimal enzymatic activity.

3. *Spe*I is a restriction site of choice, since it encodes for the Thr-Ser amino-acid sequence, which is compatible with many known linker sequences.

4. The His-tag can be placed on the N- or C-terminus of the protein. The expression of the protein could be increased in one of the positions and ease the subsequent purification step.

5. The Tm (melting temperature) is calculated as follows: 4 °C per G and C base and 2 °C per A and T base. Only the bases that are complementary to the amplified DNA are taken into account.

6. The same PCR can be performed with alternative polymerases, in which case, follow the manufacturer protocols for the PCR mix and program. However, a high-fidelity polymerase is recommended for plasmid construction, and a more inexpensive mix for the colony PCR. Similarly, different restriction enzyme brands can be used in order to digest the insert and the vector. If a different brand is chosen, follow the conditions according to the manufacturer. Nevertheless, fast digest restriction enzymes allow for a quicker, more effective, and reliable cleavage of the DNA.

7. The use of a cohesin–dockerin pair from *C. cellulovorans* and its incorporation in designer cellulosomes has also been reported in the literature [29].

8. G- and C-rich (GC) and High Fidelity (HF) buffers are provided. Use GC buffer where the PCR product content exceeds 50% GC. Use HF buffer in cases where the GC content of the PCR product is below 50%.

9. Add a linker sequence formed by a repetition of several TGS amino acids (ex: GSGGSGGSG) if one of the following situations is encountered: (a) For replacement of a native dockerin, where a natural linker sequence that connects the native dockerin with the catalytic sequence of the enzyme could not be determined. (b) For replacement of a CBM with a dockerin, where a natural linker sequence that connects the CBM with the catalytic sequence of the enzyme could not be determined. (c) For addition of a dockerin to the N-terminus of a protein (including a protein where a native CBM is retained).

10. A His-tag coding sequence should be included as well in the lower primer if the plasmid does not contain one, in the event that a His-tag is desired on the C-terminal end of the protein.

11. Thus far, the successful preparation of a recombinant chimaeric scaffoldin containing more than six cohesins has not been reported in the literature. Larger designer cellulosomes, containing multiple scaffoldins, can be designed by using the "adaptor" scaffoldin approach [16], wherein a type II dockerin can be added onto the scaffoldin to bind a second (adaptor) scaffoldin that will contain a matching type II cohesin.

12. In order to purify a scaffoldin, the affinity of the CBM to cellulose can be used by affinity purification with cellulose beads (Iontosorb®, Czech Republic or PASC [30]). However, it is recommended to add a His-tag in case a second step purification is needed or if a transformed host cell already possesses an exocellular cellulose-binding protein (*E. coli* does not).

13. If the bands are numerous, either perform a gel extraction of your specific band or repeat the PCR with different Tm conditions (an increase in annealing temperature will result in a more selective PCR reaction).

14. Colony PCR can be performed if the difference between the plasmid with the insert and the plasmid without the insert is large enough. The forward or reverse primer of the inserted sequence can be used together with a primer contained in the vector in order to avoid false-positive results.

References

1. Lynd LR, Laser MS, Bransby D et al (2008) How biotech can transform biofuels. Nat Biotechnol 26:169–172

2. Ho DP, Ngo HH, Guo W (2014) A mini review on renewable sources for biofuel. Bioresour Technol 169:742–749

3. Himmel ME, Ding S-Y, Johnson DK et al (2007) Biomass recalcitrance: engineering plants and enzymes for biofuels production. Science 315:804–807

4. Wilson DB (2004) Studies of *Thermobifida fusca* plant cell wall degrading enzymes. Chem Rec 4:72–82

5. Cantarel BL, Coutinho PM, Rancurel C et al (2009) The Carbohydrate-Active EnZymes database (CAZy): an expert resource for glycogenomics. Nucleic Acids Res 37:D233–D238

6. Himmel ME, Bayer EA (2009) Lignocellulose conversion to biofuels: current challenges, global perspectives. Curr Opin Biotechnol 20:316–317

7. Bayer EA, Lamed R, White BA, Flint HJ (2008) From cellulosomes to cellulosomics. Chem Rec 8:364–377

8. Yaron S, Morag E, Bayer EA et al (1995) Expression, purification and subunit-binding properties of cohesins 2 and 3 of the *Clostridium thermocellum* cellulosome. FEBS Lett 360:121–124

9. Pagès S, Belaich A, Belaich J-P et al (1997) Species-specificity of the cohesin-dockerin interaction between *Clostridium thermocellum* and *Clostridium cellulolyticum*: prediction of specificity determinants of the dockerin domain. Proteins 29:517–527

10. Mechaly A, Fierobe H-P, Belaich A et al (2001) Cohesin-dockerin interaction in cellulosome assembly: a single hydroxyl group of a dockerin domain distinguishes between non-recognition and high-affinity recognition. J Biol Chem 276:9883–9888. Erratum 19678

11. Arfi Y, Shamshoum M, Rogachev I et al (2014) Integration of bacterial lytic polysaccharide monooxygenases into designer cellulosomes promotes enhanced cellulose degradation. Proc Natl Acad Sci 111:9109–9114

12. Fierobe H-P, Mingardon F, Mechaly A et al (2005) Action of designer cellulosomes on homogeneous versus complex substrates: controlled incorporation of three distinct enzymes into a defined tri-functional scaffoldin. J Biol Chem 280:16325–16334

13. Stern J, Kahn A, Vazana Y et al (2015) Significance of relative position of cellulases in designer cellulosomes for optimized cellulolysis. PLoS One 10:e0127326

14. Moraïs S, Heyman A, Barak Y et al (2010) Enhanced cellulose degradation by nano-complexed enzymes: synergism between a scaffold-linked exoglucanase and a free endoglucanase. J Biotechnol 147:205–211

15. Moraïs S, Barak Y, Hadar Y et al (2011) Assembly of xylanases into designer cellulosomes promotes efficient hydrolysis of the xylan component of a natural recalcitrant cellulosic substrate. MBio 2:1–11

16. Stern J, Moraïs S, Lamed R, Bayer EA (2016) Adaptor scaffoldins: an original strategy for extended designer cellulosomes, inspired from nature. mBio 7(2):e00083–e00016

17. Bayer EA, Morag E, Lamed R (1994) The cellulosome – a treasure-trove for biotechnology. Trends Biotechnol 12:379–386

18. Fierobe HP, Pagès S, Bélaïch A et al (1999) Cellulosome from *Clostridium cellulolyticum*: molecular study of the dockerin/cohesin interaction. Biochemistry 38:12822–12832

19. Pagès S, Bélaïch A, Fierobe HP et al (1999) Sequence analysis of scaffolding protein CipC and ORFXp, a new cohesin-containing protein in *Clostridium cellulolyticum*: comparison of various cohesin domains and subcellular localization of ORFXp. J Bacteriol 181:1801–1810

20. Vazana Y, Barak Y, Unger T et al (2013) A synthetic biology approach for evaluating the functional contribution of designer cellulosome components to deconstruction of cellulosic substrates. Biotechnol Biofuels 6:182

21. Sievers F, Higgins DG (2014) Clustal omega. Accurate alignment of very large numbers of sequences. Methods Mol Biol 1079:105–116

22. Nielsen H (2017) Predicting secretory proteins with SignalP. Methods Mol Biol 1611:59–73

23. Hashimoto H (2006) Recent structural studies of carbohydrate-binding modules. Cell Mol Life Sci 63:2954–2967

24. Herve C, Rogowski A, Blake AW et al (2010) Carbohydrate-binding modules promote the enzymatic deconstruction of intact plant cell walls by targeting and proximity effects. Proc Natl Acad Sci 107:15293–15298

25. Quan J, Tian J (2014) Circular polymerase extension cloning. Methods Mol Biol 1116:103–117

26. Molinier A-L, Nouailler M, Valette O et al (2011) Synergy, structure and conformational flexibility of hybrid cellulosomes displaying various inter-cohesins linkers. J Mol Biol 405:143–157

27. Vazana Y, Moraïs S, Barak Y et al (2012) Designer cellulosomes for enhanced hydrolysis of cellulosic substrates. Methods Enzymol 510:429–452

28. Morais S, Barak Y, Caspi J et al (2010) Contribution of a xylan-binding module to the degradation of a complex cellulosic substrate by designer cellulosomes. Appl Environ Microbiol 76:3787–3796

29. Murashima K, Chen CL, Kosugi A et al (2002) Heterologous production of *Clostridium cellulovorans engB*, using protease-deficient *Bacillus subtilis*, and preparation of active recombinant cellulosomes. J Bacteriol 184:76–81

30. Stern J, Artzi L, Moraïs S et al (2017) Carbohydrate depolymerization by intricate cellulosomal systems. Methods Mol Biol 1588:93–116

Chapter 12

Evaluation of Thermal Stability of Cellulosomal Hydrolases and Their Complex Formation

Amaranta Kahn, Anastasia P. Galanopoulou, Dimitris G. Hatzinikolaou, Sarah Moraïs, and Edward A. Bayer

Abstract

Enzymatic breakdown of plant biomass is an essential step for its utilization in biorefinery applications, and the products could serve as substrates for the sustainable and environmentally friendly production of fuels and chemicals. Toward this end, the incorporation of enzymes into polyenzymatic cellulosome complexes—able to specifically bind to and hydrolyze crystalline cellulosic materials, such as plant biomass—is known to increase the efficiency and the overall hydrolysis performance of a cellulase system. Despite their relative abundance in various mesophilic anaerobic cellulolytic bacteria, there are only a few reports of cellulosomes of thermophilic origin. However, since various biorefinery processes are favored by elevated temperatures, the development of thermophilic designer cellulosomes could be of great importance. Owing to the limited number of thermophilic cellulosomes, designer cellulosomes, composed of mixtures of mesophilic and thermophilic components, have been constructed. As a result, the overall thermal profile of the individual parts and the resulting complex has to be extensively evaluated. Here, we describe a practical guide for the determination of temperature stability for cellulases in the cellulosome complexes. The approach is also appropriate for other related enzymes, notably xylanases as well as other glycoside hydrolases. We provide detailed experimental procedures for the evaluation of the thermal stability of the individual designer cellulosome components and their complexes as well as protocols for the assessment of complex integrity at elevated temperatures.

Key words Designer cellulosomes, Thermal stability, Hydrolases, Plant biomass degradation, Cohesin–dockerin pair, Scaffoldin

1 Introduction

Residual plant biomass is one of the most abundant renewable carbon sources on Earth. Its conversion into biofuels and other commodity chemicals is among the top priorities of the twenty-first century bio-based economy [1, 2]. The first and most difficult step in plant biomass utilization is the decomposition of its main carbohydrate polymers (cellulose and hemicellulose) into simple sugars that will serve as substrate for various microbial biosynthetic routes. Toward this end, one of the most promising approaches involves

Mette Lübeck (ed.), *Cellulases: Methods and Protocols*, Methods in Molecular Biology, vol. 1796, https://doi.org/10.1007/978-1-4939-7877-9_12, © Springer Science+Business Media, LLC, part of Springer Nature 2018

the enzymatic hydrolysis through designer polyenzymatic complexes, the cellulosomes [3–6].

Cellulosomes are produced from anaerobic cellulolytic bacteria and consist of a noncatalytic subunit called scaffoldin, on which enzymatic subunits are attached. This specific, intrasubunit assembly occurs through the cohesin modules of the scaffoldin and the dockerin modules of the enzymes [7]. Additionally, scaffoldins can also possess carbohydrate-binding modules (CBMs) which specifically bind to polysaccharide components of plant biomass [8]. All of these cellulosomal features enhance the rate of enzymatic hydrolysis, since in complexed form, the synergistic activity and the proximity of the enzymes with their substrate are increased [9, 10].

The designer cellulosome concept (Fig. 1) was conceived in order to study and deploy specific activities into the complex, toward the controlled hydrolysis process [11]. It is based on the species-specificity and high affinity interaction between a cohesin and a dockerin module [12–16]. Thus, designer cellulosomes comprise recombinant chimaeric scaffoldin constructs that bear multiple copies of cohesins of different specificities. The different cohesins are specifically connected with selected chimaeric enzymes that contain matching dockerin modules. This type of artificial nano-devices allows the control at the molecular level of the composition and architecture of the cellulosome, which cannot be achieved using native enzymes and/or scaffoldins.

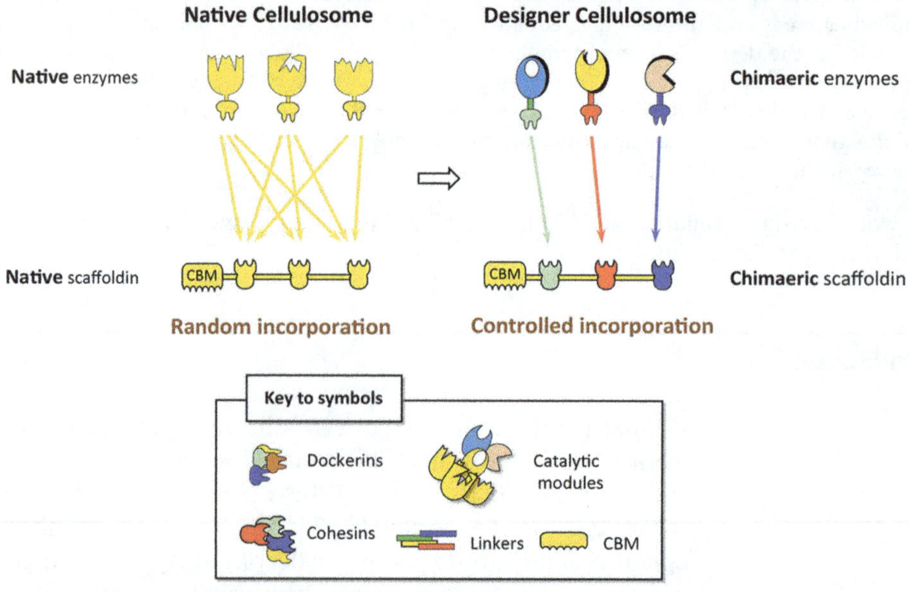

Fig. 1 Simplified schematic view of the molecular components and disposition of a trivalent (3-cohesin) cellulosome versus a trivalent designer cellulosome. In a given species, the native cohesins exhibit similar specificities for their dockerins, whereas color-coded cohesin–dockerin pairs of divergent specificity are used to fabricate designer cellulosomes. The key defines the pictograms used for the protein modules for assembly into a cellulosomal complex. Each chimaeric enzyme, as well as chimaeric scaffoldin and cohesin–dockerin pair, can be tested for thermostability as described in this chapter

Since various biorefinery processes are favored by elevated temperatures, the incorporation of stable and functional designer cellulosomes under these demanding conditions is considered of great importance [17, 18]. Such an approach also represents a great challenge, since the majority of cellulosomal components (dockerins, cohesins, and CBMs) originate from mesophilic organisms. This potential limitation can be confronted by either searching for more thermophilic and thermostable components or by improving the thermal properties of the individual components and employing cellulosomes that consist of mixtures of mesophilic and thermophilic elements. In any case, the impact of temperature on the stability and performance of the resulting designer cellulosomes must be taken into consideration and carefully evaluated.

2   Materials

2.1   Buffers and Solutions

2.1.1   Tris-Buffered Saline (TBS)—10× Stock Solution

1. Weigh 80 g NaCl, 2 g KCl, and 30 g tris (hydroxymethyl) aminomethane.

2. Transfer to a glass beaker and add 900 mL double-distilled water.

3. Mix and adjust pH to 7.4 with HCl.

4. Bring to 1 L with double-distilled water.

5. Store at 4 °C.

2.1.2   Interaction Buffer: TBS, Supplemented with 10 mM $CaCl_2$ and 0.05% Tween 20 (See Note 1)

1. Prepare a volume of 1 L of stock solution of 2 M $CaCl_2$ (*see* **Note 2**).

2. Weigh 221.96 g of $CaCl_2$, transfer to a glass beaker, and add 1 L of double-distilled water.

3. Mix and store at 4 °C.

4. Introduce 50 mL of 10× TBS, 2.5 mL of 2 M $CaCl_2$ and 250 µL of Tween 20.

5. Make up to 500 mL with double-distilled water. Store at 4 °C.

2.1.3   Pull-Down Assay Buffer

1. Acetate Buffer pH 5: mix in the following proportions, 357 mL of 0.1 M acetic acid with 643 mL of 0.1 M sodium acetate (trihydrate, 13.6 g/L).

2. Wash Buffer: 50 mM acetate buffer pH 5, supplemented with 0.05% Tween 20.

2.2   SDS-PAGE

1. Stacking Buffer: 1 M Tris pH 6.8 with SDS (500 mL). Weigh 60.5 g of tris(hydroxymethyl)aminomethane and 3.7 g SDS. Transfer to a glass beaker. Mix and adjust pH 6.8 with HCl solution. Bring to 500 mL with water. Store at 4 °C.

Table 1
SDS resolving and stacking gels preparation table

(a) Resolving gel						
	6%	8%	10%	12%	15%	20%
H_2O (mL)	10.8	9.5	8.1	6.8	4.8	1.5
Acrylamide (mL)	4	5.3	6.7	8	10	13.3
Resolving buffer (mL)	5	5	5	5	5	5
10% APS (mL)	0.2	0.2	0.2	0.2	0.2	0.2
TEMED (mL)	0.016	0.012	0.008	0.008	0.008	0.08
(b) Stacking gel						
H_2O (mL)	3.45					
Acrylamide (mL)	0.83					
Stacking buffer (mL)	0.63					
10% APS (mL)	0.05					
TEMED (mL)	0.005					

APS ammonium persulfate, TEMED tetramethylethylenediamine

2. Resolving Buffer: 1.5 M Tris pH 8.8 with sodium dodecyl sulfate (SDS). Weigh 90.75 g of tris(hydroxymethyl)aminomethane and 1.78 g SDS. Transfer to a glass beaker. Mix and adjust pH to 8.8 with HCl solution. Bring to 500 mL with double-distilled water. Store at 4 °C.

3. Running Buffer ×10 (5 L). Weigh 151.43 g tris(hydroxymethyl)aminomethane, 20.7 g glycine, and 50 g SDS. Complete to 5 L with double-distilled water. Store at room temperature.

4. Sample Buffer ×3 (100 mL). Mix 5 mL β-mercaptoethanol, 10 mg Bromophenol Blue, 3 g SDS, 10 mL glycerin, and 6.25 mL of stacking buffer. Complete to 100 mL with double-distilled water. Store at room temperature.

In order to prepare the resolving and the stacking gels, Table 1 should be followed. SDS-PAGE electrophoresis is performed for 45 min at 200 V.

2.3 Nondenaturing PAGE (Without SDS)

1. Stacking Buffer (100 mL): Weigh 5.7 g tris(hydroxymethyl)aminomethane and transfer to a glass beaker with 90 mL water. Adjust pH to 6.7 with H_3PO_4. Complete with double-distilled water up to 100 mL. Store at 4 °C.

2. Resolving Buffer (100 mL): Weigh 18.2 g tris(hydroxymethyl)aminomethane and place in a glass beaker with 90 mL water.

Adjust to pH 8.9 with HCl. Complete with double-distilled water up to 100 mL. Store at 4 °C.

3. Running Buffer ×10 (250 mL): Weigh 7.5 g tris(hydroxymethyl)aminomethane, 36 g glycine and place in glass beaker. Complete with double-distilled water up to 250 mL.

4. Sample Buffer ×3 (10 mL): Introduce 3 mL glycerol, 3 mL 10× running buffer, and 4 mL of double-distilled water. Finally, introduce 0.2 mg of Bromophenol Blue.

5. Staining Solution (1 L): Dissolve 2 g Coomassie BB R-250 in 500 mL methanol. Filter through 3 MM paper, add 100 mL 17.4 M glacial acetic acid and 400 mL of double-distilled water.

6. Destaining Solution (2 L): Add 140 mL 17.4 M glacial acetic acid, 400 mL methanol, 1460 mL of double-distilled water.

3 Methods

3.1 Components of Thermophilic Cellulosomes

In order for designer cellulosomes to be employed at elevated temperatures, it is preferable that their individual components stem from thermophilic organisms. Indeed, there is a variety of (hemi)cellulolytic enzymes of thermophilic origin extensively described. Some of them, already employed in the construction of thermostable cellulosomes, belong to the well-studied cellulosome producers *Clostridium clariflavum* [19] and *Clostridium thermocellum* [20], as well as to noncellulosomal genera, such as *Caldicellulosiruptor* [21], *Geobacillus* [22] and *Thermobifida fusca* [23–26]. Alternatively, appropriately engineered enzymes with enhanced thermostability and thermofunctionality can also be used [27, 28].

Since the only thermophilic organisms known to produce cellulosomes are three members of the genus *Clostridium* (*C. clariflavum*, *C. thermocellum*, and *C. straminisolvens*), the available thermophilic dockerins and cohesins are limited. Thus, even though several of them have already been utilized for integration into designer cellulosomes, additional dockerins, cohesins, and CBMs from mesophilic origin should also be evaluated. The formation of scaffoldins by mixing mesophilic with thermophilic cohesins and CBMs can result in complexes of increased thermal stability that may successfully be employed in applications at elevated temperatures [16, 17].

Upon selection of appropriate cellulosomal components, construction of the chimaeric scaffoldins and the addition of the dockerin modules to the free enzymes are achieved by means of standard cloning techniques, as described previously for nonthermophilic designer cellulosomes [29].

3.2 Assessment of Thermostability

The stability of the individual scaffoldins, chimaeric enzymes as well as their complexes can be assessed from their mobility in nondenaturing PAGE following incubation for different time intervals of the proteins at the desired temperature. Subsequently, the stability of the bivalent complex of each chimaeric enzyme with the matching monovalent scaffoldin as well as the full chimaeric scaffoldin should also be assessed in a similar manner.

3.2.1 Preparation of Protein Complexes

It is critical that an equimolar amount of enzyme and scaffoldin to be used for complex formation. For each enzyme and scaffoldin pair, this ratio has to be determined experimentally with nondenaturing PAGE, as the predetermined concentrations of the proteins by spectroscopic methods are not exact and might be biased due to impurities in the protein solutions. For example, Fig. 2 shows the determination of the appropriate amount of a chimaeric enzyme, (Cel9R) and its complement mini-scaffoldin (CBM-CohT). In this case, an enzyme versus scaffoldin ratio of 0.8 was shown to be the effective stoichiometric ratio for the two designated components.

The procedure for protein complex formation is as follows:

1. In 20 µL of interaction buffer, add graded amounts (4–8 µg) of each protein (*see* **Note 3**) at various ratios (usually starting from 0.4:1 and up to 1.6:1). Wherever necessary, add interaction buffer to reach a final volume of 40 µL.

2. Incubate the tubes for 1–3 h at 37 °C to allow complex formation (*see* **Note 4**).

3. Collect 20 µL of the reaction mixture for nondenaturing PAGE, and add 10 µL of nondenaturing 3× sample buffer. For a complex with an average molecular weight of

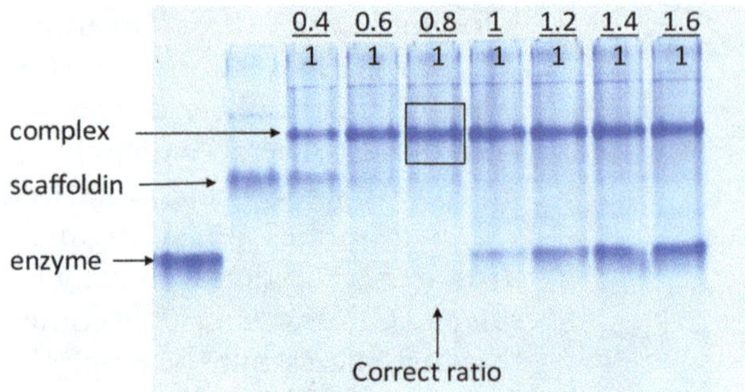

Fig. 2 Estimation of the appropriate enzyme–scaffoldin ratio for complex formation. The free enzyme and the free scaffoldin are used as markers. Different enzyme-to-scaffoldin ratios were used for complex formation (from 0.4/1 to 1.6/1). The mixtures were loaded onto a nondenaturating polyacrylamide gel. The ratio resulting in a single band, the complex, is chosen for the next steps

70–100 kDa, the stacking gel should be prepared with 3.5% final acrylamide concentration and the separating gel with 9% final acrylamide. However, for smaller complexes the stacking gel can be prepared with 4.5% polyacrylamide, and the concentration of acrylamide in the separating gel can vary accordingly (e.g., from 6% to 15%).

4. The rest of the reaction mixture can be run in parallel on 10% SDS-PAGE gels.

In some cases, the nondenaturing gel is not clear enough to determine the 1:1 ratio; if this occurs, an affinity pull down assay should be conducted (*see* Subheading 3.2.3).

3.2.2 Stability of the Scaffoldins, Chimaeric Enzymes, and Their Bivalent Complex

Upon complex formation, the protein solution is incubated at the desired temperature and for the desired duration of time. The stability of the individual components and of the complexes is inferred from their electrophoretic mobility on nondenaturing PAGE gels. The extent of denaturation can be quantified by estimating the decrease of the band density by using appropriate Image analysis software. For example, Fig. 3 illustrates the procedure followed to estimate the thermal stability of a free tetravalent scaffoldin Scaf·BTFA at 60 °C as well as the stability of the same scaffoldin in complex with a thermostable chimaeric xylanase (Xyn10-f) at the same temperature. In this case, even though 60% of the initial amount of free scaffoldin appeared denatured from the first two hours of incubation, in complex with the thermostable enzyme it remained stable for at least 12 h.

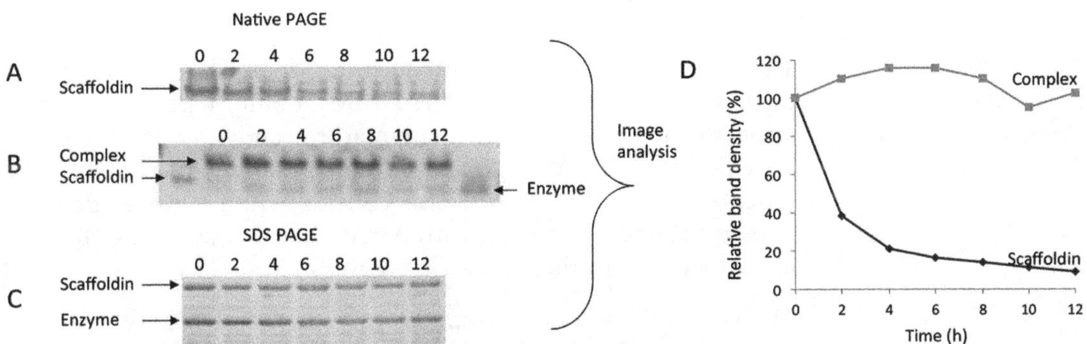

Fig. 3 Thermal stability of a free scaffoldin and the scaffoldin in complex with a thermostable chimaeric enzyme at 60 °C. (**a**) Nondenaturing PAGE of scaffoldin samples incubated at 60 °C for 0, 2, 4, 6, 8, 10, and 12 h. (**b**) Nondenaturing PAGE gel of the bivalent complex upon incubation at 60 °C for 0, 2, 4, 6, 8, 10, and 12 h. Free scaffoldin and free enzyme were also loaded on the gel as molecular weight markers. (**c**) SDS-PAGE gel of complex samples upon incubation at 60 °C for 0, 2, 4, 6, 8, 10, and 12 h. (**d**) Densitometry analysis of the thermal stability of the free scaffoldin and the bivalent complex from 0 to 12 h at 60 °C

1. For assessing the stability of the complex at different time points, prepare 250 μL of the complex to be tested using the appropriate component ratio as described above (Subheading 3.2.1).

2. Divide the mixture into 40 μL aliquots into PCR tubes (0.2 mL tubes).

3. Incubate the tubes in a preheated thermocycler (Biometra, Gottingen, Germany) at the desired temperature for the appropriate time (e.g., for 0, 2, 4, 6 8, 10, and 12 h) (*see* also **Note 5**). Samples that are not incubated can be stored at 4 °C until incubation.

4. Run the samples both in nondenaturing PAGE gels and in denaturing PAGE gels. For comparison in the nondenaturing gels, run the complex together with all individual components in the same amount used for complex formation, without prior thermal exposure. SDS-PAGE is used as a control, where completely denatured protein will not be able to penetrate the gel and will not appear in the gel.

5. Quantify the degree of denaturation of the various components and/or their complexes by estimating the decrease in the corresponding band density using an appropriate image-analysis software (*see* **Note 6**).

6. In the native gel, the free chimaeric enzyme, the free scaffoldin, the 1:1 complex without thermal denaturation is used as references. The intensity of the band after thermal denaturation either remains the same, which demonstrates stability of the complex, or is faded which indicates a thermal denaturation of the complex.

3.2.3 Stability and Functionality of CBM (Scaffoldin) at Elevated Temperatures

The effect of temperature on the ability of the CBM to bind on its carbohydrate substrate is evaluated by an affinity pull-down assay upon incubation at various temperatures and time intervals. We provide below the protocol employed to examine the stability and functionality of *C. thermocellum* CBM3a (from CipA scaffoldin), known to strongly interact with Avicel. In this example, CBM3a is incorporated into the chimaeric scaffoldin Scaf·GTV (Fig. 4).

1. To test the effect of temperature at three different time points, prepare 400 μL of 0.15–0.30 μg/μL Scaf·GTV in interaction buffer.

2. Aliquot 200 μL in three PCR tubes.

3. Incubate the tubes in a preheated thermocycler at the desired temperature and time intervals. Samples that are not incubated can be stored at 4 °C (*see* **Note 5**) until incubation.

4. In 2 mL Eppendorf tubes, weight 30 mg Avicel and add 150 μL from the above samples.

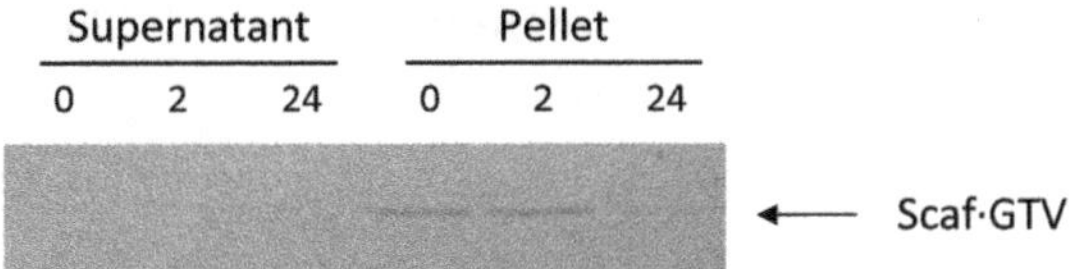

Fig. 4 Effect of temperature on the functionality of CBM3a. Samples of Scaf·GTV, which contain a CBM3a module, were incubated for 0, 2, or 24 h at 75 °C. Upon incubation, samples were allowed to interact with Avicel. Aliquots of the supernatant and the pellet from the three different time points were collected and loaded into SDS-PAGE gel. Even upon 24 h of incubation at 75 °C, Scaf·GTV was obtained but in a lower amount in the pellet fraction, indicating remaining functionality of CBM3a under these conditions that could be further quantified

5. Let the CBM interact with Avicel by incubating samples for 1 h at 4 °C with mild stirring.

6. In order to detect the unbound proteins, collect 20 µL of the supernatant from **step 5** for SDS electrophoresis.

7. For the detection of the bound proteins, centrifuge the remaining solution at 16,000 × g for 2 min and carefully discard the supernatant using a tip.

8. Add 100 µL wash buffer to the pellet and mix.

9. Centrifuge the tubes at 16,000 × g for 2 min and carefully discard the supernatant using a tip.

10. Repeat **steps 7–9**

11. Add 40 µL of 50 mM acetate buffer pH 5 to the pellet and mix.

12. Centrifuge the tubes at 16,000 × g for 2 min and remove 20 µL of the supernatant for SDS-PAGE.

3.3 Thermostability of the Whole Designer Cellulosome Complex

Upon the successful validation of the thermal properties of the individual components, the last step for evaluating cellulosome stability is to ensure the integrity of the full complex at the tested temperatures. This can be achieved either by nondenaturing PAGE or by size exclusion chromatography (SEC). The first method is simpler, faster and requires less amount of protein. Sometimes, though, because of the complexity of the cellulosome and potential impurities of the protein solutions it is not possible to get a clear image. In that case, size exclusion chromatography can be used which under appropriate calibration gives a more comprehensive result.

3.3.1 Assessment of the Cellulosome Complex Integrity via Nondenaturing Electrophoresis

The procedure to follow to test the stability of a tetravalent scaffoldin at 75 °C at 12 different time points is outlined below:

1. In 300 µL interaction buffer add equimolar amounts of the chimaeric enzymes and the scaffoldin as previously determined (*see* in Subheading 3.2.1) to prepare a 650 µL solution. If needed, fill with interaction buffer up to the final volume.

2. Prepare twelve 50 μL aliquots from the above mixture in PCR tubes.

3. Allow 1–3 h for complex formation (*see* **Note 4**).

4. Incubate samples in a preheated thermocycler at the desired temperature and time intervals. Samples that are not incubated can be stored at 4 °C (*see* **Note 5**) until incubation.

5. Collect 20 μL for nondenaturing gel electrophoresis (Acrylamide concentration at stacking gel 4% and 9% at separating gel).

6. In parallel, 20 μL of the reaction mixture can be run at a 10% SDS-PAGE gel.

3.3.2 Assessment of the Cellulosome Complex Integrity via SEC

The design of the size-exclusion chromatography (SEC) procedure should be based on the molecular weight of the individual components and their complexes. Taking this into consideration, the appropriate chromatography resin, column dimensions, and protein markers for the calibration are selected. Consequently, the adaptation of the SEC to different cellulosome complexes and their constitutive parts might require optimization of the method according to the characteristics of each individual complex. This standardization process is well reported in several textbooks and manuals [30]. In the following, we describe the assessment of the integrity of a tetravalent thermocellulosome upon incubation at 60 °C (Fig. 5).

Size-Exclusion Chromatography Column: Design, Packing, and Standardization

Since the individual components of this tetravalent cellulosome were ranging from 42 to 160 kDa, a gel filtration media such as Sephacryl S-200 with fractionation range for globular proteins ranging between 10 and 250 kDa can be employed. For the highest possible resolution, the column must have a high length-to-inner diameter ratio (approximately 100 cm height to 1 cm inner diameter would be adequate). Both packing of the chromatography column and elution of the proteins were conducted with TBS buffer and a flow rate of 0.5 mL/min.

The determination of the void volume and the calibration of the column can be conducted by injecting 100 μL of the above protein solutions into the column (0.5–5 mg/mL in TBS buffer depending on the sensitivity of the detector used).

An appropriate set of protein calibrators could be the following: blue dextran (MW > 2000 Da, void volume determination), β-amylase (200 kDa), alcohol dehydrogenase (150 kDa), albumin (66 kDa), carbonic anhydrase (29 kDa), and cytochrome c (12.4 kDa). Protein elution from the column is monitored through a UV detector (280 nm).

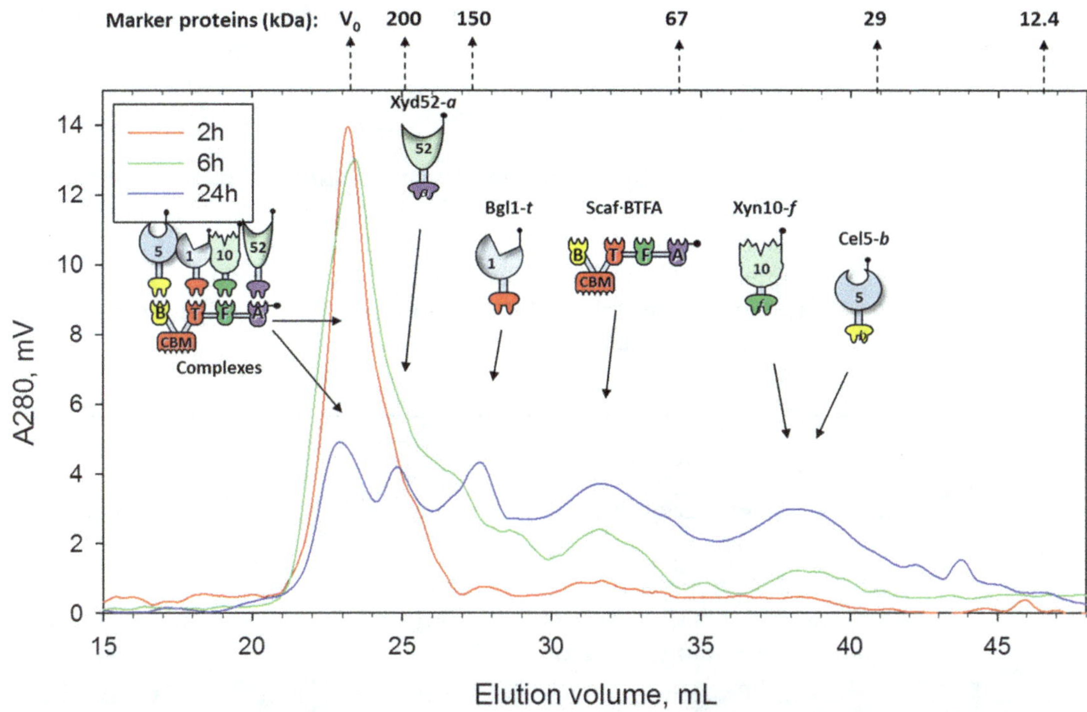

Fig. 5 Analysis of the stability of a tetravalent designer cellulosome complex by gel-filtration chromatography on a Sephacryl S-200 column. Samples of equal volume and initial concentration were introduced into the column following incubation for 2, 4, 6, and 24 h at 60 °C. The expected elution volumes and peak identification of the various cellulosomal modules are shown. The column was calibrated using the following protein markers: blue dextran (void volume, V_0), β-amylase (200 kDa), alcohol dehydrogenase (150 kDa), bovine serum albumin (66 kDa), carbonic anhydrase (29 kDa), and cytochrome c (12.4 kDa). The 0 h chromatogram coincided with the 2 h sample and is omitted for clarity. Reproduced from Anastasia P. Galanopoulou et al., 2016 with permission Springer (Berlin Heidelberg)

3.3.3 Cellulosome Sample Preparation

This protocol describes the procedure followed to test the thermal stability of the full designer cellulosome complex at four different time points:

1. Prepare 0.5 mL of complex like in Subheading 3.3.1, **step 1**.

2. Prepare four aliquots of 100 μL each.

3. Allow complex formation for 2 h (or longer for larger cellulosomes).

4. Incubate samples in a preheated thermocycler at the desired temperature and time intervals. Samples that are not incubated can be stored at 4 °C (*see* **Note 5**) until incubation.

5. In parallel prepare 100 μL samples of the individual components (scaffoldin and each individual chimaeric enzyme) in the

same concentration used for the cellulosome formation. Fill the remaining volume with interaction buffer.

6. Run all samples on SEC column consequently (Subheading 3.3.2).

Upon column and samples preparation (steps described under Subheadings 3.3.2 and 3.3.3), load the samples separately and record the individual elution volumes. The volume of the injected samples and the flow rates have to be identical with these used for the marker proteins; 100 µL and 0.5 mL/min respectively, for the present example) (Fig. 5).

4   Notes

1. For each experiment prepare 500 mL of fresh Interaction Buffer.

2. The presence of $CaCl_2$ in all protein interactions in a 5 mM-10 mM concentration is necessary for proper cohesin–dockerin binding interaction.

3. Proteins are stored in 50% glycerol at -20 °C, or filter-sterilized in TBS buffer at 4 °C. The concentration of the cellulosomal components has to be equal to or higher than 1 mg/mL. Otherwise the storage buffer of the proteins might interfere with complex formation (especially when proteins are stored in glycerol). In addition, the resulting complex solution will be of very low concentration.

4. According to the number of enzymes that are complexed on the scaffoldin the time of incubation increases from 1 to 3 h with the number of enzymes (2 h for three enzymes, 3 h for more than three enzymes).

5. For better thermal distribution during the incubation of the samples, we recommend to set both the block and lid of the thermocycler at the desired temperature. For convenience, and especially for experiments of long duration, we recommend starting with the sample with the longest incubation time and continue with the remaining samples in descending incubation times. The thermocycler apparatus can be set to drop the temperature to 4 °C after the incubation period has been completed.

6. Upon staining the gels, quantification of the bands can be evaluated by using various software such as the open source software ImageJ [31]. However, particular attention has to be paid throughout the experimental process in order to use the same amount of protein in all samples; otherwise the result will be obviously biased.

References

1. Hughes SR, Lopez-Nunez JC, Jones MA et al (2014) Sustainable conversion of coffee and other crop wastes to biofuels and bioproducts using coupled biochemical and thermochemical processes in a multi-stage biorefinery concept. Appl Microbiol Biotechnol 98 (20):8413–8431. https://doi.org/10.1007/s00253-014-5991-1

2. Lynd LR, Laser MS, Bransby D et al (2008) How biotech can transform biofuels. Nat Biotechnol 26(2):169–172. https://doi.org/10.1038/nbt0208-169

3. Artzi L, Bayer EA, Morais S (2017) Cellulosomes: bacterial nanomachines for dismantling plant polysaccharides. Nat Rev Microbiol 15 (2):83–95. https://doi.org/10.1038/nrmicro.2016.164

4. Belaich JP, Fierobe H-P, Pagès S et al (2004) From native to engineered cellulosomes. In: Ohmiya K, Sakka K, Karita S et al (eds) Genetics, biotechnology of lignocellulose degradation and biomass utilization. Uni Publishers Co., Ltd, Tokyo, pp 167–174

5. Lamed R, Setter E, Kenig R. Bayer EA (1983) The cellulosome – a discrete cell surface organelle of *Clostridium thermocellum* which exhibits separate antigenic, cellulose-binding and various cellulolytic activities. Biotech Bioeng Symp 13:163-181

6. Lynd LR, Weimer PJ, van Zyl WH, Pretorius IS (2002) Microbial cellulose utilization: fundamentals and biotechnology. Microbiol Mol Biol Rev 66(3):506–577

7. Bayer EA, Belaich J-P, Shoham Y, Lamed R (2004) The cellulosomes: multi-enzyme machines for degradation of plant cell wall polysaccharides. Ann Rev Microbiol 58:521–554

8. Shoham Y, Lamed R, Bayer EA (1999) The cellulosome concept as an efficient microbial strategy for the degradation of insoluble polysaccharides. Trends Microbiol 7:275–281

9. Fierobe HP, Bayer EA, Tardif C et al (2002) Degradation of cellulose substrates by cellulosome chimeras. Substrate targeting versus proximity of enzyme components. J Biol Chem 277(51):49621–49630. https://doi.org/10.1074/jbc.M207672200

10. You C, Zhang XZ, Sathitsuksanoh N et al (2012) Enhanced microbial utilization of recalcitrant cellulose by an ex vivo cellulosome-microbe complex. Appl Environ Microbiol 78 (5):1437–1444. https://doi.org/10.1128/AEM.07138-11

11. Bayer EA (2017) Cellulosomes and designer cellulosomes: why toy with nature? Environ Microbiol Rep 9(1):14–15. https://doi.org/10.1111/1758-2229.12489

12. Bayer EA, Morag E, Lamed R (1994) The cellulosome – a treasure-trove for biotechnology. Trends Biotechnol 12:378–386

13. Fierobe HP, Pages S, Belaich A et al (1999) Cellulosome from *Clostridium cellulolyticum*: molecular study of the Dockerin/Cohesin interaction. Biochemistry 38 (39):12822–12832

14. Mechaly A, Fierobe H-P, Belaich A et al (2001) Cohesin-dockerin interaction in cellulosome assembly: a single hydroxyl group of a dockerin domain distinguishes between non-recognition and high-affinity recognition (Erratum). J Biol Chemistry 276:19678

15. Mechaly A, Yaron S, Lamed R et al (2000) Cohesin-dockerin recognition in cellulosome assembly: experiment versus hypothesis. Proteins 39:170–177

16. Pages S, Belaich A, Tardif C et al (1996) Interaction between the endoglucanase CelA and the scaffolding protein CipC of the *Clostridium cellulolyticum* cellulosome. J Bacteriol 178 (8):2279–2286

17. Galanopoulou AP, Morais S, Georgoulis A et al (2016) Insights into the functionality and stability of designer cellulosomes at elevated temperatures. Appl Microbiol Biotechnol 100 (20):8731–8743. https://doi.org/10.1007/s00253-016-7594-5

18. Morais S, Stern J, Kahn A et al (2016) Enhancement of cellulosome-mediated deconstruction of cellulose by improving enzyme thermostability. Biotechnol Biofuels 9:164. https://doi.org/10.1186/s13068-016-0577-z

19. Artzi L, Dassa B, Borovok I et al (2014) Cellulosomics of the cellulolytic thermophile *Clostridium clariflavum*. Biotechnol Biofuels 7:100. https://doi.org/10.1186/1754-6834-7-100

20. Uttukar SM, Bayer EA, Borovok I et al (2016) Application of long sequence reads to upgrade genomes for *Clostridium thermocellum* AD2, *Clostridium thermocellum* LQRI, *and Pelosinus fermentans* R7. Genome Announc 4(5):e01043–e01016

21. Blumer-Schuette SE, Ozdemir I, Mistry D et al (2011) Complete genome sequences for the anaerobic, extremely thermophilic plant biomass-degrading bacteria *Caldicellulosiruptor hydrothermalis*, *Caldicellulosiruptor*

kristjanssonii, Caldicellulosiruptor kronotskyensis, Caldicellulosiruptor owensensis, and *Caldicellulosiruptor lactoaceticus.* J Bacteriol 193 (6):1483–1484. https://doi.org/10.1128/JB.01515-10

22. Stathopoulou PM, Galanopoulou AP, Anasontzis GE et al (2012) Assessment of the biomass hydrolysis potential in bacterial isolates from a volcanic environment: biosynthesis of the corresponding activities. World J Microbiol Biotechnol 28(9):2889–2902. https://doi.org/10.1007/s11274-012-1100-8

23. Blumer-Schuette SE, Brown SD, Sander KB et al (2014) Thermophilic lignocellulose deconstruction. FEMS Microbiol Rev 38(3):393–448. https://doi.org/10.1111/1574-6976.12044

24. Blumer-Schuette SE, Lewis DL, Kelly RM (2010) Phylogenetic, microbiological, and glycoside hydrolase diversities within the extremely thermophilic, plant biomass-degrading genus *Caldicellulosiruptor.* Appl Environ Microbiol 76(24):8084–8092. https://doi.org/10.1128/AEM.01400-10

25. Morais S, Barak Y, Hadar Y et al (2011) Assembly of xylanases into designer cellulosomes promotes efficient hydrolysis of the xylan component of a natural recalcitrant cellulosic substrate. MBio 2(6). https://doi.org/10.1128/mBio.00233-11

26. Stern J, Kahn A, Vazana Y et al (2015) Significance of relative position of cellulases in designer cellulosomes for optimized cellulolysis. PLoS One 10(5):e0127326. https://doi.org/10.1371/journal.pone.0127326

27. Anbar M, Bayer EA (2012) Approaches for improving thermostability characteristics in cellulases. Methods Enzymol 510:261–271. https://doi.org/10.1016/B978-0-12-415931-0.00014-8

28. Anbar M, Gul O, Lamed R et al (2012) Improved thermostability of *Clostridium thermocellum* endoglucanase Cel8A by using consensus-guided mutagenesis. Appl Environ Microbiol 78(9):3458–3464. https://doi.org/10.1128/AEM.07985-11

29. Vazana Y, Morais S, Barak Y et al (2012) Designer cellulosomes for enhanced hydrolysis of cellulosic substrates. Methods Enzymol 510:429–452. https://doi.org/10.1016/B978-0-12-415931-0.00023-9

30. Rhodes DG, Bossio RE, Laue TM (2009) Determination of size, molecular weight, and presence of subunits. Methods Enzymol 463:691–723. https://doi.org/10.1016/S0076-6879(09)63039-1

31. Schneider CA, Rasband WS, Eliceiri KW (2012) NIH Image to ImageJ: 25 years of image analysis. Nat Methods 9(7):671–675

Part III

Assays for Activity Testing

Chapter 13

Endoglucanase (EG) Activity Assays

M. Shafiqur Rahman, Sheran Fernando, Brian Ross, Jiangning Wu, and Wensheng Qin

Abstract

Cellulosic biomass, the most common organic compound of primary energy source on earth, is a network of interwoven biopolymers of plant cell walls. Degradation of cellulose is important for global carbon recycling. Moreover, biofuel, a renewable fuel whose energy can be derived from cellulosic biomass by enzymatic hydrolysis of cellulases. Among cellulases are endoglucases that act synergistically for subsequent hydrolytic reactions to break down the polymeric cellulose. However, in cellulolytic enzyme activity endoglucanase plays a prominent role in initiating and sustaining the hydrolytic process. Endoglucanase randomly cleaves the cellulose polymer into smaller sugar and oligomeric polysaccharides. Characterization and quantification of endoglucanase activity is important for industry and in the overall study of cellulose degradation. All assays including those for endoglucanase fall into two broad categories either qualitative or quantitative. Quantitative assays can tell if the enzyme is present, how much and its activity. Measurement can be done indirectly using a secondary colorimetric product like (1) molybdenum blue, (2) 3-amino-5-nitrosalicylic acid, (3) bicinchoninic acid (BCA), and (4) 2-cyanoacetamide or directly using an antibody in an ELISA. In this chapter, we discuss several common protocols for the measurement of endoglucanase activity.

Key words Endoglucanase, Cellulolytic enzyme, Cellulose hydrolysis, Reducing sugar, CMC

1 Introduction

Endoglucanases are a subgroup of a larger family of enzymes referred to as cellulases. Cellulases collectively refer to three broad groups of enzymes: endoglucanases, exoglucanases, and β-glucosides. Moreover, cellulases are part of a superfamily of enzymes called hydrolases, which use water to break apart larger molecules. All three types of cellulases are involved in and are required to effectively degrade cellulose, an important structural component of plant cell walls. Exoglucanases or cellobiohydrolases target cellulose chains from the ends, endoglucanases randomly cleave within the chain, and β-glucosidases hydrolyze cellobiose, the products of exoglucanases. There are many subtypes of endoglucanases (e.g., endoglucanase I, II) each with its own molecular

Mette Lübeck (ed.), *Cellulases: Methods and Protocols*, Methods in Molecular Biology, vol. 1796,
https://doi.org/10.1007/978-1-4939-7877-9_13, © Springer Science+Business Media, LLC, part of Springer Nature 2018

topology. Indeed, some organisms are capable of producing several different subtypes of endoglucanases. As a whole these endoglucanases are grouped into glycoside hydrolase families (glycoside hydrolase 5, 6, etc.) [1]. Endoglucanases vary in their substrate specificity being able to cleave differing glycosidic bonds (β-1-4, β-1-6, etc.) [2]. β-1-4-endoglucanases are the most common and are what is referred to by default. They cleave the internal β-1-4-glycosidic bonds of the cellulose chain. The exact nature and energetics of the reaction are controversial but it is believed that the mechanical sheering in the enzyme's active domain causes the breaking of the bonds [3]. Cellulose hydrolysis through enzymes occurs in two steps: (1) the physical disruption of the crystalline ultrastructure of cellulose and (2) hydrolysis of the disrupted cellulose glyosidic linkages. The hydrolysis is carried out by the two carboxylic acid groups, where one acts like an acid and the other as the base [4]. However, for the purposes of endoglucanase, the assays have relatively the same sensitivity but this is not the case for other enzymes [5]. Moreover, in this chapter, we will provide an overview of assays used to either: (1) prove or disprove the presence of the enzyme (qualitative assays) and/or (2) determine the amount of enzyme in the sample (quantitative assays).

There are many endoglucanase activity assays, and we have presented only the most popular assays. It is our hope that we have given you an idea of the general categories of assays possible and the underlying principles. Qualitative assays based on viscosity or dye decolonization will tell you whether endoglucanase is present or not. Quantitative assays based on the generation of a colorimetric product, which will indirectly measure endoglucanase levels, or ELISA assays will directly measure endoglucanase levels. For most purposes, indirect measurement is sufficient unless you are looking into a specific subtype of endoglucanase, which would require an ELISA. All the quantitative assays mentioned in the chapter have similar sensitivities for endoglucanase within a range of hydrolysis. Whichever assay you decide to use it is important to keep in mind sources of error, the assay operating conditions, and what exactly you are measuring.

1.1 Qualitative Endoglucanase Assays

Based on the methods described by Zhang et al. [6] and Hendricks et al. [7], dye decolorization assays can be used to detect endoglucanase qualitatively and semiquantitatively on agar or agarose containing media. This assay is suitable for screening a large number of samples, with differences in endoglucanase activity that are visible to the eye. In the assay, the polysaccharides (CMC in this case) are stained with a dye and forming noncovalent complexes with cellulose. Degradation of cellulose by endoglucanase results in a halo or zone of clearance caused by the simultaneous degradation of the dye. Figure 1 is an example of a dye decolorization assay. The halo can be used qualitatively and as a semiquantitative assessment using

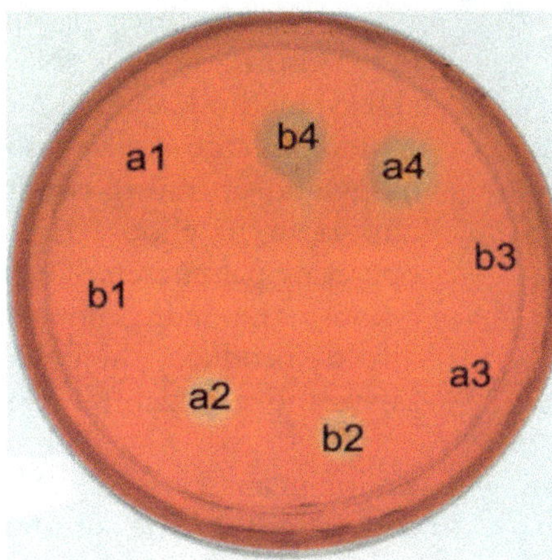

Fig. 1 Degradation of cellulose by endoglucanase results in a halo or zone of clearance caused by the degradation of the dye Congo Red [8]

the following formula, which linearly relates endonuclease activity with halo diameter:

$D = K \times \log(A) + N$, where the D is the diameter, A is the enzyme activity, and K and N are parameters determined by the constructing a standard curve with known solutions having known enzyme activities.

However, another method, enzyme-linked immunosorbent assay (ELISA) can be used for endoglucanase assays [9]. This is an ELISA assay used to measure antibody binding to antigen, in this case monoclonal antibody EG-1-2 (mAb EG-1-2) binding to endoglucanase I (EG I) produced by *T. reesei*. *T. reesei* produces many types of endoglucanases—endoglucanase II, IV, etc.; mAb EG-1-2 was made specifically for a nonconserved epitope unique to EG-1 (*see* **Note 1**). EG-1 is produced by a wide range of cellulolytic organisms. Although this mAb is highly specific to EG-1, it is possible to create a less specific mAb by targeting epitopes common to multiple endoglucanases.

1.2 Quantitative Endoglucanase Assays

1.2.1 Endoglucanase Activity Assays Using Reducing Agents

The principle behind using reducing agents such as sugars is that within a limited amount of hydrolysis, the production of these sugars such as glucose from the intact polymeric cellulose is largely controlled by endoglucanases. As endoglucanases generate reducing sugars, it results in a measurable change in the solution, either a change in viscosity or a change in color depending on the assay. Water-soluble cellulose derivatives such as Carboxymethyl Cellulose (CMC) or Hydroxyethyl Cellulose (HEC) are used instead of native cellulose because underivatized cellulose has a very low accessible fraction of β-glucosidic bonds making the reaction activity rate too slow to measure.

There are three popular colorimetric reducing sugar assays: (1) Nelson-Somogyi (NS) method, (2) 3,5-dinitrosalicylic acid (DNS) method, and (3) 2-cyanoacetamide method. All these assays achieve in large similar results but differ in their reagents, methodology and operating conditions (temp., pH etc.). Therefore, it is important to choose the correct assay for your own particular operating parameters. The BCA method is an interesting exception because the reducing agent is the endoglucanase itself, which directly produces the colorimetric product unlike the other three assays above, which rely on reducing sugars.

1.2.2 Endoglucanase Assay Using CMC by the DNS Method

Endoglucanase assay (CMCase) estimates a fixed amount of glucose (about 0.5 mg) produced by the substrate (CMC). Enzymatic activity is estimated by the DNS method. For quantitative results, enzymes must be diluted, or assay reaction time decreased until the amount of product plotted against enzyme concentration is reasonably linear. A shortcoming and important point to note is that although CMC is widely used for the assay of endoglucanase activity there are some problems with its use since the enzyme is being measured over a range of pH values. As an ionic substance like CMC, its properties will change with pH, requiring the use of mildly acid assay conditions. Nonionic substrates, such as HEC, can be used as a substitute if particular applications require more different levels of acidity or alkalinity [6, 10].

The DNS method is an estimation of reducing sugars. Reducing sugars like glucose and other oligosaccharides convert DNS (yellow) to 3-amino,5-nitrosalicylic acid (orange-red) [11]. This change in color can be measured by a spectrophotometer as an absorbance, thereby used as an estimation of endoglucanase activity, which produces the reducing sugars. Another shortcoming of this method and other methods that use reducing sugars is that citrate buffer and other substances might affect the DNS test [11]. Furthermore, the dissolved oxygen and the acidic buffer can also reduce the colorimetric agent and further confound the results [11].

1.2.3 Endoglucanase Assay Using CMC by Nelson-Somogyi Method

The principle behind this assay is similar to the CMC or BCA method except in this case it is not a peptide bond that reduces Cu^{2+} to Cu^+, instead it is a reducing sugar. Reducing sugars when heated with alkaline copper tartrate reduce Cu^{2+} to Cu^+ resulting in the formation of Cu_2O. When Cu_2O interacts with arsenomolybdic acid, the reduction of molybdic acid to molybdenum blue take place [12, 13]. For the purposes of endoglucanase, the sensitivity of this test is comparable to the DNS method [5].

1.2.4 Endoglucanase Assay Using CMC by 2-Cyanoacetamide Method

It is an accurate, rapid and nontoxic spectrophotometric quantitative assay method where 2-cyanoacetamide can detect D-glucose in a linear fashion under a wide range of pH (4.0–8.0). As sensitive as the DNS test, 2-cyanoacetamide is also detecting endoglucanase activity using CMC as a substrate [14]. This assay was developed as a safer alternative to the DNS-method since many of the reagents in the DNS method are known highly toxic, carcinogens and corrosive [14].

1.2.5 Endoglucanase Assay Using CMC and Viscosity

This is a qualitative assay based on the reduction in specific viscosity of CMC as sugars are released by endoglucanase. A viscometer should be used when preforming this assay. It is important to maintain a constant temperature in this assay, as viscosity varies greatly with temperature. Both endoglucanase and exoglucanase can reduce the specific viscosity. But within a limited degree of hydrolysis, endoglucanases activity alone can be selected for because they decrease the specific viscosity faster than exoglucanases, which decrease it more slowly [6, 15].

2   Materials

Microbial culture transfer and preparation of all the reagents should be performed under aseptic conditions (*see* **Note 2**). Prepare all reagents and buffers using deionized, double distilled or milli-Q water at room temperature and store the reagents at room temperature (unless indicated otherwise). However, for microbial media preparation you can use distilled water. Always use analytical or reagent grade chemicals.

2.1   Qualitative Assays

2.1.1   Endoglucanase Assay on Agar Medium

1. 1 g/L Congo Red solution: Dissolve 100 mg Congo Red in 99.9 mL water containing 1% ethanol.

2. 1 M NaCl solution: Dissolve 58.44 g of NaCl into distilled water in a final volume of 1 L.

3. 0.1 M sodium phosphate buffer (pH 6.5): Prepare 0.5 M sodium phosphate monobasic stock by dissolving 30.0 g of anhydrous sodium phosphate monobasic in a final volume of 500 mL of H_2O.

4. 1% w/v, low viscosity CMC agar medium: Take 0.5 g CMC, 0.1 g $NaNO_3$, 0.1 g K_2HPO_4, 0.1 g KCl, 0.05 g $MgSO_4$, 0.05 g yeast extract, 1.5 g agar, and 100 mL of distilled water in a 250 mL conical flask. Autoclave (sterilized) the medium before use to make agar plates (*see* **Note 3**).

This is an example of a possible CMC based minimal salt agar for growing bacteria, formulation can be varied as needed.

2.1.2   Endoglucanase Assay on Agarose Gel

1. 1 g/L Congo Red solution: For preparation refer to Subheading 2.1.1.

2. 1 M NaCl solution: For preparation refer to Subheading 2.1.1.

3. 0.1 M sodium phosphate buffer pH 6.5: For preparation refer to Subheading 2.1.1.

4. 1% w/v, low viscosity CMC in 0.8% agarose: For preparation refer to Subheading 2.1.1. The only difference is to use 0.8 g agar.

2.1.3 Endoglucanase Assay on Polyacrylamide Gel

This method can separate the protein components of a sample by electrophoresis and then detects endoglucanase activity on polyacrylamide gels via dye staining and zones of clearance. Refer to Fig. 2 for an example of how the assay looks.

1. 1 g/L Congo Red solution: For preparation refer to Subheading 2.1.1.

2. 1 M NaCl solution: For preparation refer to Subheading 2.1.1.

3. 0.1 M sodium phosphate buffer pH 6.5: For preparation refer to Subheading 2.1.1.

4. 1% w/v CMC in sodium phosphate buffer whose pH is chosen depending on the specific cellulase: For preparation refer to Subheading 2.1.1.

2.1.4 Enzyme-Linked Immunosorbent Assay (ELISA)

1. Monoclonal Antibody (mAb) EG-I against Endoglucanase I.

2. Phosphate buffered saline with Tween-20 (PBST): 20 mM sodium phosphate buffer (pH 7.2) containing 150 mM sodium chloride, 0.02% sodium azide, and 0.0 5% Tween 20. Or, mix 3.2 mM Na_2HPO_4, 0.5 mM KH_2PO_4, 1.3 mM KCl, 150 mM NaCl, 0.05% Tween® 20, pH 7.4. Sterilize by autoclaving, and store in room temperature.

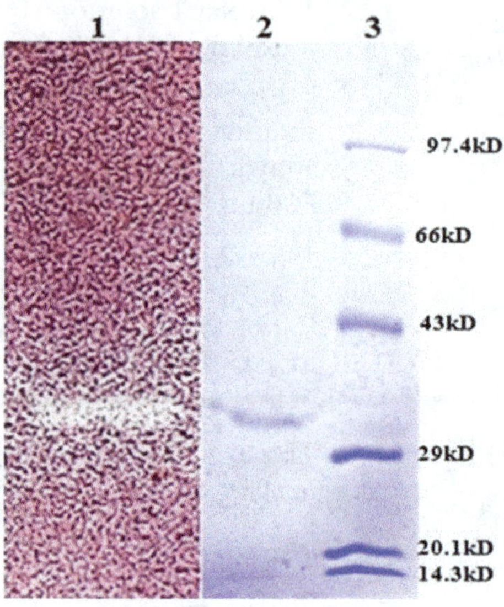

Fig. 2 Separation of protein mixtures by SDS-PAGE [16]

3. 1.5 M Tris–HCL buffer (pH 8.8): Dissolve 18.15 g of Tris base in 80 mL distilled water. Adjust pH to 8.8 using 6 M HCl. Afterward add 100 mL distilled water to bring volume up to 150 mL.

4. Bovine serum albumin (BSA) from Sigma-Aldrich.

5. 200 μL of a 1-mg/mL *p*-nitrophenyl phosphate (PNPP) solution: Mix 1 mg PNPP salt with 1 mL of distilled water.

6. 2 M sodium hydroxide: Dissolve 79.99 g NaOH in 1 L of distilled water.

7. Alkaline phosphatase-conjugated anti-rabbit immunoglobulin G produced in swine and alkaline phosphatase-conjugated anti-mouse IgG (heavy and light chains) produced in rabbits such as that made by Orion Diagnostics, Espoo, Finland and other companies.

2.2 Quantitative Assays

2.2.1 Endoglucanase Assay Using CMC by the DNS Method

1. 50 mM Citrate buffer pH 4.8: Add 210 g citric acid monohydrate to 750 mL distilled water. Then add 1.0 M NaOH (39.997 g/L distilled water) until pH equals 4.3 (around 50–60 g). Dilute the solution to 1000 mL and check pH. If necessary add 1.0 M NaOH until pH = 4.8.

2. 2% w/v CMC in 0.05 M citrate buffer: Add 2.0 g of CMC per 100 mL of distilled water. Heat and dissolve, centrifuge to remove any residue if needed.

3. 500 mL DNS reagent: Add 3.15 g DNS and 10.48 g NaOH into 250 mL of distilled water. Dissolve the above ingredients, then add 91 g Rochelle salts (Na-K tartrate), 2.5 g phenol, and 2.5 g sodium metabisulfite. Add distilled water up to 500 mL.

4. Glucose standards (GS):

   Preparation of known glucose standards. For example:

   GS1–0.125 mL of 2 mg/mL glucose + 0.875 mL of 0.05 M citrate buffer.

   GS2–0.250 mL of 2 mg/mL of glucose + 0.750 mL of 0.05 M citrate buffer.

   GS3–0.330 mL of 2 mg/mL of glucose + 0.670 mL of 0.05 M citrate buffer.

   GS4–0.500 mL of 2 mg/mL of glucose + 0.500 mL of 0.05 M citrate buffer.

   GS5–1.000 mL of 2 mg/mL of glucose.

   Note: concentrations and preparation volumes can vary as appropriate.

5. Prepare enzyme dilution series/standards:

   EZ1–0.1 mL enzyme/crude enzyme extract + 0.9 mL 0.05 M citrate buffer.

   EZ2–0.1 mL of EZ1 + 0.9 mL 0.05 M citrate buffer.

2.2.2 Endoglucanase Assay Using CMC and Bicinchoninic Acid (BCA)

1. 0.05 M Citrate buffer pH 4.8: Prepare 0.05 M solutions of both 10.51 g/L citric acid and 14.71 g/L sodium citrate in double distilled water. Adjust the pH of the 0.05 M citrate solution to 4.8 with the 0.05 M citric acid solution (may require about 667 mL of citric acid solution per 1 L of sodium citrate solution). Store in refrigerator.

2. 0.05% w/v CMC solution in 0.05 M citrate buffer: Add 0.05 g CMC per 100 mL of 0.05 M citrate buffer.

3. Working BCA reagent. Mix equal volumes of following reagents A and B. The reagent should be used immediately after being made.

   (a) BCA reagent A: Dissolve 97.1 mg disodium 2,2-bicinchoninate in a solution of 2.714 g of Na_2CO_3 and 1.21 g of $NaHCO_3$ with a final volume of 50 mL. Solution A will remain stable for 4 weeks at 4 °C in darkness.

   (b) BCA Solution B: Dissolve 62.4 mg $CuSO_4 \cdot 5H_2O$ and 63.1 mg L-serine in 50 mL of water. Solution B will remain stable for 4 weeks at 4 °C in darkness.

4. Glucose standard (GS) solution 1 mL of 5 mM glucose diluted to 50 mM using citrate buffer. Prepare the sugar standards (GS) as below:

   GS1–0.4 mL of 50 mM glucose + 1.6 mL of buffer.

   GS2–0.8 mL of 50 mM glucose + 1.2 mL of buffer.

   GS3–1.2 mL of 50 mM glucose + 0.8 mL of buffer.

   GS4–1.6 mL of 50 mM glucose + 0.4 mL of buffer.

   GS5–2.0 mL of 50 mM glucose.

5. Prepare enzyme blank and substrate blank:
   Substrate blank: 1.8 mL of CMC solution + 0.2 mL of citrate buffer.

   Enzyme blank: 1.8 mL of CMC solution + 0.2 mL of dilute enzyme solutions. Treat blanks identically as the experimental samples.

2.2.3 Endoglucanase Assay Using CMC by Nelson-Somogyi Method

1. Alkaline copper Tartrate:
   Solution A: Dissolve 2.5 g anhydrous sodium carbonate, 2.0 g sodium bicarbonate, 2.5 g potassium sodium tartrate and 20.0 g anhydrous sodium sulfate in 60 mL distilled water and dilute to 100.0 mL.
   Solution B: Dissolve 15.0 g copper sulfate in 80.0 mL distilled water. Add one drop of sulfuric acid and dilute to 100.0 mL.
   Mix 4 mL of Solution B and 96 mL of solution A before use.

2. Arsenomolybdate reagent: Dissolve 2.5 g ammonium molybdate in 45 mL water. Add 2.5 mL sulfuric acid and mix well.

Then add 0.3 g disodium hydrogen arsenate dissolved in 25 mL water. Mix well and incubate at 37 °C for 24–48 h.

3. Standard glucose solution (Stock): Dissolve 100 mg glucose in 100 mL distilled water. From this stock solution, make seven working standards. For example, 10 mL of stock diluted to 100 mL with distilled water (100 µg/mL).

2.2.4 Endoglucanase Assay Using CMC by 2-Cyanoacetamide Method

1. 1% 2-cyanoacetamide (w/v): Dissolve 0.1 g of 2-cyanoacetamide in 10.0 mL of distilled water.

2. 100 mM borate (pH 9.0): Dissolve about 6.2 g of boric acid in 800 mL distilled water, adjust pH by adding 1.0 M NaOH (39.997 g/L distilled water), and then top up to 1 L. Adjust pH as needed. Heat a little to dissolve all ingredients.

2.2.5 Endoglucanase Assay Using CMC/Viscosity

1. 50 mM sodium acetate buffer pH 5.0: Mix 3 g of acetic acid and 0.5 g of sodium acetate in 1 L of distilled water. Adjust to desired pH with 1.0 M NaOH or 1.0 M HCl (*see* **Note 4**).

2. 0.05% CMC solution in acetate buffer: Add 0.05 g of CMC to 100 mL of acetate buffer.

3 Methods

3.1 Qualitative Assays

3.1.1 Endoglucanase Assay Using Agar Medium

1. Inoculate a NanoDrop (6 µL) of fungal spore or bacterial suspension (*see* **Note 5**) or a tiny amount of fungal mycelial growth onto a CMC agar plate (*see* **Note 6**), incubate for 2–3 days.

2. Stain the culture medium with the microorganism by adding 5–10 mL of Congo Red solution at room temperature for 20–30 min.

3. Rinse the residual dye on the plate using distilled water.

4. Destain Congo Red with 5–10 mL of 1 M NaCl for 20–30 min.

    (a) This will kill the bacteria, so make sure that you do not need them anymore.

    (b) If the halos are not clear, destain again using another 5–10 mL of 1 M NaCl solution.

5. Measure the clear or halo zone diameter in millimeters, and yellow halos for endoglucanase activity with the red background (*see* **Note 7**).

3.1.2 Endoglucanase Assay on Agarose Gel

1. Pour CMC agarose solution into a Petri dish to cover and let it set.

2. Make several wells at equidistance using a sterilized cork borer (*see* **Note 8**) in the solidified agarose gel, and remove the agarose particles from the wells by a sterilized forceps (*see* **Note 8**).

3. Add 0.1–0.2 mL of the enzyme solution into the holes.

4. Put the plate at room temperature for several hours or even overnight (*see* **Note 9**).

5. Wash the plate with distilled water.

6. Add 5–10 mL of the Congo Red solution and incubate at room temperature for 30 min.

7. Wash the residual dye on the plate by distilled water.

8. Destain the dye by using 5–10 mL of 1 M NaCl solution at room temperature for 20–30 min, and pour away the destain solution.

9. Measure the clear yellow halo zone in mm in diameter with the red background.

3.1.3 Endoglucanase Assay on Polyacrylamide Gel

1. Separate the protein mixtures by native PAGE or SDS-PAGE: Perform electrophoresis at room temperature at 60 V until samples migrate into resolving gel and then at 80 V until the dye front reaches toward end of the resolving gel.

2. Rinse the gel in distilled water for 5 min.

3. Soak the gel in the sodium phosphate buffer with gentle shaking for 20 min, and repeat the washing procedure three times to remove the SDS. (If using native PAGE one soak is enough because the sodium will precipitate.)

4. Transfer the gel into the CMC/phosphate buffer for 30 min.

5. Rinse the gel with distilled water.

6. Incubate the gel in 0.1 M sodium phosphate buffer at 40 °C for 1 h.

7. Stain the gel with the Congo Red solution at room temperature for 30 min.

8. Wash the gel with distilled water, and destain the gel in 1 M NaCl solution at room temperature for 30 min.

9. Detect the clear yellow halo with the red background.

3.1.4 Enzyme-Linked Immunosorbent Assay (ELISA)

1. Dilute 100 μL of 1 μg/mL antigen in 30 mM Tris–HCl pH 8.8. Incubate overnight at 4 °C.

2. Wash the plates with PBST by emptying and filling the plates three times.

3. Incubate with 200 μL of a 1% BSA solution in PBST per well for 1 h at 37 °C to block the remaining binding sites on the plastic surface.

4. Empty plates. Then add 100 μL of antiserum, hybridoma culture supernatant or another solution containing the mAb diluted in 1% BSA in PBST, incubate for 2 h at 37 °C and then wash three times with PBST.

5. To detect bound polyclonal antibodies, incubate plates for 2 h at 37 °C with 100 µL of anti-rabbit IgG-alkaline phosphatase conjugate diluted in PBST. To detect bound monoclonal antibodies, incubate the plates with anti-mouse IgG-alkaline phosphatase conjugate.

6. After three final washes with PBST, add 200 µL of a 1-mg/mL *p*-nitrophenyl phosphate solution in substrate buffer to the wells, and develop the color for 30 min at room temperature. Stop the reaction by the addition of 50 µL of 2 M sodium hydroxide, and measure the A405 nm.

7. The detection limit is defined as the concentration at the intersection of a standard dilution curve with the mean blank A405 plus 4 standard deviations. There are six blank values per plate and one standard curve.

3.2 Quantitative Assays

3.2.1 Endoglucanase Assay Using CMC by the DNS Method

1. Prepare enzyme dilution series, with the aim that at least one dilution will release slightly more than 0.5 mg of glucose and one releasing slightly less than 0.5 mg of glucose.

2. Mix 0.5 mL of CMC solution and 0.5 mL of each diluted enzyme solution (DES) from the dilution series in a test tube (s). Maintain the temperature of the solution at approximately 50 °C.

3. Incubate at 50 °C for 30 min.

4. Add 3.0 mL of DNS reagent and mix.

5. Heat in boiling water for 5 min. Then put into ice or in ice water bath to stop the reaction.

6. Add 20 mL of distilled water and mix homogenously.

7. Prepare an enzyme blank without CMC, and contains 0.1 mL enzyme solution + 0.9 mL 0.05 M citrate buffer, and substrate blank without DNS containing 0.1 mL CMC 2% w/v + 0.9 mL 0.05 M citrate buffer. These blanks should also be boiled for 5 min and then cooled.

8. Take about 200 µL from each sample plus blanks and put into the microplate/96 well plate in triplicate.

9. Read absorbance at 540 nm.

10. Subtract the absorbance values of enzyme blank from substrate blank to adjust for background and analyze the enzyme activity.

11. Draw the relationship between the glucose concentration and their respective EDRs. Figure 3 below is an example of such a graph.

12. Use standard glucose curve (standard calibration graph) for known concentrations of glucose to determine reducing sugars. Whereas the graph of glucose is used to estimate the total sugars. Graph the relationship between absorbance, glucose concentration and enzyme dilution rates (EDRs).

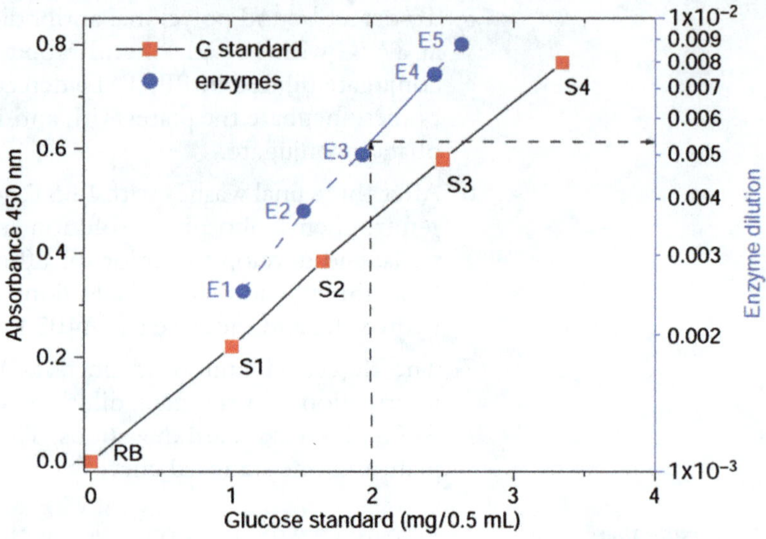

The relationship of absorbance at 540 nm for the DNS assay and EDRs in terms of glucose concentration.

Fig. 3 Standard curve for conversion of absorbance to glucose concentration [6]

13. The linear relationship between sugar released and enzyme dilution help to estimate the enzyme activity. One unit (IU) of CMCase corresponds to the amount of enzyme consuming or forming 1 μmol substrate or 1 μmol product per *min* under standard conditions. Therefore, one unit of enzyme corresponds to the release of 1 μM of glucose equivalent per minute from the substrate.

14. Calculate the CMCase activity of the original concentrated enzyme solution in terms of IU/mL: CMCase = 0.85/EDR.

3.2.2 Endoglucanase Assay Using CMC and Bicinochoninic Acid (BCA)

1. Dilute enzyme solution (e.g., 1000×) using 50 mM citrate buffer and prepare DES.

2. Add 1.8 mL of CMC solution into the test tubes.

3. Heat to 50 °C in a water bath.

4. Add 0.2 mL of diluted enzyme solution (DES) and mix well.

5. Incubate at 50 °C for 10 min.

6. Add 2 mL of working BCA reagents and mix well.

7. Incubate at 75 °C for 30 min.

8. Place samples + enzyme blank + substrate blank into 96-well plate. Remember to place samples in triplicate.

9. Read absorbance at 560 nm.

10. Calculate the enzyme activity based on a linear relationship between reducing sugar concentrations and enzyme concentrations. Refer to Fig. 3 for an example graph.

<table>
<tr><td valign="top">

3.2.3 Endoglucanase Assay Using CMC by Nelson-Somogyi Method

</td><td valign="top">

1. Dilute enzyme solution (e.g. 1000×) using 50 mM citrate buffer and prepare DES.
2. Add 1.8 mL of CMC solution into the test tubes.
3. Heat to 50 °C in a water bath.
4. Add 0.2 mL of diluted enzyme solution (DES) and mix well.
5. Incubate at 50 °C for 10 min.
6. Add 8 mL distilled water.
7. Pipette out aliquots of 0.1 or 0.2 mL to separate test tubes.
8. Pipette out 0.2, 0.4, 0.6, 0.8 and 1 mL of the working standard solution into a series of test tubes.
9. Make up the volume in both sample and standard tubes to 2 mL with distilled water.
10. Pipette out 2 mL distilled water in a separate tube to set a blank.
11. Add 1 mL of alkaline copper tartrate reagent to each tube.
12. Place the tubes in boiling water for 10 min.
13. Cool the tubes and add 1 mL of arsenomolybolic acid reagent to all the tubes.
14. Make up the volume in each tube to 10 mL with water.
15. Read the absorbance of blue color at 620 nm after 10 min.
16. From the graph drawn, Calculate the amount of reducing sugars present in the sample (*see* **Note 10** for calculation).

</td></tr>
<tr><td valign="top">

3.2.4 Endoglucanase Assay Using CMC by 2-Cyanoacetamide Method

</td><td valign="top">

1. Dilute enzyme solution (e.g., 1000×) using 50 mM sodium acetate buffer pH 5.0 and prepare DES.
2. Add 1.8 mL of CMC solution into the test tubes.
3. Heat to 50 °C in a water bath.
4. Add 0.2 mL of diluted enzyme solution (DES) and mix well.
5. Incubate at 50 °C for 10 min.
6. Hundred microliters of glucose solution is added to 1 mL of 100 mM borate pH 9.0 and 200 μL of 1% 2-cyanoacetamide.
7. The solution is mixed by vortexing.
8. Heat to boil for 10 min and cool to room temperature.
9. Measure absorbance at 276 nm.
10. Calculate the enzyme activity based on a linear relationship between reducing sugar concentrations and enzyme concentrations. Refer to Fig. 3 for an example graph.

</td></tr>
<tr><td valign="top">

3.2.5 S Viscometer (Water Flow Time of 15 s at 30 °C) or a Similar Device

</td><td valign="top">

1. Add 1.0 mL of the prewarmed DES—keep temperature constant.
2. Determine the flow rates every 5 or 10 min.

</td></tr>
</table>

3. Calculate specific viscosity ($h$sp): $Np = t - t0/to$, where t is the effluent time of the buffer (s) and $t0$ is the efflux time of the buffer (s).

4. Plot the increasing rate of the reciprocal of the specific viscosity against the enzyme concentration; a linear relation should be obtained.

5. Calculate unit of activity from the linear relationship between enzyme concentration/rate of increase of reciprocal of the viscosity of the CMC solution.

4   Notes

1. Aseptic technique is a set of principles and practices used for microbial culture preparation, and transferring culture to reduce the undesirable microbes. All the culture transfer work should be done under biosafety cabinet (laminar air flow) using sterilized materials, media, or reagents.

2. Autoclave (121 °C at 15 psi pressure for 20 min) the medium. Gently shake the flask to mix the agar homogenously throughout the mixture, and pour the cooled and melted (Temp. 45–50 °C) agar medium into sterilized petri discs. Use biosafety cabinet (laminar air flow) for preparing the agar plates as well as transfer the microbial culture to prevent contamination. Leave the plates for 10–20 min under laminar air flow until solidified the medium. The plate sizes may vary with its availability and researcher's interest.

3. When producing buffer solution there are many ways to prepare it. A general approach would be: say you needed a 50 mM buffer—start with 50 mM of an appropriate base and 50 mM of appropriate acid.

4. Transfer one loop full of 20 h bacterial or 72–120 h fungal spore culture into 10 mL sterilized distilled water, mix thoroughly, diluted to 10^6 cell or spore per mL by adding sterilized water.

5. Transfer the same diameter of fungal mycelial growth from 72 to 120 h culture on Potato Dextrose or Sabouraud agar medium for each experiment.

6. Optional: In order to increase halo contrast, add 5–10 mL of 5% acetate acid or 1 M HCl to the plate at room temperature for 10 min, and pour off. The background of the plate will turn blue.

7. Make wells in to agar plate using 4–6 mm in diameter cork borer. Use flame-sterilized cork borer before applying to make hole. Use flame sterilized forceps for removing medium from the wells.

8. Keep the plate at room temperature for several hours to diffuse the enzyme into gel.

9. Please note when you use this assay, if you which to detect all endoglucanases you will need created your own antibody using a conserved epitope.

10. Calculation.

Absorbance corresponds to 0.1 mL of test = "x" mg of glucose.

10 mL contains = "x"/0.1 × 10 mg of glucose = % of reducing sugars.

References

1. Poidevin L, Feliu J, Doan A et al (2013) Insights into exo- and endoglucanase activities of family 6 glycoside hydrolases from *Podospora anserina*. Appl Environ Microbiol 79 (14):4220–4229

2. Wang D, Kim DH, Seo N et al (2016) A novel glycoside hydrolase family 5 β-1,3-1,6 endoglucanase from *Saccharophagus degradans* 2-40T and its transglycosylase activity. Appl Environ Microbiol 82(14):4340–4349

3. Zhang YH, Lynd LR (2004) Toward an aggregated understanding of enzymatic hydrolysis of cellulose: noncomplexed cellulase systems. Biotechnol Bioeng 88(7):797–824

4. Bhaumik P, Dhepe PL (2015) Conversion of biomass into sugars. In: Murzin D, Simakova O (eds) Biomass sugars for non-fuel applications. Green Chem series. Royal Society of Chemistry, London, pp 1–53

5. Gusakov AV, Kondratyeva EG, Sinitsyn AP (2011) Comparison of two methods for assaying reducing sugars in the determination of carbohydrase activities. Int J Anal Chem 2011:1–4

6. Zhang YH, Hong J, Ye X, Mielenz JR (2009) Biofuels: methods and protocols. Molecular Biology. Humana, New York, p 581

7. Hendricks CW, Doyle JD, Hugley B (1995) A new solid medium for enumerating cellulose-utilizing bacteria in soil. Appl Environ Microbiol 61(5):2016–2019

8. Wang L, Yang Y, Cai B et al (2014) Coexpression and secretion of endoglucanase and phytase genes in *Lactobacillus reuteri*. Int J Mol Sci 15(7):12842–12860

9. Bühler R (1991) Double-antibody sandwich enzyme-linked immunosorbent assay for quantitation of endoglucanase I of *Trichoderma reesei*. Appl Environ Microbiol 57 (11):3317–3321

10. Ghose TK (1987) Measurement of cellulase activities. Pure Appl Chem 59(2):257–268

11. Marsden WL, Gray PP, Nippard GJ, Quinlan MR (1983) Evalution of the DNS method for analysing lignocellulosic hydrolysates. J Chem Tech Biotechnol 32:1016–1022

12. Somogyi M (1952) Estimation of sugars by colorimetric method. J Biol Chem 200:245

13. Krishnaveni S, Balasubramanian T, Sadasivam S (1984) Sugar distribution in sweet stalk sorghum. Food Chem 15(3):229–232

14. Jurick WM, Whitaker BD, Gaskins VL, Janisiewicz WJ (2012) Application of the 2-Cyanoacetamide method for spectrophotometric assay of cellulase enzyme activity. Plant Pathol J 11(1):38–41

15. Zhang YH, Lynd LR (2006) A functionally-based model for hydrolysis of cellulose by fungal cellulase. Biotechnol Bioeng 94:888–898

16. Dave BR, Sudhir AP, Subramanian RB (2015) Purification and properties of an endoglucanase from *Thermoascus aurantiacus*. Biotechnol Rep 6:85–90

Chapter 14

Cellobiohydrolase (CBH) Activity Assays

Hem Kanta Sharma, Wensheng Qin, and Chunbao (Charles) Xu

Abstract

Cellulosic biomass is the most abundant biopolymer on the earth. It has great potential to quench the thirst of liquid energy by producing biofuels and thus help to mitigate human reliance on fossil fuels. Although several cellulase activity assay methods have been used to disintegrate the glycosidic bonds, the appropriate selection of substrates and synergistic involvement of multiple enzymes in hydrolytic activity is not yet fully understood. The proper quantification of hydrolytic enzymes and hydrolysates is challenging because of the heterogeneity of cellulose, changes in enzyme-substrate ratio and the presence of some inhibitory compounds like cellobiose and cellodextran. In the glycosyl hydrolase (GH) family, cellobiohydrolase (CBH) is expected to disrupt the crystalline cellulose and release the sugar molecules. Several methods have been proposed for CBH assay with slight modification in substrate and quantification of hydrolysates. However, the Avicel method is still considered as the most promising and efficient hydrolytic technique so far. The most commonly used CBH assays including Avicel and other recent methods for proper quantification are outlined in this chapter. Also a qualitative screening of CBH producing bacteria using carboxymethyl cellulose (CMC) agar plates is described.

Key words Cellobiohydrolase, Cellulose, Enzyme assay, Glucose, Avicel, ELISA

1 Introduction

Cellulose is the most abundant natural biopolymer, composed of D-glucose subunits linked by β-1,4 glycosidic bonds. The crystalline nature of cellulose and the protective covering of lignin in the lignocellulosic biomass are the main hurdles for its efficient hydrolysis [1–3]. The crystallinity of cellulose can be degraded into monomeric sugar units by synergistic action of hydrolytic enzymes collectively called as cellulase. Cellulase consists of 1,4-β-endoglucanase, 1,4-β-exoglucanase or cellobiohydrolase (CBH), and 1,4-β-glucosidase, all these enzymes belong to glycosyl hydrolase (GH) family [4]. Among the 128 GH families, the CBH can be found in GH families 5, 6, 7, 9, 48, and 74 [5]. Two major types of cellobiohydrolase are CBHI and CBHII, which effectively degrade the crystalline cellulose, presumably by peeling the microcrystalline structure of cellulose chain, whereas endoglucanase

Mette Lübeck (ed.), *Cellulases: Methods and Protocols*, Methods in Molecular Biology, vol. 1796, https://doi.org/10.1007/978-1-4939-7877-9_14, © Springer Science+Business Media, LLC, part of Springer Nature 2018

typically acts on more soluble amorphous region of cellulose, showing high degree of synergism and thus releasing the sugar molecules [5, 6]. There has been increasing interest in the hydrolysis of cellulose for sustainable and renewable biofuels production, a promising alternative to fossil fuels. Various bacteria and fungi are known to secrete endo or exo-acting cellulases that act on cellulose, resulting in the release of glucose and cellobiose. There have been extensive studies into the cellulolytic system of *Trichoderma reesei*, which is composed of two cellobiohydrolases (CBHI and CBHII) [7, 8]. Miettinen-Oinonen and Suominen [7] quantified the CBHI and CBHII in *T. reesei* by double-antibody sandwich enzyme-linked immunosorbent assay (ELISA). However, Brook et al. [9] isolated three cellobiohydrolases named CBHI-A, CBHI-B, and CBHII from crude extracts of *Talaromyces emersonii* liquid cultures. Recently, four cellobiohydrolase I enzymes named as CBHI-A, CBHI-B, CBHI-C, and CBHI-D have been purified from the growth of *Penicillium decumbens* JU-A10. The enzyme activity was tested against p-nitrophenyl-β-D-cellobioside (pNPC) [10]. The cellodextrin and cellobiose have their inhibitory activities during cellulose hydrolysis, thus β-glucosidase is essential to break the final glycosidic bonds of cellobiose so as to produce sufficient glucose molecules and reduce or eliminate cellobiose inhibition [11, 12]. The CBH assay is more difficult than endoglucanase and β-glucosidase assays due to lack of proper substrates and hindrances from cellulase components [13]. Although there is no such single standard assay method for CBH activity, the Avicel method [12, 13] has been repeatedly used. Some other substrates and quantitative methodologies are also described in this chapter along with a qualitative screening method.

1.1 Qualitative Assay

Qualitative screening of CBH producing bacteria can be carried out by comparing the relative diameter of cellulase hydrolytic activity using carboxymethyl cellulose (CMC) agar plates.

1.2 Quantitative Assays

1.2.1 CBH Assay Using Avicel as a Substrate

Avicel is a microcrystalline substrate and in comparison, to carboxymethyl cellulose (CMC), it is easily hydrolyzed by CBH showing high enzyme activity. Thus, Avicel has been routinely used as a suitable substrate in CBH assays. Most of the CBHs release cellobiose and a small amount of glucose from the cellulose. The reducing sugars produced after hydrolysis can be estimated by dinitrosalicylic acid (DNS) and Nelson Somogyi method whereas the total sugar is estimated by phenol-sulfuric acid method [14]. Centrifugation (14,000 $\times$ g for 3 min) of host culture is a simple step to collect the crude enzyme as a supernatant. The appropriate dilution of enzyme may be required.

1.2.2   CBH Assay Using Regenerated Amorphous Cellulose as a Substrate

Native cellulose has crystalline structure, comparatively more stable than regenerated cellulose due to network of intramolecular and intermolecular hydrogen bonds. Regenerated amorphous cellulose (RAC) is made from the chemical modification of microcrystalline cellulose or Avicel. Several solvents such as SO_2-amine solvent system using SO_2, diethylamine, and dimethylsulfoxide [15, 16] and acidic treatment using sulfuric acid [17], phosphoric acid [18–21] are being used for preparation of RAC. The treatment on cellulose crystals is used to disrupts the hydrogen bonds and thus provide larger (about 20-fold) surface area for enzymatic hydrolysis [20, 21].

1.2.3   CBH Assay Using p-Hydroxybenzoic Acid Hydrazide Method

The cellobiohydrolase assay on different polysaccharides were measured by Takahashi et al. [22], following the *p*-hydroxybenzoic acid hydrazide (PAHBAH) method discovered by Lever, M [23]. The activity of *Magnaporthe oryzae* GH-6 family cellobiohydrolase (MoCel6A) catalyzed the hydrolysis of amorphous and water-soluble polysaccharides. Since the acid hydrazides react with reducing sugars in alkaline solution, the PAHBAH can be used to detect less than 1 μg of sugars [22, 23]. There are different choices for polysaccharide selection. It can be cellulose, Avicel, CMC, hydroxyethyl cellulose, or other polysaccharides. However, very high concentration of protein and calcium are the impeding factors [23], and a strong agitation is also required if water-insoluble polysaccharides are selected as substrates [22].

1.2.4   CBH Assay Using Lytic Polysaccharide Monooxygenase

Lytic polysaccharide monooxygenase (LPMO) is a relatively newly developed method for depolymerization of recalcitrant polysaccharide chains in their crystalline regions based on the principle of oxidative disintegration so as to release oxidized oligosaccharides [24, 25]. The LPMO was initially discovered for its activity on chitin degradation [2, 25] however it also degrades cellulose [2]. The LPMO belongs to the auxiliary activities (AA) enzyme class. It is a copper-dependent monooxygenase [26, 27] and works slowly in association with hydrolytic enzymes. LPMO can strongly boost up the saccharification process and enhance the soluble sugar yield from lignocellulosic biomass [28].

1.2.5   CBH Assay Using Double-Antibody Sandwich Enzyme-Linked Immunosorbent Assay

Enzyme-linked immunosorbent assay (ELISA) technique is aimed at detecting and quantifying substances such as antigen, antibodies, peptides, proteins, and hormones. The quantification of CBHI in crude cellulase enzyme can be done by using double-antibody sandwich enzyme-linked immunosorbent assay. At least two antibodies (either monoclonal or polyclonal antibodies) are required to act in the sandwich. The polyclonal antibodies (PAb) have been used in characterization of CBHI and CBHII. This is a highly specific and direct procedure, used to quantify the CBHI at its range of 0.1–0.8 μg/mL [29].

2 Materials

Use analytical grade chemicals, sterilized/autoclaved instruments and protective wearing (gloves/protective clothing/eye protection/face protection) throughout all the experiments. Use laminar air flow/biosafety cabinet during any microbial handling and transfer with high level of caution to prevent possible contamination. Prepare all reagents in autoclaved distilled water and store at room temperature unless otherwise indicated. Autoclave (at 121 °C for 30 min) all the contaminated growth medium and other used materials such as petri discs, centrifuge tubes, pipette tips, glass wares, gloves etc. before its effective disposal following the regulation.

2.1 Qualitative Assay

1. Gram's iodine solution (300 mL): Take 300 mL distilled water into a 500-mL glass bottle. Add 2 g potassium iodide and 1 g iodine (*see* **Note 1**) into it by continuous steering with magnetic bar.

2. LB broth (100 mL): Take 100 mL distilled water into a 250-mL conical flask. Add 1 g peptone, 0.5 g yeast extract, and 0.5 g NaCl into the flask with continuous steering (*see* **Note 2**).

3. CMC Agar plate (100 mL): Take 100 mL distilled water into a 250-mL conical flask. Add 0.5 g CMC, 0.1 g $NaNO_3$, 0.1 g K_2HPO_4, 0.1 g KCl, 0.05 g $MgSO_4$, 0.05 g yeast extract, and 1.5 g agar into the flask with continuous steering. Autoclave the mixture before use to make agar plate (*see* **Note 3**).

2.2 Quantitative Assays

2.2.1 CBH Assay Using Avicel as a Substrate

1. Avicel suspension in sodium acetate buffer (pH 4.8): Prepare 0.1 M sodium acetate buffer pH 4.8 (*see* **Note 4**). Add 1.25% (w/v) Avicel and mixed well to make the Avicel suspension in sodium acetate buffer.

2. Enzyme dilution: Dilute a series of enzyme solutions (one of which release less than and another release more than 0.5 mg of glucose) by using 0.1 M sodium acetate buffer (optional) or collect the crude enzyme from supernatant (*see* **Note 5**) of bacterial culture.

3. Reaction mixture (2 mL): Prepare the reaction mixture by adding 1.6 mL of Avicel suspension in 0.1 M sodium acetate buffer (pH 4.8) and 0.4 mL of diluted enzyme or crude enzyme samples in a 5 mL glass test tube.

4. Other reagents: Avicel, 5% phenol solution (*see* **Note 6**), ~98% sulfuric acid (*see* **Note 7**), 0.1 M sodium acetate buffer (pH 4.8).

2.2.2 CBH Assay Using RAC as a Substrate

1. Reaction mixture (1 mL): Prepare a reaction mixture containing 0.5 mL of 1% (w/v) RAC (*see* preparation of RAC in Subheading 3.2.2), 0.05 mL of 1 M citrate buffer pH 4.5 (*see* **Note 8**), 0.25 mL of distilled water, 0.2 mL of diluted enzyme solution or crude enzyme from host culture.

2. Enzyme dilution (optional): Dilute a series of enzyme solutions by using 50 mM acetate buffer.

3. Other reagents: Regenerated amorphous cellulose (RAC) 1% (w/v), phenol solution (5%), sulfuric acid (~98%), phosphoric acid solution (86%), 2 M sodium carbonate, 0.2% (w/v) sodium azide solution (*see* **Note 9**).

2.2.3 CBH Assay Using PAHBAH

1. Reaction mixture (100 μL) composition: Prepare a reaction mixture of 100 μL final volume by adding 100 mM sodium phosphate pH 6.0, 0.1 μg of CBH, and 0.5–5 mg of polysaccharide (*see* **Note 10**) in distilled water.

2. 1% w/v of *p*-hydroxybenzoic acid hydrazide (PAHBAH).

2.2.4 CBH Assay Using LPMO

1. Cellulosic substrate: Prepare the substrate containing 0.1% (w/v) cellulose in 50 mM sodium phosphate buffer pH 6.0 (*see* **Note 11**).

2. Other reagents: Purified LPMO (9 μg LPMO/mg of cellulosic substrate), 0.1% (w/v) Avicel or microcrystalline cellulose, 0.1% (w/v) amorphous cellulose (or, follow the RAC preparation Subheading 3.2.2), 7.5 μM ascorbic acid, hydrolytic enzymes (see below).

3. Hydrolytic enzymes (μg of protein/mg cellulosic substrate):

    (a) Complete cellulase: 25 μg cellulase/mg of cellulosic substrate.

    (b) CBHI: 100 μg CBHI/mg of cellulosic substrate.

    (c) CBHII: 100 μg CBHII/mg of cellulosic substrate.

    (d) Endoglucanase: 100 μg endoglucanase/mg of cellulosic substrate.

    (e) β-glucosidase: 5 μg β-glucosidase/mg of cellulosic substrate.

    (f) The crude enzymes extracted from host culture can also be used.

2.2.5 CBH Assay Using ELISA

1. Immunoglobulin (IgG): Rabbit anti-mouse IgG or goat anti-rabbit IgG.

2. Other reagents: 1 μg/mL polyclonal antibodies (PAb), 5–10 μg/mL monoclonal antibodies (MAb), 50 mM carbonate buffer pH 9.5 (*see* **Note 12**), 0.1 M phosphate-buffered saline (PBS) pH 7.4 (*see* **Note 13**), CBHI, 1× PBS containing

0.1% Tween 20 (1× PBS Tween 20) (*see* **Note 13**), 2% bovine serum albumin, 2,2′-azino-bis-3-ethylbenzthiazoline-6-sulfonic acid (ABTS).

---

3   Methods

3.1 Qualitative Assay of CBH

The bacterial strain grown overnight in 1.5 mL Luria–Bertani liquid nutrient medium (LB broth) at 30 °C could be a better choice to inoculate in CMC agar plates, however the slower growing strains may require 2–3 days of incubation in plate along with positive (*Cellulomonas xylanilyticus*) and negative controls (*E. coli*) to measure and compare the clear zone or halo size, representing the cellulase activity [30, 31]. The qualitative assay can be done with following procedures.

1. Transfer 1.5 mL of LB broth (culture medium) into a 5 mL test tube.

2. Inoculate a pure single strain bacterial colony into a test tube with LB broth (*see* **Note 14**).

3. Keep the tube in shaking incubator at 30 °C and 200 rpm for 24 h (*see* **Note 15**).

4. After incubation, inoculate 5 μL of bacterial culture on center of the CMC agar plate and incubate at 30 °C for 48 h.

5. Gently pour the Gram's iodine solution on the CMC agar plates until the plates are covered. Observe the halo or zone size of clearance after 1 min.

6. Measure the diameter of halo regions to judge the hydrolytic ability. The larger the size of the halo the higher the hydrolytic activity of bacterial strains.

3.2 Quantitative Assay of CBH

Following are some of the widely used quantitative methods for CBH assay.

1. CBH assay using Avicel as substrate [12–14, 22].

2. CBH assay using amorphous cellulose as substrate [13].

3. CBH assay using *p*-hydroxybenzoic acid hydrazide method [22].

4. CBH assay using lytic polysaccharide monooxygenase (LPMO) for oxidative degradation of cellulosic substrate [2, 24].

5. CBH assay using double-antibody sandwich enzyme-linked immunosorbent assay (ELISA) [29].

3.2.1 CBH Assay Using Avicel as a Substrate

1. Transfer 2 mL of reaction mixture (*see* Subheading 2.2.1) into the microcentrifuge tubes.

2. Incubate the reaction mixture at 50 °C for 2 h.

3. Quench the reaction by placing the tubes in ice-cold water for 10 min.

4. Centrifuge the mixture at $13,000 \times g$ for 3 min and collect the supernatant.

5. Analyze the supernatant by DNS method (see below) or Nelson Somogyi method for reducing sugars [14, 32]; and phenol–sulfuric acid method for total sugar [13] (make sure to have triplicates of each sample).

6. Prepare the blanks:

    (a) Enzyme blank: 0.4 mL of diluted enzymes or crude enzyme + 1.6 mL of 0.1 M acetate buffer.

    (b) Substrate blank: 0.4 mL of 0.1 M acetate buffer + 1.6 mL of 1.25% (w/v) Avicel suspension buffer.

Procedure for DNS method: The following procedures for DNS method consist of larger volume of sample solution and other chemicals. However, it can be scaled down by using minimal volume (*see* **Note 16**).

1. Take 1 mL of supernatant in a 25 mL test tube.

2. Add 3 mL of DNS solution (*see* **Note 17**) into the tube and mix well by gently vortex for 3–5 s.

3. Put the tube in boiling water bath for 5 min.

4. Remove the tube from hot water bath and cool down the tube by keeping it in ice-cold water for 10 min.

5. Add 20 mL of distilled water into the tube and mix it properly.

6. Measure the absorbance at 540 nm in spectrophotometer with three replicates for each sample.

7. Subtract the absorbance values of enzyme blank from substrate blank and analyze the enzyme activity.

8. The standard calibration graph for known amount of cellobiose is used to determine reducing sugars, whereas the graph of glucose is used to estimate the total sugars.

9. The linear relationship between sugar released and enzyme dilution helps to estimate the enzyme activity. One unit of CBH corresponds to the release of 1 µM of glucose equivalent per minute from Avicel.

3.2.2 CBH Assay Using RAC as a Substrate

Preparation of RAC [13, 20].

1. Add 0.2 g of microcrystalline cellulose or Avicel and 0.6 mL distilled water into a regular 50 mL centrifuge tube (cellulose slurry will form).

2. Gently pour 10 mL of ice-cold (86%) phosphoric acid into the tube with continuous stirring so that the cellulose suspension solution will thoroughly mixed and turns transparent.

3. Add another 40 mL of ice-cold water (10 mL per addition for four times) with vigorous stirring produce a white cloudy precipitate.

4. Centrifuge the precipitated cellulose at $5000 \times g$ for 1 min and keep it in water bath at 4 °C for 20 min.

5. Discard the supernatant and keep the pellet for further wash with ice-cold water followed by centrifugation.

6. Suspend the cellulose pellet by adding 45 mL of ice-cold water in the tube and centrifuge at $5000 \times g$ for 1 min to remove the supernatant containing phosphoric acid (repeat this step four times).

7. Add 0.5 mL of 2 M Na_2CO_3 to neutralize the residual phosphoric acid.

8. Suspend the cellulose pellet in 45 mL of ice-cold distilled water and centrifuge at $5000 \times g$ for 1 min (repeat this step two times).

9. Collect the cellulose pellet and make 1% (w/v) of RAC suspension ready to use and store (*see* **Note 18**).

Assay procedures

1. Take 1 mL of reaction mixture (*see* Subheading 2.2.2) in 1 mL centrifuge tubes.

2. Incubate the reaction mixture at 50 °C for 30 min.

3. Quench the reaction by submerging the tubes in ice-cooled water bath.

4. Centrifuge the mixture at $10,000 \times g$ for 3 min and the collect the supernatant.

5. Prepare the blanks:

   (a) Enzyme blank: 0.2 mL of diluted enzymes or crude enzyme + 0.05 mL of 1 M citrate buffer + 0.75 mL of distilled water.

   (b) Substrate blank: 0.5 mL of 1 M citrate buffer + 0.5 mL of 1% (w/v) RAC + 0.45 mL of distilled water.

6. Analyze the total soluble sugars in the supernatants by the phenol–sulfuric acid method (*see* below) at 490 nm absorbance.

7. Subtract the absorbance values of enzyme blank from substrate blank and analyze the enzyme activity.

Procedure for phenol–sulfuric acid method

1. Transfer 0.7 mL of supernatant into a 5 mL glass tubes.

2. Add 0.7 mL of 5% phenol solution and 3.5 mL of sulfuric acid with continuous mixing (*see* **Note 19**).

3. Allow 20–30 min time to cool down the exothermic reaction to room temperature and measure the absorbance at 490 nm.

4. Use the standard calibration graph of known amount of glucose to determine the total sugars.

5. A linear range gives 0.005–0.1 g/L sugars in the samples. It helps to analyze the enzyme activity which is one unit of exoglucanase or CBH corresponds to the release of 1 μmol of glucose equivalent per minute from Avicel.

3.2.3 CBH Assay Using PAHBAH Method

1. Transfer 100 μL of reaction mixture (*see* Subheading 2.2.3) into 1 mL centrifuge tubes.

2. Incubate the tubes at 30 °C for up to 18 h.

3. Centrifuge the tubes at 22,000 × *g* for 5 min.

4. Take 50 μL of supernatants into microcentrifuge tubes and mix with 150 μL of PAHBAH solution (1% w/v).

5. Keep the microcentrifuge tubes in a boiling water bath for 5 min.

6. Remove the tubes from water bath and wait till cooled down to room temperature.

7. Measure the absorbance at 410 nm in spectrophotometer.

8. Use standard calibration graph of glucose to estimate the reducing sugars.

9. The linear relationship between sugar released and enzyme dilution help to estimate the enzyme activity.

10. The liquid chromatography-mass spectrometry (LC-MS) or high pressure liquid chromatography-mass spectrometry (HPLC-MS) of the hydrolysates is useful in product analysis.

3.2.4 CBH Assay Using LPMO

Method I (involving pretreatment of substrate)

1. For the pretreatment in the initial step, transfer 450 μL of cellulosic substrate into Eppendorf tubes (make triplicate of each sample).

2. Add LPMO (9 μg LPMO/mg of cellulosic substrate) into the tubes.

3. Cover each tube with oxygen-permeable parafilm and incubate at 25 °C for 5 h.

4. After 5 h of pretreatment, add 50 μL of one of the hydrolytic enzyme (from the list of hydrolytic enzymes, Subheading 2.2.4). Again, incubate the tube at 50 °C for 12 h.

5. Remove one of reaction tube (for sampling) from incubation in every 15 min for up to 1 h (total four tubes).

6. Stop the enzyme action by keeping the tubes at 95 °C for 10 min.

7. Lower the temperature to 4 °C by keeping in cold (4 °C) water

8. Centrifuge the tubes at 9000 × g for 1 min.

9. Collect the supernatant for final product analysis using high performance anion exchange chromatography (HPAEC).

10. For control (*see* **Note 20**).

Method II (without pretreatment of substrate)

1. Transfer 500 μL of cellulosic substrate plus LPMO, CBHI and β-Glucosidase into the Eppendorf tubes (*see* **Note 21**)

2. Cover the tubes with oxygen-permeable parafilm.

3. Incubate the tubes at 50 °C for up to 96 h.

4. Remove one reaction tube (for sampling) from incubation in 5 h and another tube in 96 h for product analysis.

5. Stop the enzyme reaction by keeping tubes at 95 °C for 10 min.

6. Lower the temperature to 4 °C by keeping the tubes in 4 °C cold water.

7. Centrifuge the tubes at 9000 × g for 1 min and collect the supernatant.

8. Use the supernatant for final product analysis by using HPAEC.

9. For control (*see* **Note 20**).

<table>
<tr><td>

3.2.5 CBH Assay Using ELISA

</td><td>

1. Pour 100 μL of PAb (1 μg/mL) in 50 mM carbonate buffer (pH 9.5) or purified MAb (5–10 μg/mL) in 1× PBS for coating into the ELISA plates wells.

2. Incubate the plates at 4 °C for 12 h.

3. Wash the wells with 1× PBS Tween-20.

4. Block the nonspecific spots with 2% bovine serum albumin at 37 °C for 2 h.

5. Add 100 μL of CBHI or diluted crude cellulase (from host culture) into the wells.

6. Incubate the plates at 37 °C for 1 h.

7. Wash the wells with 1× PBS Tween 20 and then add the second antibody. It may be MAb or PAb (*see* **Note 22**).

8. Incubate the wells at 37 °C for 30 min to 1 h (*see* **Note 23**).

9. If unlabeled antibody is used in second time, then a peroxidase-labeled third antibody is used in detection. It may include either rabbit anti-mouse or goat anti-rabbit immunoglobulin (IgG) and incubate at 37 °C for 30 min.

</td></tr>
</table>

10. Wash the plate with 1× PBS Tween-20 and add 100 μL of ABTS substrate into the wells.

11. Keep the plates in room temperature for 30 min to observe the color.

12. Measure the absorbance values in microplate reader at 410 nm.

13. Analyze the absorbance values by plotting standard curve against the concentration of CBHI standards.

3.3 Optimization of Cellobiohydrolase Assay

Higher yield of cellobiohydrolase is feasible only after its optimization. Generally, optimization is to analyze the hydrolytic activities of enzyme at various temperatures, pH, agitation speed, incubation time, the source of carbon/polysaccharides, source of nitrogen, and other inorganic salts' concentration. Replication of reaction mixtures for various substrates is required for the optimization. It may vary with different methods, however generally the following "one at a time" factors are taken into consideration.

1. The optimization of temperature for hydrolysis can be analyzed by incubating the reaction mixture at different temperatures (20–60 °C).

2. The optimization of incubation time for hydrolysis can be analyzed by varying the time of incubation in hot water bath. It may range from 1 to 25 h.

3. The optimization of pH on the hydrolytic activity of CBH can be measured by equilibrating the reaction mixture at different pH (ranges from pH 3.5 to 11.0), for example, with sodium acetate (pH 3.5–5.5), sodium phosphate (pH 5.5–7.5), Tris–HCl (pH 7.5–9.0), or CAPS (N-cyclohexyl-3-aminopropanesulfonic acid; pH 9.0–11.0) [22].

4. Determination of the optimum agitation speed for hydrolysis can be measured by incubating in shaking incubator at 120–220 rpm.

5. The optimum CBH production can be analysed by using various sources of carbon (such as starch, CMC, Avicel, RAC, xylose, and glucose), nitrogen (such as yeast extract, peptone, urea, and ammonium sulfate) and metal ions (such as Ca^{2+}, Co^{2+}, Mn^{2+}, Mg^{2+}, and Zn^{2+}).

3.4 Concluding Remarks

There are several methods developed for cellulase assays. The CBH plays a significant role in production of cellobiose and some other sugars from crystalline cellulose. However, the specific substrate and suitable CBH enzyme assay protocols need to be further developed. Recent CBH assays are slightly varied on the types of substrate used and detection methods for hydrolyzed products. Some of the direct methods like high pressure liquid chromatography (HPLC) and cellobiose oxidoreductase assay are complicated due to production of cellobiose during CBH activities [29]. The ELISA is gaining its popularity in better quantification of CBH. However, Avicel and RAC are widely used for CBH enzyme assay because these can act as a substrate for exoglucanases and endoglucanases due to some amorphous cellulose and soluble cellodextrans [14, 33].

4   Notes

1. Iodine takes a long time to dissolve in distilled water. So, continuous 1–2 h of stirring may require. Keep in mind that iodine also has volatile tendency and should be kept in air tight bottle and better to cover with aluminum foil.

2. Autoclave (121 °C for 30 min) the LB broth before use and it can be stored at 4 °C for few days. LB broth is a good nutrient medium for bacterial growth. So, precaution should be made while transferring the bacterial strain to avoid the contamination. Use biosafety cabinet during all bacterial inoculation.

3. Autoclave (121 °C for 30 min) the agar mixture. Gently shake the flask and pour the agar mixture into the sterilized/disposable plastic petri discs. Use biosafety cabinet to make the agar plates. Leave the plate for 20 min in the cabinet to become solidified. The agar will be solidified at room temperature. The plate sizes can vary with its availability and researcher's interest.

4. Take 0.82 g of sodium acetate in a beaker and add distilled water till up to 100 mL. Adjust the desired pH of sodium acetate buffer by using 5 N NaOH and 1 N HCl.

5. Take 1 mL of overnight cultured broth medium in a 1.5 mL microcentrifuge tube. Centrifuge the tube at $12,000 \times g$ for 3 min and collect the crude enzyme from supernatant.

6. Phenol is a strongly corrosive and combustible solid. It can severely affect the skin, eyes, and mucosal membrane. Use personal protective wear/equipment while handling. Since it is also toxic to plants and animals, do not release into environment and always follow the local, regional, and or national laws and regulations for disposal. Store in a cool, dry, and well-ventilated room.

7. The 98% sulfuric acid is highly corrosive in nature. It can cause serious damage to the skin upon contact by chemical burning and dehydration, can lead to blindness if splashed onto eyes. Use of protective wearing is highly recommended.

8. Dissolve 19.21 g of citric acid in a final volume of 100 mL distilled water. Adjust the pH 4.5 by dissolving solid NaOH.

9. Sodium azide is soluble in water, and very toxic (comparable to alkali cyanides) and has a severe poisonous effect. Even an exposure to small quantities through skin contact or swallowing may be fatal. The lethal dose for an adult human is about 0.7 g.

10. The amount of polysaccharide depends upon its types, such as 5 mg for Avicel, 5 mg for cellulose, or 0.5 mg for other polysaccharides.

11. Prepare the 50 mM sodium phosphate buffer pH 6.0 by dissolving 0.599 g of monosodium phosphate to the final volume of 100 mL distilled water. Adjust the pH by using NaOH and phosphoric acid.

12. 50 mM carbonate buffer pH 9.5 by dissolving 1.59 g of Na_2CO_3 and 2.93 g of $NaHCO_3$ to a final volume of 1000 mL deionized water. Adjust the pH 9.5.

13. Prepare 0.1 M phosphate-buffered saline (PBS) pH 7.4 by using 10.9 g anhydrous Na_2HPO_4, 3.2 g anhydrous NaH_2PO_4, and 90 g NaCl in a final volume of 1000 mL deionized water. Make sure to adjust the pH 7.4. Add 1 mL of Tween 20 to make $1\times$ PBS Tween-20.

14. A separate culture of positive control (*Cellulomonas xylanilyticus*) and negative control (*E. coli*) is highly desirable for comparative qualitative assay with other bacterial strains.

15. The hours of incubation vary according to bacterial growth rate. The slower growing bacterial strain may require an additional 24–48 h of incubation to get comparatively equivalent biomass growth as that of faster growing strains.

16. The DNS method can be scale down with minimal volume of enzymes and reagents. Say for example 50 or 100 µL of diluted enzyme or crude enzyme from supernatant and 100 or 200 µL of DNS can use in 1 mL microcentrifuge tube for incubation. However, make sure to use equivalent proportion of volume and concentration of reagents during experiment and estimation/calculation of enzyme activity.

17. Preparation of DNS reagent (500 mL): Take a 250 mL conical flask. Put 3.15 g of DNS and 10.48 g of sodium hydroxide into the flask containing 250 mL of distilled water and make a homogeneous mixture with the help of magnetic steering. Take another 250 mL conical flask. Put 91 g sodium-potassium tartrate, 2.5 g phenol and 2.5 g sodium-metabisulfite into the flask containing 250 mL of distilled water and make a homogeneous mixture with the help of magnetic steering. Gently mix the reagent solution (from first and second flask) into a separate 500 mL glass bottle, wrap it with aluminum foil and store at 4 °C.

18. RAC suspension can be stored for about 1 year at 4 °C when mixed with 0.2% (w/v) sodium azide solution (*see* **Note 8**).

19. The phenol and sulfuric acid reaction is highly exothermic. So be cautious and use all protective wearing while handling.

20. The negative control does not contain LPMO or contains LPMO lacking ascorbic acid.

21. Make triplicate of sample and put all three enzymes in each tube.

22. Use of antibodies MAb or PAb depends on which antibody was used as the coating antibody. If PAb is used first as coating antibody then MAb would serve as the second antibody and vice versa.

23. Incubation time depends on the types of second antibody. Incubation time for labeled antibody (horseradish peroxidase) is 1 h, whereas if unlabeled antibody is used the incubation time is 30 min.

References

1. Chen H, Zhao X, Liu D (2016) Relative significance of the negative impacts of hemicelluloses on enzymatic cellulose hydrolysis is dependent on lignin content: evidence from substrate structural features and protein adsorption. ACS Sustain Chem Eng 4(12):6668–6679

2. Horn S, Vaaje-Kolstad G, Westereng B, Eijsink VG (2012) Novel enzymes for the degradation of cellulose. Biotechnol Biofuels 5(1):45

3. Zhang Y, Ding SY, Mielenz JR et al (2007) Fractionating recalcitrant lignocellulose at modest reaction conditions. Biotechnol Bioeng 97(2):214–223

4. Henrissat B, Davies G (1997) Structural and sequence-based classification of glycoside hydrolases. Curr Opin Struct Biol 7(5):637–644

5. Annamalai N, Rajeswari MV, Sivakumar N (2016) Cellobiohydrolases: role, mechanism, and recent developments. In: Vijai Kumar G (ed) Microbial enzymes in bioconversions of biomass. Springer, New York, pp 29–35

6. Divne C, Ståhlberg J, Teeri TT, Jones TA (1998) High-resolution crystal structures reveal how a cellulose chain is bound in the 50 Å long tunnel of cellobiohydrolase I from *Trichoderma reesei*. J Mol Biol 275(2):309–325

7. Miettinen-Oinonen A, Suominen P (2002) Enhanced production of *Trichoderma reesei* endoglucanases and use of the new cellulase preparations in producing the stonewashed effect on denim fabric. Appl Environ Microbiol 68(8):3956–3964

8. Geng A (2014) Genetic transformation and engineering of *Trichoderma reesei* for enhanced enzyme production. In: Geng A (ed) Biotechnology and biology of Trichoderma. Elsevier, Amsterdam, pp 193–200

9. Brooks MM, Tuohy MG, Savage AV et al (1992) The stereochemical course of reactions catalysed by the cellobiohydrolases produced by *Talaromyces emersonii*. Biochem J 283:31–34

10. Gao L, Gao F, Wang L et al (2012) N-glycoform diversity of cellobiohydrolase I from *Penicillium decumbens* and synergism of nonhydrolytic glycoform in cellulose degradation. J Biol Chem 287(19):15906–15915

11. Maki M, Leung K, Qin W (2009) The prospects of cellulase-producing bacteria for the bioconversion of lignocellulosic biomass. Int J Biol Sci 5(5):500–516

12. Dashtban M, Maki M, Leung KT et al (2010) Cellulase activities in biomass conversion: measurement methods and comparison. Crit Rev Biotechnol 30(4):302–309

13. Zhang YHP, Hong J, Ye X (2009) Cellulase assays. In: Mielenz JR (ed) Biofuels: methods and protocols. Springer, New York, pp 213–231

14. Wood TM, Bhat KM (1988) Methods for measuring cellulase activities. Methods Enzymol 160:87–112

15. Ciolacu D, Ciolacu F, Popa VI (2011) Amorphous cellulose – structure and characterization. Cellul Chem Technol 45(12):13–21

16. Isogai A, Atalla R (1991) Amorphous celluloses stable in aqueous media: regeneration from SO_2-amine solvent systems. J Polym Sci Part A Polym Chem 29(1):113–119

17. Ioelovich M (2012) Study of cellulose interaction with concentrated solutions of sulfuric acid. ISRN Chem Eng 2012:428974

18. Zhang Y, Cui J, Lynd L, Kuang L (2006) A transition from cellulose swelling to cellulose dissolution by o-phosphoric acid: evidence from enzymatic hydrolysis and supramolecular structure. Biomacromolecules 7(2):644–648

19. Hao X, Shen W, Chen Z et al (2015) Self-assembled nanostructured cellulose prepared by a dissolution and regeneration process using phosphoric acid as a solvent. Carbohydr Polym 123:297–304

20. Hong J, Wang Y, Ye X, Zhang YHP (2008) Simple protein purification through affinity adsorption on regenerated amorphous cellulose followed by intein self-cleavage. J Chromatogr A 1194(2):150–154

21. Jia X, Chen Y, Shi C et al (2013) Preparation and characterization of cellulose regenerated from phosphoric acid. J Agric Food Chem 61 (50):12405–12414

22. Takahashi M, Takahashi H, Nakano Y et al (2010) Characterization of a cellobiohydrolase (MoCel6A) produced by *Magnaporthe oryzae*. Appl Environ Microbiol 76(19):6583–6590

23. Lever M (1972) A new reaction for colorimetric determination of carbohydrates. Anal Biochem 47(1):273–279

24. Eibinger M, Ganner T, Bubner P et al (2014) Cellulose surface degradation by a lytic polysaccharide monooxygenase and its effect on cellulase hydrolytic efficiency. J Biol Chem 289(52):35929–35938

25. Vaaje-Kolstad G, Westereng B, Horn SJ et al (2010) An oxidative enzyme boosting the enzymatic conversion of recalcitrant polysaccharides. Science 80(330):6001

26. Hemsworth GR, Taylor EJ, Kim RQ et al (2013) The copper active site of CBM33 polysaccharide oxygenases. J Am Chem Soc 135 (16):6069–6077

27. Aachmann FL, Sørlie M, Skjåk-Bræk G et al (2012) NMR structure of a lytic polysaccharide monooxygenase provides insight into copper binding, protein dynamics, and substrate interactions. Proc Natl Acad Sci U S A 109 (46):18779–18784

28. Patel I, Kracher D, Su M et al (2016) Salt-responsive lytic polysaccharide monooxygenases from the mangrove fungus *Pestalotiopsis* sp. NCI6. Biotechnol Biofuels 9(1):108

29. Riske FJ, Eveleigh DE, Macmillan JD (1990) Double-antibody sandwich enzyme-linked immunosorbent assay for cellobiohydrolase I. Appl Environ Microbiol 56(11):3261–3265

30. Paudel YP, Qin W (2015) Characterization of novel cellulase-producing bacteria isolated from rotting wood samples. Appl Biochem Biotechnol 177(5):1186–1198

31. Maki M, Broere M, Leung K, Qin W (2011) Characterization of some efficient cellulase producing bacteria isolated from paper mill sludges and organic fertilizers. Int J Biochem Mol Biol 2(2):146–154

32. Miller GL (1959) Use of dinitrosalicylic acid reagent for determination of reducing sugar. Anal Chem 31(3):426–428

33. Ozioko PC, Ikeyi Adachukwu IP, Ugwu OPC (2013) Review article cellulases their substrates activity and assay methods. Experiment 12 (2):778–785

Two-Dimensional High-Throughput *Endo*-Enzyme Screening Assays Based on Chromogenic Polysaccharide Hydrogel and Complex Biomass Substrates

Julia Schückel and Stjepan Krešimir Kračun

Abstract

In this chapter, we present a two-dimensional approach for high-throughput screening of *endo*-cellulases as well as other *endo*-acting enzymes. The method is based on chromogenic substrates, produced either from purified or complex material, providing valuable information about enzyme activity toward its target as well as that same target in a context of complex natural material normally encountered in bioindustrial settings. The enzymes that can be tested using this assay can be from virtually any source: in purified form, directly from microbial cultures or even from raw materials, enabling study of the interplay between enzyme mixtures such as synergistic or inhibitory effects. By using the method of analysis described in this chapter, enzymes can be screened and evaluated quickly and information pertinent to both the inherent properties of the enzyme itself as well as predictions about its performance on complex biomass samples can be obtained.

Key words High-throughput enzyme assays, Chromogenic substrates, *Endo*-acting enzymes, Glycosyl hydrolases, Lyases, Lytic polysaccharide monooxygenases

1 Introduction

1.1 Importance of High-Throughput Enzyme Discovery and Biochemical Characterization

Advances in genomics, proteomics, and associated bioinformatics techniques have enabled the identification of extensive numbers of putative carbohydrate-active enzymes that have currently, in the carbohydrate active enzyme (CAZy) database, been assigned to 145 glycosyl hydrolase (GH) families, 27 polysaccharide lyase families, and 13 families classified as auxiliary activities covering redox enzymes [1–4] such as lytic polysaccharide monooxygenases (LPMOs). However, it is estimated that the activity of no more than 20% of the enzymes in CAZy can be reliably predicted with confidence [5–9], and the speed at which these enzymes can be biochemically characterized is increasingly lagging behind the speed at which new enzymes can be identified.

Mette Lübeck (ed.), *Cellulases: Methods and Protocols*, Methods in Molecular Biology, vol. 1796,
https://doi.org/10.1007/978-1-4939-7877-9_15, © Springer Science+Business Media, LLC, part of Springer Nature 2018

Many of these enzymes are pivotal in efforts toward green fuels, such as bioethanol and biogas, based on efficient bioconversion of plant-based material into fermentable carbohydrates [10–12] that is hindered by the inherent natural complexity of the raw starting materials used [13, 14]. A bioconversion process that is not optimized effectively becomes inefficient and results in recalcitrant material [15] that remains both an obstacle and a challenge for our efforts toward biosustainability and efficient biofuel production [16–18]. In particular, polysaccharide degrading enzymes including GHs, glycosyl lyases, and LPMOs are abundant in nature and of major biotechnological importance [19–26].

In order to utilize these enzymes for industrial purposes or otherwise, they need to be biochemically characterized in order to, firstly, confirm their predicted activity and secondly, determine their biochemical properties such as reaction condition optima among other parameters [27]. Only after they have been biochemically characterized, they can be utilized in industrial processes, such as the bioconversion industry, or find uses in research or other areas.

The number of techniques with which activities of these enzymes can be characterized is abundant and it can account for all of the necessary aspects of enzyme activity from the properties of the enzyme itself to the properties and characteristics of its products. These techniques include high performance chromatography (HPLC) often coupled to mass spectrometry (MS) where oligosaccharide products can be analyzed [28]. This is a powerful and quantitative approach, but it suffers from low throughput. Another very powerful technique is nuclear magnetic resonance (NMR) which is capable of producing an astonishing level of detail [29–32] and has seen significant progress in the past three decades [33, 34], however, it too suffers from low throughput as well as the need for relatively large amounts of relatively pure material [35]. Polysaccharide degrading enzyme activities can also be monitored by measuring the generation of reducing ends as a result of degradation [36], for example using the Nelson-Somogyi [37], the 3,5-dinitrosalicylic acid [38], *p*-hydroxybenzoic acid hydrazide (PAHBAH) [39], and 3-methyl-2-benzothiazoninone hydrazine (MBTH) [40] methods. However, these reducing-end assays also have limited throughput and can be prone to side-reactions [41]. Chromogenic substrates, such as *para*-nitrophenyl (pNP) glycosides or related compounds, are a convenient way of screening enzyme activities however they are available for screening only some activities [25, 42]. For example, *para*-nitrophenyl (pNP) glycosides can be useful for rapidly assaying GH activities, even though they are more suitable for studies of *exo*-acting enzymes since these small artificial compounds are of limited use for study of high molecular weight and sometimes crystalline substrates such as chitin, cellulose, and some arabinoxylans [25, 42].

With regard to high-throughput enzyme screening, it is important to recognize a necessary and unavoidable compromise between the accuracy of information obtained and the speed at which information of any kind can be obtained. For in-depth studies of enzymes, one will have to compromise on throughput, however, before enzyme candidates deemed important enough for that level of detail are selected—they first need to be identified and selected by more high-throughput methods that compromise on detail but have the ability to screen large numbers of enzymes in a short amount of time. This is why high-throughput screening of enzyme activity is a recommended first choice when dealing with such a vast number of enzymes while low-throughout methods are then a logical subsequent choice once the most interesting candidates have been identified [27, 43, 44].

1.2 Enzyme Activity and Enzyme Accessibility—Addressing Both Questions with One Assay

When dealing with degradation of polysaccharide components of complex biological materials one has to take into account enzyme activity against its target in a direct fashion and its activity against its target when it is in the context of complex biological material. While the overall specificity, activity, efficiency, and reaction condition optima most likely do not change regardless of the scenario, some very significant circumstances arising from the complexity of natural biomaterial [45] can easily outcompete and severely affect the efficiency of an enzyme as recorded by direct enzyme-to-target assays.

1.3 Introduction to Chromogenic Substrates

Chromogenic substrates are, by definition, substrates that produce color upon enzymatic degradation. As such, they differ in the mode by which they produce color which is usually directly correlated to their chemical structure. With focus on substrates intended for screening polysaccharide degradation, most chromogenic substrates can be roughly divided into three groups: (1) substrates with chromophores covalently attached to polysaccharides [27, 44], (2) substrates with chromophores passively (noncovalently) adsorbed to polysaccharides [46, 47], and (3) substrates with chromophore molecules attached to the reducing end of polysaccharides [36, 48, 49]. All three groups have their advantages and disadvantages but generally achieve the same purpose. Assays conducted with group (1) are usually based on color generation as a result of enzyme action [27] while assays conducted by group (2) are, inversely, based on disappearance of color, such as the appearance of "clear zones" in agar plate assays [46]. Assay conducted with group (3) are usually more suitable for *exo*-acting enzymes even though they have been incorporated into agar plate assays as well [36].

The method presented in this chapter is based on substrates belonging to group (1) with chromophores covalently attached to polysaccharide chains. As expected, they are only suitable for screening of *endo*-acting enzymes such as glycosyl hydrolases, lyases, and lytic polysaccharide monooxygenases among others as *exo*-acting enzymes are sterically hindered by the large chromophore molecules and do not produce a colorimetric response [27].

The chemical nature of the chromophore also determines the type of assay that these substrates are suitable for. For example, water-soluble chromophores are more suited for in-solution assays, while water-insoluble chromophores are more suited for assays where chromogenic polysaccharides are immobilized on some type of solid support such as microscope slides [50, 51] or agar plates [52–54].

1.4 Chromogenic Polysaccharide Hydrogel (CPH) Substrates

Chromogenic polymer hydrogel (CPH) substrates are synthesized from purified polysaccharides and then chemically dyed and cross-linked rendering them chromogenic and insoluble. As such, their flexibility in terms of different assay layouts such as 96-well plates and agar plates has already been demonstrated [27, 44, 55]. Additionally, the possibility of dyeing the substrates with dyes of different colors has also been demonstrated in abovementioned work giving the assay an additional level of flexibility and high-throughput nature. Cross-linking makes the substrates insoluble facilitating easier separation of the substrate and the resulting dyed oligosaccharide products that are a result of enzyme degradation. This is an important advantage as it enables incorporation of CPH substrates into a variety of assay layouts such as 96-well microtiter plates and agar plates as well as others and makes them flexible toward screening a variety of samples containing enzymes such as single purified enzymes, enzyme cocktails, and microbial broths relevant for biomass deconstruction [27].

The ability to produce CPH substrates in four different colors is an important feature that can be used to increase the throughput of CPH substrate assays as substrate mixtures also enable multiple enzyme activities to be detected simultaneously.

The main principle upon which the CPH substrates work is based on the fact that upon enzymatic degradation of the insoluble matrix—small, dyed oligomers are released into solution rendered soluble by the virtue of their small molecular weight (Fig. 1).

The increasing color response is directly correlated to increasing enzyme concentration as shown in Fig. 2 and the absorbance values can easily be converted to activity units (U/mL) by using positive controls and that is discussed later in the practical part of this chapter.

1.5 Insoluble Chromogenic Biomaterial (ICB) Substrates

Enzymes that degrade biomass feedstocks seldom encounter single polysaccharides in isolation when used for bioconversion purposes. Instead, they are challenged with highly complex mixtures of polysaccharides that are physically intertwined and sometimes covalently linked [56]. Ideally, assays for screening enzymes intended

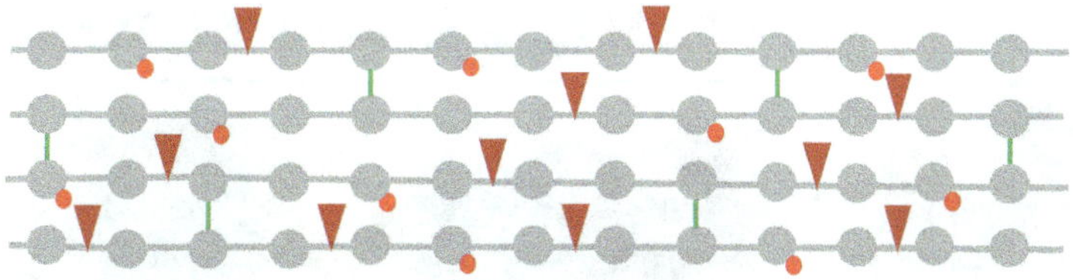

Fig. 1 Scheme of the CPH substrate insoluble matrix with polysaccharide chains (grey) with covalently linked chromophores (bright red) and covalently cross-linked with a bifunctional cross-linker (green). Upon enzymatic digestion (dark red arrowheads), the insoluble matrix disintegrates into smaller fragments that diffuse into solution

for biomass deconstruction should address this complexity and heterogeneity and chromogenic versions of biomass materials typically used as biofuel/biorefinery feedstocks provide such a substrate [27]. Insoluble Chromogenic Biomass (ICB) substrates are not intended to resolve individual enzyme activities since multiple polymers are dyed with the same color within a single ICB substrate type. However, they provide information about the ability of an enzyme, cocktail or microbial broth to release oligomeric products from a complex biomass matrix and enable assessment of enzyme efficiency on feedstocks normally used in biorefineries [27].

An example of profiling different biomass feedstocks is shown in Fig. 3 where different plant materials were synthetically converted into ICB substrates and assayed with a panel of commercial enzymes including a microbial broth [27].

The enzymes used and the formulation of the microbial broth are specified in Table 1.

ICB substrates are based on raw biomass samples or alcohol-insoluble residue (AIR) preparations, sometimes used to remove pigments and/or other chemical compounds which may interfere with the chemical dyeing process. AIR preparations are standard crude preparations of polysaccharides widely used as the starting point for biomass analysis, for example saccharification assays [27, 28] and microarray-based polysaccharide composition analysis [57].

On a molecular level, the mode of response to enzyme degradation for ICB substrates is identical to CPH substrates as depicted in Fig. 1.

CPH substrates provide a valuable tool for assessing the specific activity of an enzyme while ICB substrates are used to evaluate the capacity of an enzyme to digest a component within the context of complex substrate mixtures that enzymes usually encounter within biomass. Although ICB substrates do not provide information about individual enzyme specificities, they are nonetheless useful tools for assessing the commercial performance of enzymes, cocktails, or broths [58].

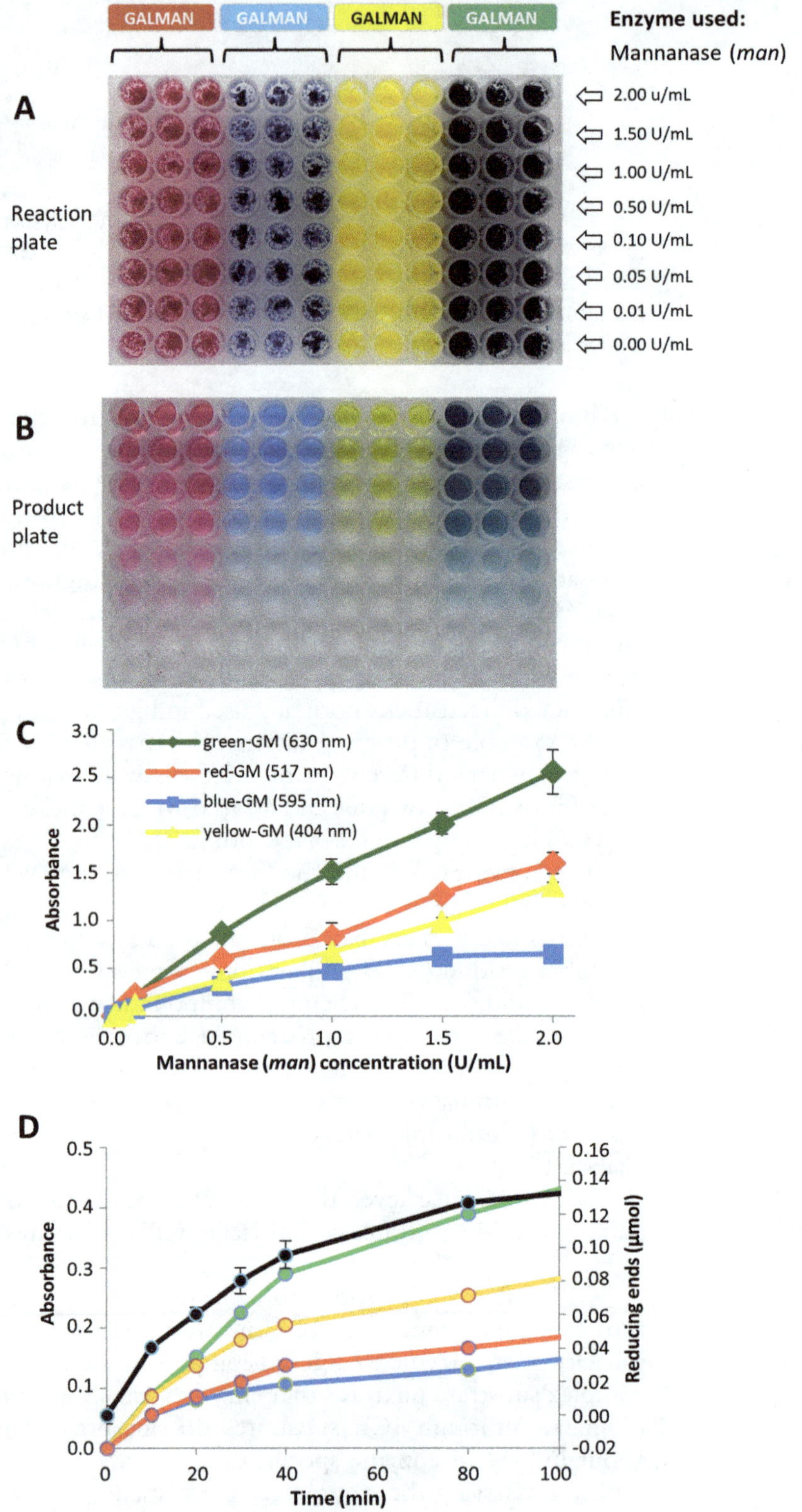

Fig. 2 Dose and time responses of CPH substrates. (**a**) Reaction plate containing four different colored versions of CPH-galactomannan (GALMAN)—red, blue, yellow, and green. (**b**) Product plate containing the products of

1.6 Two-Dimensional Analysis of Endo-Enzyme Activity and Biomass Degradability Using CPH and ICB Substrates

CPH and ICB substrates together facilitate high-throughput multiplexed assays where multiple enzymes, side-activities of a single enzyme or microbial culture broths. When combined, they have the ability to provide information about enzyme activity against a single target (CPH substrates) and then that same target within the context of complex biomaterial such as biomass (ICB substrates) adding a second dimension to enzyme analysis. This enables evaluation of enzyme efficiency both in the discovery/screening stage [43, 55, 58, 59], comprising collecting data about enzyme reaction condition optima using CPH substrates, and in the application stage by evaluating their efficiency against complex biomass samples, the ultimate target in bioindustrial applications, using ICB substrates [58] (Fig. 4).

An example study using this methodology, as shown by Jimenez et al. (2016), evaluated a metagenomics and metasecretomics based approach for developing new strategies for lignocellulose saccharification. Enzymes belonging to several GH families were identified (GH3, GH43, GH13, GH10, GH29, GH28, GH16, GH4, and GH92) in addition to several AA families (AA2, AA6, and AA10) and evaluated on CPH substrates: arabinan, arabinoxylan, xylan, galactomannan, rhamnogalacturonan, β-glucan from barley and ICB substrates: sugarcane bagasse, wheat straw, and willow demonstrating a seamless process starting with enzyme discovery/characterization and ending in efficient evaluation of their bioindustrial potential [58].

2 Materials

1. Assay kit plates containing chromogenic substrates (*see* **Note 1**).

2. Activation solution (*see* **Note 1**).

3. Micropipettes.

4. Deionized water.

5. 96-well plates for activating/washing the substrates.

6. Centrifuge with rotor for plates (recommended speed 2500–2700 × *g*) or vacuum manifold with 350 mL receiver plate space block and a diaphragm vacuum pump (maximum negative pressure of −60 kPa).

Fig. 2 (continued) the digestion of the substrates shown in (**a**) with mannanase (*man*) used in the range of 0 to 2 U/mL. (**d**) Graph showing the absorbance (*y*-axis to the left) of products released over a time period (from 0 to 100 min) by treatment of four differently colored versions of CPH-xyloglucan with xyloglucanase (*xg*) at 0.25 U/mL in 100 mM sodium acetate buffer, pH 5.5. Also shown on this graph (*y*-axis to the right) is the production of reducing ends over the same time period from undyed xyloglucan, as measured using the 3-methyl-2-benzothiazolinone hydrazone (MBTH) method, shown as a black line [40]. Note that an undyed xyloglucan hydrogel was used since the dye would likely interfere with the MBTH method. *See* Table 1 for details about the enzymes used. Figure reproduced from [27] with kind permission from Springer

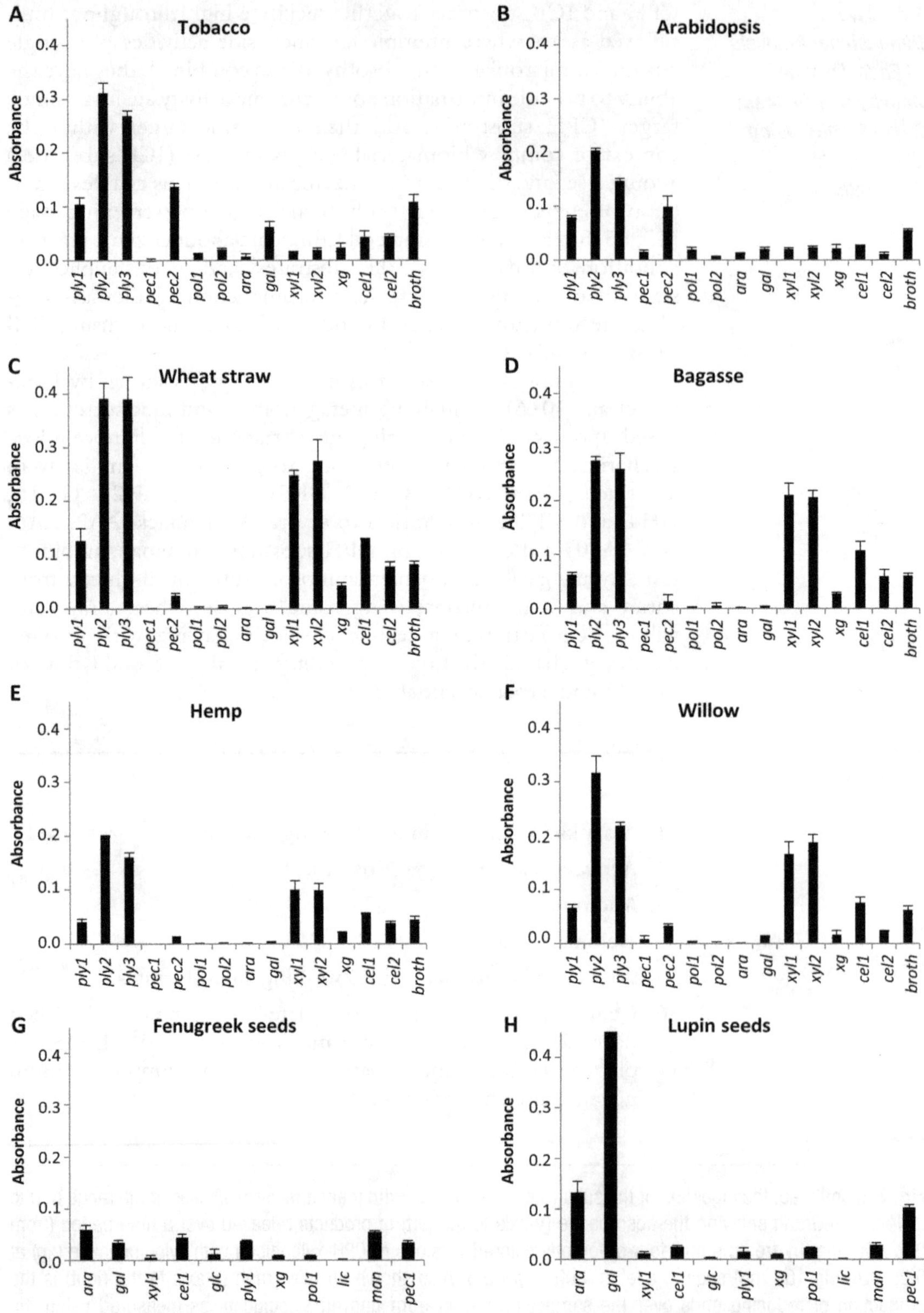

Fig. 3 A range of insoluble chromogenic biomass substrates from different plants and feedstocks: (**a**) tobacco, (**b**) Arabidopsis, (**c**) wheat straw, (**d**) sugarcane bagasse, (**e**) hemp, (**f**) willow, (**g**) fenugreek seeds, and (**h**)

7. Buffer suitable for the target enzyme.

8. 96-well polypropylene clear-well microplates as collection plates.

9. Incubator capable of shaking and heating.

10. Microtiter plate reader (*see* **Note 2**).

3 Methods

3.1 General Notes About the Assay Procedure

The assay procedure is identical regardless whether the substrates tested are CPH or ICB substrates.

3.2 Activation of the Assay Kit Plate

1. The plate comes covered with a plastic lid and silicone mat as shown in Fig. 5. Remove the plastic lid and silicone mat on top of the 96-well plate (*see* **Note 3**).

2. Activate the assay kit plate (96-well filter plates containing cellulase substrate) by adding 200 µL activation solution into each well, followed by 10 min incubation at room temperature without agitation. This removes the stabilizer compounds from the substrate.

3. Centrifuge (recommended speed 2500–2700 × g) for 10 min or apply vacuum using a vacuum manifold (with the spacer block inside and any standard, transparent 96-well plate as a collection plate) to remove extant activation solution.

4. Wash the substrates by adding 100 µL sterile water and centrifuge or apply vacuum to remove the stabilizer. Repeat this step two more times and the plates are now ready to use.

3.3 Enzyme Reaction

1. Add 145 µL of an appropriate buffer (*see* **Note 4**) and 5 µL of enzyme solution (*see* **Note 5**) to each well of the assay kit plate.

2. Put the product plate (a clear 96-well plate compatible with your microtiter plate reader) underneath the assay kit plate (*see* **Note 6**).

3. Seal the assay kit plate with an adhesive film to prevent the liquid from spilling during agitation as shown in Fig. 6.

Fig. 3 (continued) lupin seeds treated with a range of enzymes (shown on *x*-axes). The absorbance values are means from three wells. Enzymes were used at 10 U/mL for 24 h at room temperature. The assay was performed using 100 mM sodium acetate buffer pH 4.5, except for the enzymes *ply1*, *ply2*, *ply3*, *pol1*, and *xg* was used pH 5.5, for *cel2* pH 6.0, for *pec3* 100 mM sodium phosphate pH 7.0, and for *pec1* sodium carbonate pH 10.0. *See* Table 1 for details about the enzymes used. Figure reproduced from [27] with kind permission from Springer

Table 1
Specifications about the enzymes used in Figs. 2 and 3

Code name	Description
ara	*Endo*-arabinase (*Aspergillus niger*)
cel1	*Endo*-cellulase (*endo*-β-1,4-glucanase) (*Trichoderma longibrachiatum*)
cel2	Cellulase (*endo*-β-1,4-glucanase) (*Bacillus amyloliquefaciens*)
gal	*Endo*-β-1,4-D-galactanase (*Aspergillus niger*)
ply1	Macerase™ Pectinase (*Rhizopus* sp.)
ply2	Pectolyase Y-23 (*Aspergillus japonicus*)
ply3	Pectolyase (*Aspergillus japonicus*)
pec1	Pectate lyase (*Cellvibrio japonicus*)
pec2	Pectate lyase (*Aspergillus* sp.)
pol1	*Endo*-polygalacturonase M2 (*Aspergillus niger*)
pol2	*Endo*-polygalacturonase M1 (*Aspergillus niger*)
xg	Xyloglucanase (*Paenibacillus* sp.)
xyl1	β-xylanase, M4 (*Aspergillus niger*)
xyl2	*Endo*-β-1,4-xylanase M1 (*Trichoderma viride*)
broth	Culture broth from *Phanerochaete chrysosporium* (3 days after inoculation)

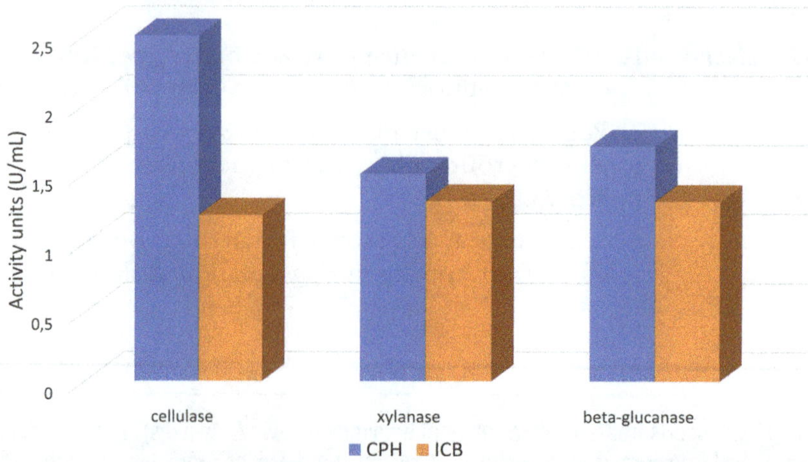

Fig. 4 A schematic example of two-dimensional analysis of enzymes against CPH (blue) and ICB (orange) substrates reflecting their activity against purified targets (CPH) and complex biomass material (ICB) with respective absorbance unit values stacked on top of each other. Graphs are for illustration purposes only and not based on real experimental data

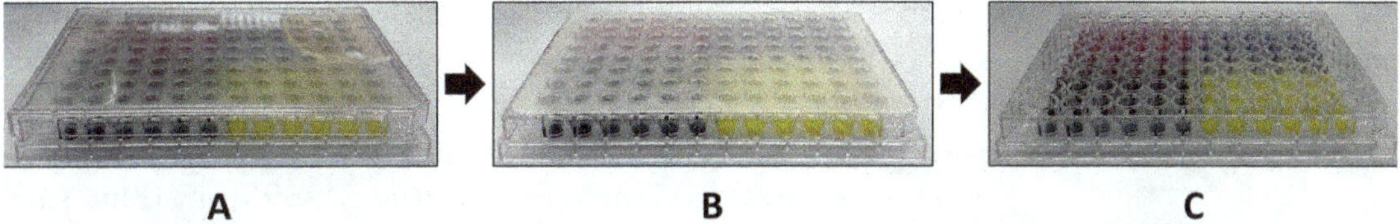

A　　　　　　　　　**B**　　　　　　　　　**C**

Fig. 5 The plate comes covered with a plastic lid and a silicone mat. Remove the plastic lid (**a**) and the silicone mat (**b**) to expose the ready-to-use plate (**c**)

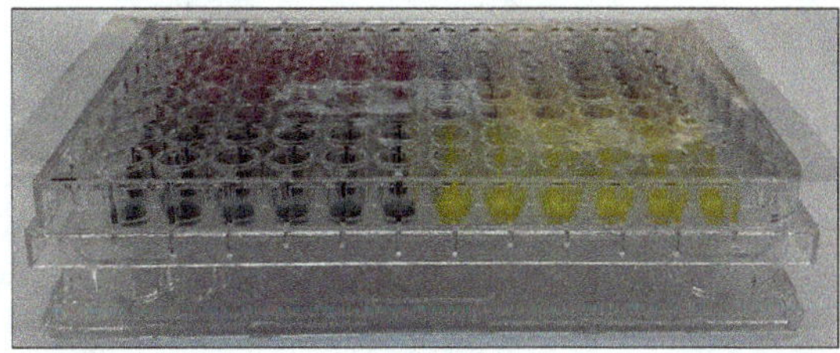

Fig. 6 The assay plate from Fig. 3 sealed with an adhesive film and placed atop a 96-well clear-well plate

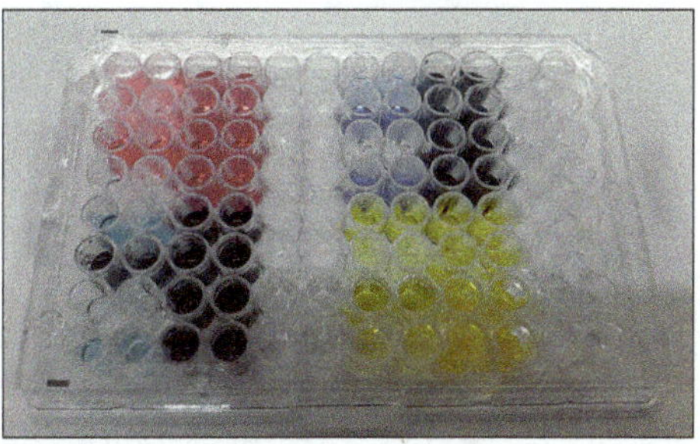

Fig. 7 An example of the product plate (still sealed with an adhesive film) with different color intensities corresponding to different enzyme activities

4. Incubate the reaction at an appropriate temperature for a desired time with agitation (*see* **Note 7**).

5. Place the assay plate with the product plate in the centrifuge and spin down the reaction product from the assay plate into the product plate for 10 min. It is also possible to use the vacuum manifold by placing the assay kit plate on top of the manifold with the product plate underneath and applying vacuum (maximum negative pressure of −60 kPa).

6. The filtrate containing the colored oligosaccharides as reaction products is now in the product plate and can be analyzed further (*see* **Note 8**). An example of how the product plate can look like is shown in Fig. 7.

3.4 Detection and Quantification

1. Check that the volume of liquid in each well of the product plate is the approximately the same by visual inspection (*see* **Note 9**).

2. Read the absorbance of the reaction products at 404 nm (yellow substrates), 517 nm (red substrates), 595 nm (blue substrate), and 630 nm (green substrate) using a plate reader.

3. Subtract the value of the negative control from the sample and the positive control.

4. Based on the number of replicates, calculate the average value and standard error.

5. Plot the corrected absorbance values (*see* **Note 10**).

4   Notes

1. The assay kit plates, including the activation solution, can be obtained from GlycoSpot IVS. Cellulase degrading substrates available are for example CPH-2-hydroxylethylcellulose, CPH-carboxymethylcellulose, ICB-filter paper, and substrates from virtually any biomass feedstock which can be combined with other substrates related to hemicelluloses (e.g., xylan, arabinoxylan, mixed-linkage glucans from different sources, xyloglucan among many others), pectins (e.g., citrus peel pectin, rhamnogalacturonan), starch-related polysaccharides (amylose and amylopectin), and other biopolymers (e.g., polysaccharides and proteins from other organisms). The substrates are available in different colors (yellow, red, green, and blue) and can be combined in one well enable simultaneous measurement of different enzyme activities possible. Additionally, GlycoSpot IVS produces tailor-made substrates from raw materials supplied by customers themselves, both in a form of CPH (purified polysaccharides) and ICB (complex biomaterial) substrates, enabling researchers to screen their own enzymes on their own raw materials. Examples of such raw materials that can be supplied to GlycoSpot and chromogenic substrates can be produced are: modified cellulose derivatives (cellulose with different degrees of substitution, e.g., alkylated celluloses), cellulose model materials (e.g., Avicel, filter paper, cotton fiber) and biomass feedstocks (e.g., sugarcane bagasse, sugarcane straw, wheat straw, softwoods, hardwoods, cereals, and generally any type of complex biological material including not only plant-based but also any type of material containing mixtures of polysaccharides and/or proteins).

2. The spectrophotometer should have a program for endpoint absorbance measurements with the possibility to measure at the following wavelengths: 404 nm for yellow, 517 nm for red, 595 nm for blue, and 630 nm for green.

3. The silicone mat on top of the plate is intended for protection of the substrates during shipment and packaging and should be removed and discarded. There is a silicone filter underneath the plate is intended for keeping the liquid inside the plate during the assay procedure and should not be removed.

4. An appropriate buffer is a buffer with the pH optimum of the target enzyme or simulates the conditions in a process. The substrates are stable between pH 3.0 and 10.0. Additional components required for the reaction such as metal cations or reducing agents needed by the target enzyme should also be included. Oxidizing agents such as hydrogen peroxide, sodium hypochlorite and similar compounds should be avoided in excess amount as they can oxidize and destroy the dye molecules, however they do not interfere with the assay in catalytic amounts.

5. The final concentration of the enzyme should be around 100 μg/mL or 1 U/mL, when purified enzyme is used. In cases where the enzyme is not purified, and/or the enzyme concentration is unknown, starting with the highest possible sample concentration is recommended. Plant tissue and raw material extracts or culture broths can also be used in addition to purified enzyme solutions. Always include buffer alone as a negative control and, if possible, previously characterized enzymes as positive controls. In the case of complex samples such as microbial cultures, the growth medium buffer can be used as a negative control.

6. The product plate will collect any potential leakage from the reaction plate during shaking. This can happen in the range of 1 μL, if any, when the reaction is incubated at high temperatures and for longer periods of time.

7. The substrates are stable up to 90 °C. Mixing the reaction in the assay kit plate during the incubation is crucial for achieving a consistent and reproducible result: shaking at a minimum rate of 100 rpm is recommended. The incubation time should be increased when testing ICB substrates and up to 24 h when testing unknown enzyme concentrations (e.g., culture supernatants) with CPH substrates. For enzyme activities in the range of 1 U/mL, 30 min are sufficient for CPH substrates, but 24 h are still recommended for ICB substrates because of the complexity of the material they are produced from. Note that appropriate incubation times depend on the activity of the enzyme(s), but in general if there is no detectable activity within 24 h, it is likely that the enzyme(s) present are not able to degrade the tested substrate.

8. Sometimes the enzymatic degradation products can block the filter in the bottom of the assay plate and after centrifugation or

vacuum suction all of the reaction solution does not get transferred to the product plate. One possibility is to increase the centrifugation time, the negative pressure in the vacuum manifold or transfer the solution carefully with a pipette. Make sure that no substrate is transferred and that the volume is equal in each well in the product plate. This may result in uneven levels of liquid in the product plate. If the levels of liquid are very different upon visual inspection, the wells should be supplemented with small amounts of reaction buffer in order to even out the volume. Another, more accurate solution is to use the path-check option available in many microplate readers that normalize the liquid levels during measurements by detecting the level of liquid in each well automatically.

9. If you see a significant dispersion of the absorbance values for the different replicas of the same sample, make sure that the volume is equal in each well and that there are no bubbles in the wells.

10. Data can be plotted in a number of ways such as bar graphs [27] or heat maps [43]. By using positive controls in a form of enzymes with defined activity and specificity, and based on the linear response of color release to enzyme activity [27, 44], the absorbance values can easily be recalculated into units per milliliter (U/mL) providing a more accurate result such as the one depicted in Fig. 2.

References

1. Cantarel BL, Coutinho PM, Rancurel C et al (2009) The Carbohydrate-Active EnZymes database (CAZy): an expert resource for glycogenomics. Nucleic Acids Res 37:D233–D238. https://doi.org/10.1093/nar/gkn663

2. Lombard V, Ramulu HG, Drula E et al (2014) The carbohydrate-active enzymes database (CAZy) in 2013. Nucleic Acids Res 42: D490–D495. https://doi.org/10.1093/ Nar/Gkt1178

3. Lombard V, Bernard T, Rancurel C et al (2010) A hierarchical classification of polysaccharide lyases for glycogenomics. Biochem J 432:437–444. https://doi.org/10.1042/ BJ20101185

4. Levasseur A, Drula E, Lombard V et al (2013) Expansion of the enzymatic repertoire of the CAZy database to integrate auxiliary redox enzymes. Biotechnol Biofuels 6:41. https:// doi.org/10.1186/1754-6834-6-41

5. Davies G, Henrissat B (1995) Structures and mechanisms of glycosyl hydrolases. Structure 3 (9):853. https://doi.org/10.1016/S0969-2126(01)00220-9

6. Henrissat B (1991) A classification of glycosyl hydrolases based on amino-acid-sequence similarities. Biochem J 280:309–316

7. Henrissat B, Bairoch A (1993) New families in the classification of glycosyl hydrolases based on amino-acid-sequence similarities. Biochem J 293:781–788

8. Henrissat B, Bairoch A (1996) Updating the sequence-based classification of glycosyl hydrolases. Biochem J 316:695–696

9. Henrissat B, Davies G (1997) Structural and sequence-based classification of glycoside hydrolases. Curr Opin Struct Biol 7:637–644. https://doi.org/10.1016/S0959-440x(97) 80072-3

10. Canilha L, Chandel AK, Milessi TSDS et al (2012) Bioconversion of sugarcane biomass into ethanol: an overview about composition, pretreatment methods, detoxification of hydrolysates, enzymatic saccharification, and ethanol fermentation. J Biomed Biotechnol 7:989572. https://doi.org/10.1155/2012/ 989572

11. Liguori R, Ventorino V, Pepe O, Faraco V (2016) Bioreactors for lignocellulose conversion into fermentable sugars for production of high added value products. Appl Microbiol Biotechnol 100:597–611. https://doi.org/10.1007/s00253-015-7125-9

12. Maurya DP, Singla A, Negi S (2015) An overview of key pretreatment processes for biological conversion of lignocellulosic biomass to bioethanol. 3 Biotech 5:597–609. https://doi.org/10.1007/s13205-015-0279-4

13. Scott F, Quintero J, Morales M et al (2013) Process design and sustainability in the production of bioethanol from lignocellulosic materials. Electron J Biotechnol 16:1–16. https://doi.org/10.2225/vol16-issue3-fulltext-7

14. Kumar P, Barrett DM, Delwiche MJ, Stroeve P (2009) Methods for pretreatment of lignocellulosic biomass for efficient hydrolysis and biofuel production. Ind Eng Chem Res 48:3713–3729. https://doi.org/10.1021/ie801542g

15. Pu Y, Hu F, Huang F et al (2013) Assessing the molecular structure basis for biomass recalcitrance during dilute acid and hydrothermal pretreatments. Biotechnol Biofuels 6:15. https://doi.org/10.1186/1754-6834-6-15

16. Madadi M (2017) Recent status on enzymatic saccharification of lignocellulosic biomass for bioethanol production. Electron J Biol 13:135–143

17. Singh R, Kapoor V, Kumar V (2012) Utilization of agro-industrial wastes for the simultaneous production of amylase and xylanase by thermophilic actinomycetes. Brazil J Microbiol 43:1545–1552. https://doi.org/10.1590/S1517-83822012000400039

18. McCann MC, Carpita NC (2015) Biomass recalcitrance: a multi-scale, multi-factor, and conversion-specific property. J Exp Bot 66:4109–4118. https://doi.org/10.1093/jxb/erv267.

19. Tamayo-Ramos JA, Orejas M (2014) Enhanced glycosyl hydrolase production in Aspergillus nidulans using transcription factor engineering approaches. Biotechnol Biofuels 7:103. https://doi.org/10.1186/1754-6834-7-103

20. Li LL, Taghavi S, McCorkle SM et al (2011) Bioprospecting metagenomics of decaying wood: mining for new glycoside hydrolases. Biotechnol Biofuels 4:23. https://doi.org/10.1186/1754-6834-4-23

21. Dimarogona M, Topakas E, Christakopoulos P (2013) Recalcitrant polysaccharide degradation by novel oxidative biocatalysts. Appl Microbiol Biotechnol 97:8455–8465. https://doi.org/10.1007/s00253-013-5197-y

22. Agger JW, Isaksen T, Varnai A et al (2014) Discovery of LPMO activity on hemicelluloses shows the importance of oxidative processes in plant cell wall degradation. Proc Natl Acad Sci U S A 111:6287–6292. https://doi.org/10.1073/pnas.1323629111

23. Giordano RJ, Cardo-Vila M, Lahdenranta J (2001) Biopanning and rapid analysis of selective interactive ligands. Nat Med 7:1249–1253. https://doi.org/10.1038/nm1101-1249

24. Sánchez ÓJ, Cardona CA (2008) Trends in biotechnological production of fuel ethanol from different feedstocks. Bioresour Technol 99:5270–5295. https://doi.org/10.1016/j.biortech.2007.11.013

25. Sweeney MD, Xu F (2012) Biomass converting enzymes as industrial biocatalysts for fuels and chemicals: recent developments. Catalysts 2:244–263. https://doi.org/10.3390/Catal2020244

26. Sims REH, Mabee W, Saddler JN, Taylor M (2010) An overview of second generation biofuel technologies. Bioresour Technol 101:1570–1580. https://doi.org/10.1016/j.biortech.2009.11.046

27. Kračun SK, Schückel J, Westereng B et al (2015) A new generation of versatile chromogenic substrates for high-throughput analysis of biomass-degrading enzymes. Biotechnol Biofuels 8:70. https://doi.org/10.1186/s13068-015-0250-y

28. Pena MJ, Tuomivaara ST, Urbanowicz BR et al (2012) Methods for structural characterization of the products of cellulose- and xyloglucan-hydrolyzing enzymes. Cell 510:121–139. https://doi.org/10.1016/B978-0-12-415931-0.00007-0

29. Simmons TJ, Mortimer JC, Bernardinelli OD et al (2016) Folding of xylan onto cellulose fibrils in plant cell walls revealed by solid-state NMR. Nat Commun 7:13902. https://doi.org/10.1038/ncomms13902

30. Grantham NJ, Wurman-Rodrich J, Terrett OM et al (2017) An even pattern of xylan substitution is critical for interaction with cellulose in plant cell walls. Nat Plants 3:859–865. https://doi.org/10.1038/s41477-017-0030-8.

31. Mansfield SD, Kim H, Lu FC, Ralph J (2012) Whole plant cell wall characterization using solution-state 2D NMR. Nat Protoc 7:1579–1589. https://doi.org/10.1038/nprot.2012.064

32. Foston M, Samuel R, He J, Ragauskas AJ (2016) A review of whole cell wall NMR by

the direct-dissolution of biomass. Green Chem 18:608–621. https://doi.org/10.1039/C5GC02828K

33. Peters T, Jiménez-Barbero J (2002) NMR spectroscopy of glycoconjugates. Wiley, Hoboken. https://doi.org/10.1002/352760071X

34. Hounsell EF (1995) 1H NMR in the structural and conformational analysis of oligosaccharides and glycoconjugates. Prog Nucl Magn Reson Spectrosc 27:445–474. https://doi.org/10.1016/0079-6565(95)01012-2

35. von der Lieth C-W, Lütteke T, Frank M (2009) Bioinformatics for glycobiology and glycomics: an introduction. Wiley-Blackwell, Hoboken

36. Sadhu S, Maiti TK (2013) Cellulase production by bacteria: a review. Br Microbiol Res J 3:235–258. https://doi.org/10.9734/BMRJ/2013/2367.

37. Nelson N (1944) A photometric adaptation of the Somogyi method for the determination of glucose. J Biol Chem 153:375–380

38. Miller GL (1959) Use of dinitrosalicylic acid reagent for determination of reducing sugar. Anal Chem 31:426–428. https://doi.org/10.1021/Ac60147a030

39. Lever M (1973) Colorimetric and fluorometric carbohydrate determination with para-hydroxybenzoic acid hydrazide. Biochem Med 7:274–281. https://doi.org/10.1016/0006-2944(73)90083-5

40. Anthon GE, Barrett DM (2002) Determination of reducing sugars with 3-methyl-2-benzothiazolinonehydrazone. Anal Biochem 305:287–289. https://doi.org/10.1006/abio.2002.5644

41. Gusakov AV, Kondratyeva EG, Sinitsyn AP (2011) Comparison of two methods for assaying reducing sugars in the determination of carbohydrase activities. Int J Anal Chem 2011:283658. https://doi.org/10.1155/2011/283658

42. Biely P, Mastihubova M, Tenkanen M et al (2011) Action of xylan deacetylating enzymes on monoacetyl derivatives of 4-nitrophenyl glycosides of beta-D-xylopyranose and alpha-L-arabinofuranose. J Biotechnol 151:137–142. https://doi.org/10.1016/j.jbiotec.2010.10.074

43. Schückel J, Kračun SK, Lausen TF et al (2017) High-throughput analysis of endogenous fruit glycosyl hydrolases using a novel chromogenic hydrogel substrate assay. Anal Methods 9:1242–1247. https://doi.org/10.1039/C6AY03431D

44. Schückel J, Kračun SK, Willats WGT (2016) High-throughput screening of carbohydrate-degrading enzymes using novel insoluble chromogenic substrate assay kits. J Vis Exp 115:e54286. https://doi.org/10.3791/54286.

45. Yang B, Dai Z, Ding S-Y, Wyman CE (2011) Enzymatic hydrolysis of cellulosic biomass. Biofuels 2:421–449. https://doi.org/10.4155/bfs.11.116

46. Yoon JH, Park JE, Suh DY et al (2007) Comparison of dyes for easy detection of extracellular cellulases in fungi. Mycobiology 35:21. https://doi.org/10.4489/MYCO.2007.35.1.021

47. Meddeb-Mouelhi F, Moisan JK, Beauregard M (2014) A comparison of plate assay methods for detecting extracellular cellulase and xylanase activity. Enzym Microb Technol 66:16–19. https://doi.org/10.1016/j.enzmictec.2014.07.004

48. Dashtban M, Maki M, Leung KT (2010) Cellulase activities in biomass conversion: measurement methods and comparison. Crit Rev Biotechnol 30:302–309. https://doi.org/10.3109/07388551.2010.490938

49. Zverlov VV, Velikodvorskaya GA, Schwarz WH (2002) A newly described cellulosomal cellobiohydrolase, CelO, from Clostridium thermocellum: Investigation of the exo-mode of hydrolysis, and binding capacity to crystalline cellulose. Microbiology 148:247–255. https://doi.org/10.1099/00221287-148-1-247

50. Broome AM, Bhavsar N, Ramamurthy G et al (2010) Expanding the utility of β-galactosidase complementation: Piece by piece. Mol Pharm 7:60–74. https://doi.org/10.1021/mp900188e

51. Seymour PA, Sander M (2007) Immunohistochemical detection of beta-galactosidase or green fluorescent protein on tissue sections. Methods Mol Biol 411:13–23

52. Gadd GM, Sariaslani S (2013) Advances in applied microbiology, vol 83. Elsevier, Amsterdam

53. Browne NK, Huang Z, Dockrell M et al (2010) Evaluation of new chromogenic substrates for the detection of coliforms. J Appl Microbiol 108:1828–1838. https://doi.org/10.1111/j.1365-2672.2009.04588.x

54. Uchiyama T, Yaoi K, Miyazaki K (2015) Glucose-tolerant β-glucosidase retrieved from a Kusaya gravy metagenome. Front Microbiol 6:548. https://doi.org/10.3389/fmicb.2015.00548

55. Mravec J, Guo X, Hansen AR, Schückel J et al (2017) Pea border cell maturation and release involve complex cell wall structural dynamics.

Plant Physiol 174:00097. https://doi.org/10.1104/pp.16.00097

56. Jin Z, Katsumata KS, Lam TBT, Iiyama K (2006) Covalent linkages between cellulose and lignin in cell walls of coniferous and non-coniferous woods. Biopolymers 83:103–110. https://doi.org/10.1002/bip.20533

57. Kračun SK, Fangel JU, Rydahl MG et al (2017) Carbohydrate microarray technology applied to high-throughput mapping of plant cell wall glycans using comprehensive microarray polymer profiling (CoMPP). Methods Mol Biol 1503:147–165. https://doi.org/10.1007/978-1-4939-6493-2_12

58. Jiménez DJ, de Lima Brossi MJ et al (2016) Characterization of three plant biomass-degrading microbial consortia by metagenomics- and metasecretomics-based approaches. Appl Microbiol Biotechnol 100 (24):10463–10477. https://doi.org/10.1007/s00253-016-7713-3.

59. Mackenzie AK, Naas AE, Kracun SK et al (2015) A polysaccharide utilization locus from an uncultured bacteroidetes phylotype suggests ecological adaptation and substrate versatility. Appl Environ Microbiol 81:187–195. https://doi.org/10.1128/Aem.02858-14

Analytical Tools for Characterizing Cellulose-Active Lytic Polysaccharide Monooxygenases (LPMOs)

Bjørge Westereng, Jennifer S. M. Loose, Gustav Vaaje-Kolstad, Finn L. Aachmann, Morten Sørlie, and Vincent G. H. Eijsink

Abstract

Lytic polysaccharide monooxygenases are copper-dependent enzymes that perform oxidative cleavage of glycosidic bonds in cellulose and various other polysaccharides. LPMOs acting on cellulose use a reactive oxygen species to abstract a hydrogen from the C1 or C4, followed by hydroxylation of the resulting substrate radical. The resulting hydroxylated species is unstable, resulting in glycoside bond scission and formation of an oxidized new chain end. These oxidized chain ends are spontaneously hydrated at neutral pH, leading to formation of an aldonic acid or a gemdiol, respectively. LPMO activity may be characterized using a variety of analytic tools, the most common of which are high-performance anion exchange chromatography system with pulsed amperometric detection (HPAEC-PAD) and MALDI-TOF mass spectrometry (MALDI-MS). NMR may be used to increase the certainty of product identifications, in particular the site of oxidation. Kinetic studies of LPMOs have several pitfalls and to avoid these, it is important to secure copper saturation, avoid the presence of free transition metals in solution, and control the amount of reductant (i.e., electron supply to the LPMO). Further insight into LPMO properties may be obtained by determining the redox potential and by determining the affinity for copper. In some cases, substrate affinity can be assessed using isothermal titration calorimetry. These methods are described in this chapter.

Key words Lytic polysaccharide monooxygenase, High-performance anion-exchange chromatography, MALDI-TOF mass spectrometry, Copper, Isothermal titration calorimetry

1 Introduction

1.1 A Short History of LPMOs

The discovery of LPMOs in 2010 by Vaaje-Kolstad et al. [1] has led to major improvements in enzymatic depolymerization of cellulose [2]. Using molecular oxygen, externally supplied electrons and a single copper ion as co-factor, LPMOs carry out oxidative cleavage of glycosidic bonds. In contrast to cellulases, which to some extent need to extract single polysaccharide chains from their crystalline context in order to productively bind their substrates, LPMOs can act directly on crystalline material [1, 3]. By introducing chain breaks, LPMOs make the crystalline substrates more accessible to

Mette Lübeck (ed.), *Cellulases: Methods and Protocols*, Methods in Molecular Biology, vol. 1796,
https://doi.org/10.1007/978-1-4939-7877-9_16, © Springer Science+Business Media, LLC, part of Springer Nature 2018

the action of classical cellulases, explaining their considerable impact on the overall efficiency of the enzymatic degradation process. There is an ongoing discussion on the catalytic mechanism of these enzymes [4–6] but their products are well characterized and the nature of the expected products does not seem to depend on which catalytic mechanism is valid.

1.2 Mapping and Quantifying LPMO Activity on Cellulose

LPMOs acting on cellulose abstract a hydrogen atom from the C1 or the C4 in the scissile glycosidic bond, followed by hydroxylation of the resulting substrate-radical. The resulting hydroxylated species is unstable, resulting in glycoside bond scission and formation of an oxidized new chain end [7]. Glycosidic bond cleavage with C1 oxidation leads to formation of one chain end that is a lactone and a regular nonreducing end; the lactone is in equilibrium with the aldonic acid form, which dominates at neutral pH. Glycosidic bond cleavage with C4 oxidation leads to formation of a new regular reducing end, whereas the nonreducing end is a 4-ketosugar. The 4-keto sugar is hydrated to the corresponding gemdiol form in aqueous conditions. The lactone and the 4-keto form have identical masses and the same applies to their hydrated forms, the aldonic acid and the gemdiol. The two types of products may nevertheless be discriminated in MALDI-TOF MS because (1) the hydrated gemdiol is more easily dehydrated as a result of sample preparation for MALDI-ToF MS analysis than the aldonic acids, and (2) the aldonic acid form is charged and thus tends to form double adducts with for example sodium (Fig. 1) [8, 9].

There are several chromatographic methods for analyzing the native and oxidized products, generated when an LPMO acts on cellulose [10]. The most common method, accessible in most laboratories, is high-performance anion-exchange chromatography (HPAEC) combined with pulsed amperometric detection (PAD) (Fig. 2). At the alkaline pH during the HPAEC analysis, the equilibrium between the lactone and acid is strongly shifted toward the aldonic acid, and this makes HPAEC ideal for analysis of C1-oxidized products, which are stable and negatively charged (the pK_a of cellobionic acid is 3.5 [10]), and are readily separated from native products. C4 oxidized products are less stable and undergo on-column decomposition during HPAEC [10], leading to products with additional oxidations and, importantly, native cello-oligosaccharides that have one less glucose than the original C4-oxidized product. Still, C4-oxidized products yield characteristic and diagnostic signals in HPAEC-PAD, albeit with less signal intensity due to diversion into multiple peaks and appearance later in the gradient where signal suppression by acetate becomes more prominent. Importantly, some LPMOs oxidize both C1 and C4, meaning that they have a mixed product profile that includes double-oxidized products. Products that are diagnostic for double oxidized products elute late from the column with little separation

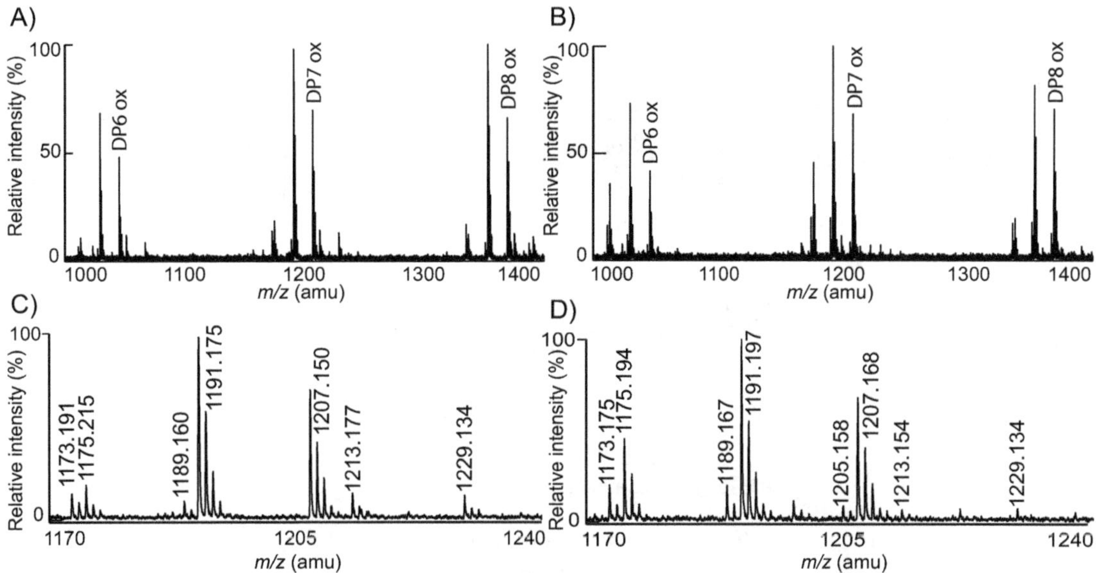

Fig. 1 Comparison of products generated by two cellulose active LPMOs from *Thermobifida fusca*. Panels **a** and **c** show products generated by the LPMO domain of C1-oxidizing E8 (or TfLPMO10B), whereas panels **b** and **d** show products generated by C1&C4-oxidizing E7 (or TfLPMO10A). Panels **c** and **d** are zoom-in views relative to panels **a** and **b**, respectively, focusing on the heptamer cluster. The *m/z* values in panels **c** and **d** correspond to: 1173, sodium adduct of lactone or ketoaldose; 1175, sodium adduct of native Glc$_7$; 1189, potassium adduct of lactone or ketoaldose or sodium adduct of double oxidized Glc$_7$; 1191, sodium adduct of aldonic acid or potassium adduct of native Glc$_7$ or sodium adduct of gemdiol (4-ketoaldose + water); 1205, potassium adduct of double oxidized sugar; 1207, potassium adduct of aldonic acid or gemdiol form of the 1189 species; 1213, sodium adduct of the aldonic acid sodium salt; 1229, sodium adduct of the aldonic acid potassium salt. One would expect the signal at 1173 to be relatively larger in panel **d** than in panel **c**, because the 4-ketoaldose generated by C4 oxidation is less readily hydrated that the lactone generated by C1 oxidation. In the case of a C1–C4 oxidizer this is somewhat hidden because these mixed oxidzers produce more native oligomers meaning that the adjacent peak (1175) becomes stronger. The relatively higher signals at 1189 (relatively stable 4-keto form and/or sodium adduct of the double oxidized form) and at 1205 (potassium adduct of the double oxidized form) in panel D confirm the mixed C1 and C4-oxidizing activity of *Tf*LPMO10A. This picture is reproduced from the supplementary information of [43]

(Fig. 2; [11]) and their identity is not well resolved. A beautiful series of chromatograms, presenting many variants of product profiles is shown in Vu et al. [12].

Originally, there was some doubt in the field as to the nature of the oxidized products. In particular, the possibility of C6 oxidation was considered. Resolving these issues by mass spectrometry is challenging because of overlapping masses. For example, although the hydrated form of oxidized C6 is an uronic acid with an *m/z* difference of −2 compared with gemdiols and aldonic acids, unambiguous identification by MS is complicated by complex adduct formation patterns. Therefore, efforts have been made to use NMR for reliable product identification (Fig. 3) [11, 13]. The C1-oxidized products can be recognized by the absence of the

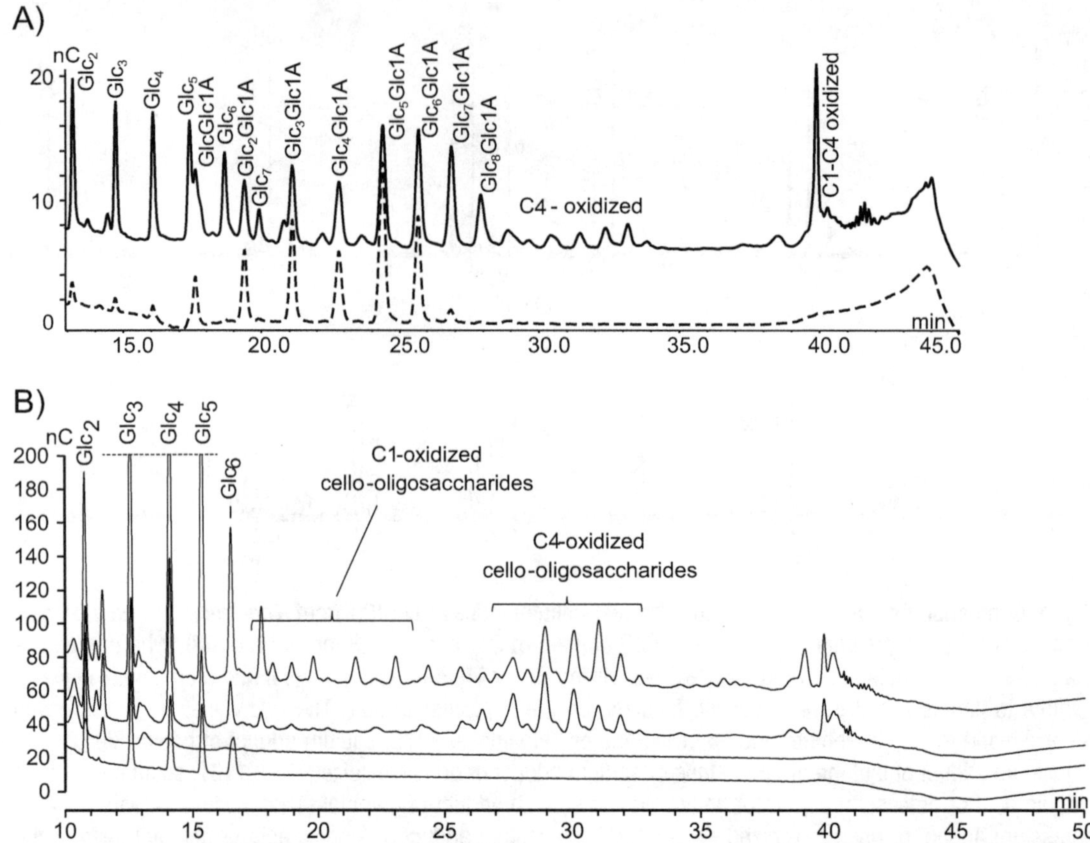

Fig. 2 (**a**) HPAEC product profile for the LPMO domain of CelS2 (or *Sc*LPMO10C), a bacterial C1-oxidizer (dotted chromatogram) and *Sc*LPMO10B, a bacterial C1-C4 oxidizer (solid chromatogram). Native cello-oligosaccharides elute first followed by the aldonic acids. There is a slight overlap between the two product clusters, which implies that the C1-oxidized monomer and dimer elute among the late eluting native oligosaccharides. Peaks representing C4-oxidized products elute after the C1-oxidized products and tend to be lower (although this varies with the LPMO and the conditions used; for example see panel **b** and Fig. S4 in Vu et al. [12]). C4-oxidizing LPMOs tend to show higher amounts of native products (as observed here and in Fig. S4 in Vu et al. [12]), for two reasons: (1) C4-oxidized cello-oligosaccharides undergo on-column degradation reactions that lead to the formation of native cello-oligomer that are one sugar shorter than the original C4-oxidized product [10]; (2) If the LPMO can oxidize C4 and C1, as in the case of *Sc*LPMO10B, its action will lead to the eventual formation of native products. The chemical identity of the compound or compounds eluting around 40 minutes in the chromatogram for the C1&C4 oxidizer only is not known, but it has been shown that peaks eluting in this area are associated with the formation of double oxidized products [11]. (**b**) Products generated from cellulose by a strictly C4 oxidizing fungal LPMO, *Nc*LPMO9C (upper chromatogram), and a C1&C4 oxidizing fungal LPMO, *Gt*LPMO9A-2 (second chromatogram from top). This panel includes chromatograms for a negative control (no enzyme) and a standard of native oligomers (lower two chromatograms). Panel **a** is reproduced from [43] and panel **b** is reproduced from [44]

reducing end signals (usually present at H1 α~5.22; β~4.66 ppm and C1 α~94.7; β~98.6 ppm) and more deshielded chemical shifts, especially for protons at carbons two (C2) and three (C3). On the other hand, C4-oxidized products are harder to identify as they lack

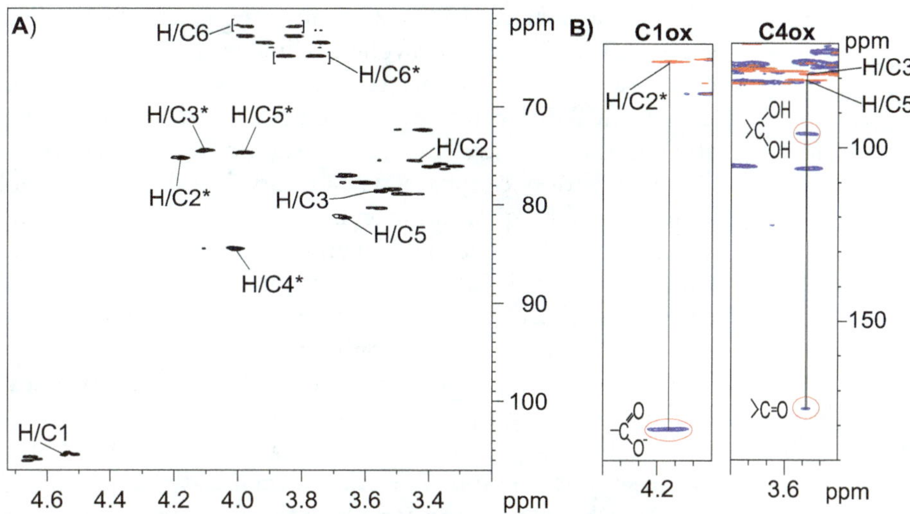

Fig. 3 NMR analysis of the C1 and C4 oxidized LPMO products. (a) ^{13}C HSQC spectrum of the products generated by treating 0.9 mg/mL cellopentaose with 2.9 μM *Nc*LPMO9C in the presence of 0.9 μM of cellobiose dehydrogenase *(Mt*CDH). Under these conditions, the reaction will result in both C1 oxidized and C4 oxidized oligosaccharide ends. The sample was in 99.996% D$_2$O with 5 mM sodium acetate pD 6.0 and spectra were recorded at 25 °C. Peaks for the proton/carbon signals of the C4 oxidized monosaccharide residue are marked by H/C#, where # refers to the ring carbon number for the oxidized monosaccharide residue. Peaks for the proton/carbon signals of the C1 oxidized monosaccharide residue are marked by H/C#*. Brackets indicate pairs of proton/carbon signals attached to the same C6 (the primary alcohol group) in the oxidized monosaccharide residue. For the sake of simplicity peaks related to nonoxidized monosaccharide residues are not marked (a full assignment of chemical shifts is provided in [11]). Panel **b** shows regions of a ^{13}C HSQC spectrum overlaid by a ^{13}C HMBC spectrum recorded for products obtained in a reaction with *Nc*LPMO9C and *Mt*CDH. The left panel (C1ox) shows a correlation (indicated by a vertical line) from the H/C2* peak in HSQC (red) to a peak with a carbon chemical shift of 181.1 ppm in HMBC (blue), corresponding well to a carboxylate group at position C1, which is a hallmark of a C1 oxidized product. The right panel (C4ox) shows correlations from both the H/C3 and H/C5 peaks in HSQC (red) to two carbon peaks with carbon chemical shifts of 95.9 ppm and 175.2 ppm in HMBC (blue), corresponding well to the presence of a geminal diol and a keto group at C4, respectively. The signal intensity of the geminal diol is about 4 times bigger than for the keto group (approximately 80% and 20% of the signal intensity, respectively). The formed chemical groups are drawn next to the diagnostic peaks. This figure and its legend were modified from Isaksen [11]

signals for the proton directly attached to carbon four (C4), and show minimal changes in chemical shifts for the rest of the protons, as compared to the nonoxidized monosaccharide residues. The C1 or C4 signals (in C1 or C4-oxidized products, respectively) may be directly observed in a 1D ^{13}C NMR spectrum, but due to lack of directly attached protons they are difficult to detect. Therefore, it is recommendable to use heteronuclear multi bond correlation (HMBC), which takes advantage of heteronuclear polarization transfer, which enhances the signal by a factor of 32 and enables the detection of correlations from ^{1}H to ^{13}C that are mainly separated by 2 and 3 bonds (Fig. 3b).

Cellulose-active LPMOs act on a variety of cellulose substrates, such as phosphoric-acid swollen cellulose (PASC), more crystalline cellulose (Avicel) and pretreated biomass. It seems evident that LPMOs will differ in terms of their preferred substrate but systematic studies to unravel and explain such differences are lacking. For characterization purposes, PASC and Avicel are preferable substrates. Product quantification is not straightforward and opportunities and methods needed depend on the reaction setup. Reactions with commercial LPMO- and β-glucosidase-containing enzyme cocktails such as Cellic Ctec2® from Novozymes will yield gluconic acid and 4-keto-cellobiose only [14, 15], due to the fact that all longer oxidized products are degraded by the enzymes in the cocktail. Although this has not yet been investigated in-depth, it seems safe to assume in this case that almost all oxidized sites end up in the soluble fraction. There are several ways to quantify gluconic acid, which is stable, and for which there is a commercially available standard. Quantification of 4-keto-cellobiose is not straightforward, because there are no standards and because the product is unstable during the HPAEC-PAD. Nevertheless, quantification has been achieved by exploiting the ability of certain C4 oxidizing LPMOs to cleave cellodextrins, thus generating two products, a C4-oxidized oligomer and an equal amount of a native oligomer, where the latter can easily be quantified [15].

Notably, when assessing the activity of LPMOs acting alone, only soluble products are usually measured and quantified. For substrates with high degrees of polymerization, such soluble products result from the LPMO cleaving twice in the same polysaccharide chain. Thus, clearly, LPMO activities will be underestimated, especially in the start of a reaction, when almost every LPMO reaction will occur in individual polymeric chains that do not become soluble upon one single cleavage. Solubilization of LPMO-generated chain ends is promoted if also cellulases are present in the reaction. Some approaches for quantification of insoluble oxidized chain ends are shortly discussed in Subheading 1.4.

The progress of LPMO-reactions depends on many factors. The characteristics of the insoluble substrate like crystallinity, particle size, pretreatment, and the presence of lignin affect LPMO-activity. Notably, while commonly used substrate concentrations tend to seem saturating, this may not always be the case if the LPMO in question only binds to a small subfraction of the potentially heterogeneous material.

Another important factor is the choice of reductant. LPMOs are capable of using a wide array of reductants that can be either chemical reductants or another redox enzyme [1, 16–19]. The most commonly used types of reductant are small chemical reductants such as ascorbic acid, gallic acid, or reduced glutathione. The advantage of chemical reductants over the best-known protein electron donor, cellobiose dehydrogenase (CDH; [7, 20]), is that neither the cellulosic substrate nor the liberated oligosaccharides will be oxidized by the reductant. In reactions containing CDH, all soluble cellooligosaccharides will be oxidized by CDH yielding aldonic acids, i.e., products identical to products formed by C1 oxidizing LPMOs. On the other hand, it is more difficult to control chemical electron donors due to their instability in the presence of metal ions, their oxidation by the LPMO, and, often, the dependence of their redox potential on pH. Another potential complication lies in the fact that certain reductants may give interfering signals during product analysis. Furthermore, some chemical reductants may cause side reactions resulting in even more complex reaction kinetics and product mixtures. The benefit of protein electron donors is that they can be controlled and maintain electron supply stable over a long time. Moreover, it may be possible to chromatographically monitor simultaneously the activity of both the LPMO and the electron donating enzyme (e.g., [21]), which may lead to increased insight into reaction kinetics (e.g., [21]). Considering that CDH is active on cello-oligosaccharides, we expect increased use of other enzymatic electron supplying systems ([16, 22] in the near future.

Literature data show that LPMOs often display nonlinear kinetics during typical LPMO reactions (Fig. 4; [6, 21, 23], the

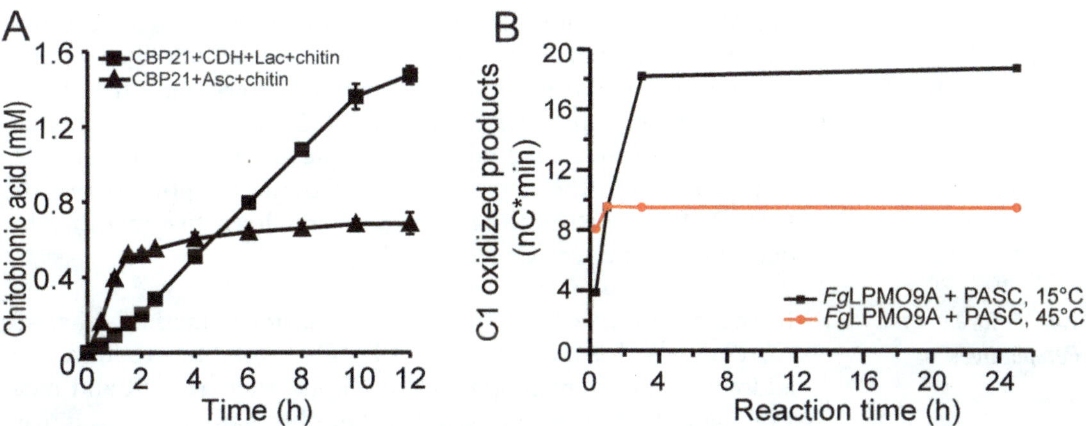

Fig. 4 Nonlinearity of LPMO kinetics. Panel **a** shows an experiment with a chitin-active LPMO which is fueled by either ascorbic acid or CDH/lactose. With ascorbic acid, the reaction is fast but terminates early. With CDH/lactose, the reaction is slower but the system is much more stable. It is worth noting that if single time point experiments would have been done, using the same conditions, the conclusions of a 1 h and a 10 h experiment regarding reductant efficiency would have been totally different. Panel **b** shows the activity of a fungal LPMO, *Fg*LPMO9A on cellulose at two different temperatures. In both cases, the reactions terminate early, long before substrate is depleted. The speed of the reaction shows an expected temperature dependency, but the outcome of a 24 h incubation would be totally dominated by stability effects. Notably, control experiments showed that the inactivation of *Fg*LPMO9A at 45 °C is not due to temperature-induced unfolding. Panel **a** is reproduced from [21], whereas panel **b** is reproduced from [23]

reasons of which are often unclear, although recently, some insights have been obtained (*see* ref. 6, and below). Depletion of reductant, molecular oxygen, and productive binding sites on the substrates may all contribute to the explanation of these observations. Notably, recent data indicate that autocatalytic oxidative self-inactivation of the LPMO, brought about by an imbalance between the amount of reducing power and substrate availability [6] could be a major reason for nonlinear reaction kinetics. This recent work also indicates that H_2O_2, rather than O_2 is the preferred cosubstrate of LPMOs, shedding new light on the role of the reductant [6]. While a detailed discussion of these recent developments is beyond the scope of this review, the key point to make is that, to achieve linear kinetics, reaction conditions, including the type and concentration of the reductant need to be carefully optimized.

1.3 Redox Potential and the Affinity for Ligands and Co-Factor

A deeper understanding of LPMO functionalities requires insight into enzyme-substrate interactions and the chemical and electronic structure of the catalytic center. Several biophysical methods (such as EPR; e.g., [17]) and computational techniques (such as density functional theory calculation; e.g., [24]) are available to obtain such insight. Recently, the first crystal structure of an LPMO in complex with a soluble substrate has become available [25]. The pK_a values of the catalytic histidines may be determined by NMR [26].

Biochemical characterization of LPMO properties is at the basis of these advanced studies of LPMO functionality and may also help in assessing the validity of the results of computations approaches. Below, several such characterization methods are described; (1) the use of isothermal titration calorimetry for determining the affinity of the copper binding site (Fig. 5a); (2) the use of isothermal titration calorimetry for determining the affinity for the substrate (Fig. 5b–d); (3) the determination of the redox potential of the LPMO. It is worth noting that there are alternative methods for determining the redox potential [16].

1.4 Future Perspectives

In this chapter, we have outlined several of the methods that are currently used to characterize LPMOs. Notably, we have not addressed important biophysical methods, such as EPR and measurement of electron transfer rates. The chromatographic and mass spectrometric methods discussed shortly above have been reviewed in more detail in a previous chapter of this series [27] and in-depth discussions of the HPAEC methods and several alternatives may be found in two previous publications [10, 13]. Generally, these methods are well established, but the (in) stability of C4-oxidized products remains a challenge, which likely will be met by introducing C4-specific product modifications [28, 29].

Quantification of LPMO activity remains a challenge because of stability issues concerning both the reductant and the enzyme itself [6, 16, 21], and because it is not straightforward to monitor oxidations on the insoluble material. The stability issues must be

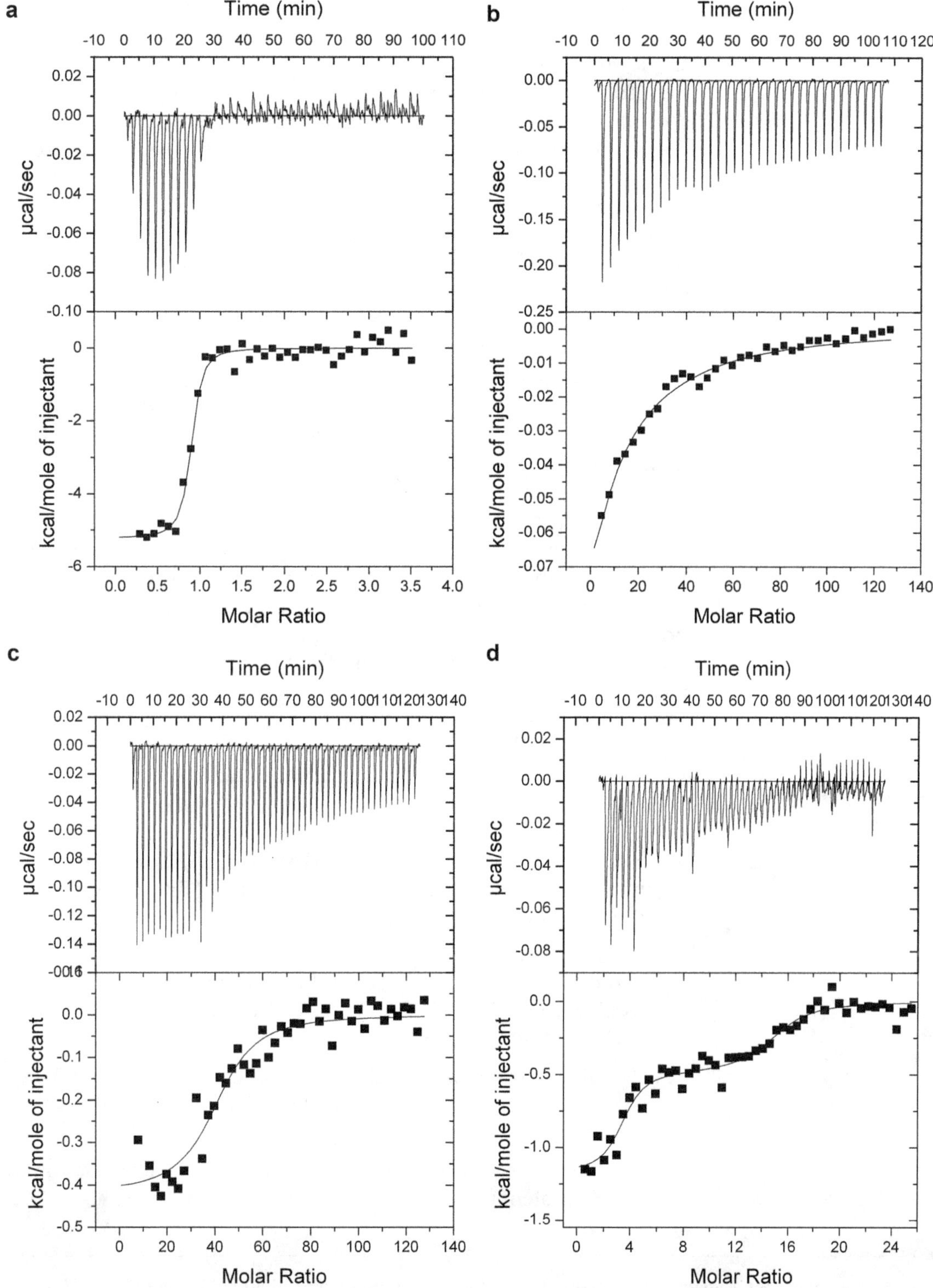

Fig. 5 Thermograms (upper parts) and binding isotherms with theoretical fits (lower parts). (**a**) Binding of 150 μM of Cu^{2+} to 5 μM apo *Nc*LPMO9*C* at pH 5.5, 10 °C. (**b**) Binding of a 11 mM 4-β-D-cellohexaose to 30 μM of *Nc*LPMO9C at pH 5.5, 25 °C. (**c**) Binding of 500 μM of *Nc*LPMO9C-Cu^{2+} to xyloglucan from tamarind

taken very seriously to the extent that quantitative statements about LPMO activity can only be based on progress curves (and not on single time point measurements; Fig. 4a). Insight into oxidations on insoluble products may in some cases be obtained by completely solubilizing LPMO-treated material with hydrolases and then analyze soluble oxidized products (e.g., [21]). Furthermore, the use of size exclusion chromatography in ionic liquid mode for analyzing molecular distributions in cellulose [30] as well as for studying the molecular distribution of product mixtures after enzymatic treatments has a large potential to broaden our understanding of the effects of LPMO treatments. Labeling techniques are of interest, to visualize, and perhaps quantify the occurrence of oxidized chain ends in insoluble material [3, 31].

Despite their obvious importance in Nature and the biorefinery, and despite major research efforts since their discovery in 2010, several aspects of LPMOs remain enigmatic. One recent development concerns the involvement of H_2O_2 in LPMO catalysis. In particular, Bissaro et al. have shown that LPMO activity can be boosted by providing the reduced enzyme continuously with low amounts of H_2O_2, whereas supply of higher dosages of H_2O_2 leads to rapid enzyme inactivation [6]. As noted above, quantification of LPMO activity is challenging and these recent findings may provide both an explanation and a solution to the challenge. Clearly, there is much exciting research ahead in the LPMO field. The analytical tools described below may help in further unraveling of LPMO function in nature and in the biorefinery.

2   Materials

2.1 *MALDI-TOF MS*

1. Equipment: Bruker Ultraflex MALDI-ToF/ToF instrument with a Nitrogen 337-nm laser beam (Bruker Daltonics GmbH, Bremen, Germany).

2. (Optional) Lithium chloride solution (the LiCl concentration should be approximately twice the concentration of the buffer used in the LPMO reaction). Dissolve the desired amount of LiCl in Milli-Q water.

3. 2,5-dihydroxybenzoic acid (DHB) solution: dissolve 4.5 mg DHB (Bruker Daltonics) in 150 µL acetonitrile and 350 µL water.

4. MTP 384 target plate ground steel TF from Bruker Daltonics (or equivalent).

Fig. 5 (continued) seeds (estimated at 0.9 µM), at pH 5.5, 10 °C (**d**) Binding of 500 µM of *Nc*LPMO9C-Cu^{2+} to PASC (estimated 4.5 µM) at pH 5.5, 25 °C. All data and figures are from [41], except panel A, which has not been published previously

2.2 HPAEC-PAD

1. Equipment: Ion exchange chromatography system with pulsed amperometric detection (PAD) (e.g., ICS3000, Dionex).

2. Columns: CarboPac PA1 (2 × 250 mm) and a CarboPac PA1 guard (2 × 50 mm) columns (Dionex, Thermo).

3. MilliQ water. Measure the desired volume of Milli-Q water (Type I, 18.2 MΩ·cm) directly in a dedicated HPAEC mobile phase bottle. Sonicate for 20 min to remove dissolved carbon dioxide and transfer immediately hereafter to the HPAEC system and store under N_2-saturated headspace.

4. Sodium hydroxide (0.1 M). Measure exactly 2 L of Milli-Q water (Type I, 18.2 MΩ·cm) directly in a dedicated HPAEC mobile phase bottle. Sonicate for 20 min to remove dissolved carbon dioxide and transfer immediately hereafter to the HPAEC system and store under N_2-saturated headspace. Add 10.4 mL of NaOH from a 50% (w/w) solution. Do not use NaOH pellets. Close the mobile phase bottle and swirl gently to ensure proper mixing. Maintain N_2-saturated headspace until the mobile phase is discarded.

5. Sodium acetate (1 M in 0.1 M NaOH). Dissolve 82.03 g of anhydrous sodium acetate (≥99% purity) in 1 L of Milli-Q water (Type I, 18.2 MΩ·cm). Filter the solution through no less than a 0.45 μm filter directly into a dedicated HPAEC mobile phase bottle. Sonicate for 20 min to remove dissolved carbon dioxide and transfer immediately hereafter to the HPAEC system and store under N_2-saturated headspace. Add 5.2 mL of NaOH from a 50% (w/w) solution. Do not use NaOH pellets. Close the mobile phase bottle and swirl gently to ensure proper mixing. Maintain N_2-saturated headspace until the mobile phase is discarded.

2.3 Product Identification by NMR

1. Equipment: Bruker Avance 600 MHz spectrometer equipped with a 5-mm cryogenic CP-TCI z-gradient probe.

2. High quality NMR tube 3 or 5 mm (e.g., Schott professional, Norell 509UP8, Wilmad 535-PP-7, Shigemi susceptibility matched to D_2O).

3. Deuterium oxide. 99.9% or 99.96% D_2O (Cambridge Isotope Laboratories, Andover, MA).

4. TSP stock solution (1% w/v; chemical shift reference for proton and carbon): Dissolve 0.1 g of 3-(trimethylsilyl)-propionic-2,2,3,3-d4 acid sodium salt (Aldrich, Milwaukee, WI, USA) in 10 mL of 99.9% D_2O.

5. Buffer. Sodium acetate (5 mM, pD 6.0): Dissolve 0.041 g of anhydrous CH_3COONa in ~80 mL of H_2O, adjust the pH and hereafter adjust the volume to 100 mL with H_2O. To reduce the interference of the water signal transfer 10 mL of the

acetate buffer to a 33 mm 50 mL conical centrifuge tube, lyophilize and redissolve the powder in 10 mL 99.96% D_2O.

6. Cellopentaose stock solution (0.1% w/v): Dissolve cellopentaose (1.0 mg/mL; Megazyme) in sodium acetate (5 mM, pD 6.0) and add 10 µL TSP stock solution (1‰ w/v) per mL of cellopentaose stock solution as chemical shift reference for proton and carbon.

7. Hydroquinone solution (50 mM): Dissolve 0.055 g of hydroquinone in 10 mL of 99.96% D_2O.

8. Cellobiose dehydrogenase (CDH) solution: Make an enzyme solution with a concentration of 15–20 µM CDH, e.g., CDH from *Myriococcum thermophilum* (*Mt*CDH; [32]), using 99.96% D_2O for the dilution (*see* **Note 1**).

9. LPMO solution: Make an enzyme solution with a concentration of 45–60 µM LPMO using 99.96% D_2O for the dilution (*see* **Note 2**).

10. Oxygen gas (100%).

2.4 Copper Saturation

Copper saturation of the LPMO should be performed in the buffer that is planned used in downstream experiments and may thus vary considerably based on the experiment (for example, sodium phosphate is optimal for EPR, but incompatible with MS). Commonly used buffers are Bis-Tris pH 6.0, MES pH 6.0, sodium phosphate pH 6.0, HEPES pH 7.0 and Tris–HCl pH 8.0 in concentrations ranging from 1 to 50 mM. It should be noted that some buffers, including Tris, have the ability to bind/chelate copper and could interfere with LPMO activity, but it is so far not known whether this property of Tris influences LPMO activity under commonly uses reaction conditions. However, it is advised to resaturate LPMOs with copper before experiments if they have been stored in Tris-containing buffers. A copper saturation protocol using 25 mM Bis-Tris pH 6.0 is outlined below.

1. A PD midiTrap G-25 (3.5 mL) column (GE Healthcare; *see* **Note 3**).

2. 25 mM BisTris buffer pH 6.0. Dissolve 0.52 g in 80 mL of Milli-Q water (Type I, 18.2 MΩ cm). Adjust to pH 6.0 with HCl and to 100 mL final volume with Milli-Q water (Type I, 18.2 MΩ cm).

3. 50 mM $CuSO_4$. Dissolve 0.62 g $CuSO_4 \cdot 5H_2O$ in a final volume of 50 mL Milli-Q water (Type I, 18.2 MΩ cm).

2.5 Quantitative Activity Assays

1. 2.0 mL Eppendorf tubes.

2. Substrate (e.g., Avicel or PASC) or lignin-containing substrates such as steam-exploded lignocellulosic material.

3. Buffer of choice (*see* Subheading 2.4).

4. Purified LPMO, copper-saturated as described in Subheading 3.4.

5. Cellulases, for example Cel5A from *Thermobifida fusca* or an LPMO-poor Cellulase mixture such as Celluclast (optional, if quantification of the oxidations in the soluble material or the total sample is required, respectively).

6. Reducing agent; either a small molecule reductant such as ascorbic acid or reduced glutathione (100 mM stock solution in water), gallic acid (100 mM in $EtOH_{absolute}$), or a protein electron donor and its substrate (e.g., cellobiose dehydrogenase and a suitable substrate, such as cellobiose or lactose).

7. Milli-Q water (Type I, 18.2 MΩ cm).

8. Thermomixer C with a thermoblock suitable for 2.0 mL tubes and a ThermoTop (Eppendorf).

9. Water bath or high-temperature heat block that can heat up to 100 °C.

10. 200 mM NaOH. Dilute 0.53 mL NaOH 50% (w/w) solution (as in Subheading 2.2) in a final volume of 50 mL Milli-Q water (Type I, 18.2 MΩ cm).

11. A MultiScreen 96-well plate filter (0.45 μm) operated by a vacuum manifold (Millipore) and a 96-well plate.

12. Standards: native cello-oligosaccharides and cellobiose dehydrogenase, for C1-oxidized standards, or *Nc*LPMO9C, for C4 oxidized standards.

13. 20 mM Tris–HCl pH 8.0. Dissolve 0.24 g Tris base in 80 mL Milli-Q water (Type I, 18.2 MΩ cm) and adjust the pH to 8.0 with HCl. Fill up to a final volume of 100 mL with Milli-Q water (Type I, 18.2 MΩ cm).

2.6 Redox Potential

1. Equipment: Ultraviolet-visible spectrophotometer (i.e., Hitachi U-1900 or Agilent Technologies Cary 8454 UV-Vis spectrophotometer). Cuvettes allowing emission at $\lambda = 610$ nm with a recommended total volume of 100 μL to save protein solution.

2. Milli-Q water and a buffer that does not chelate copper ions (e.g., Chelex-treated 20 mM MES buffer, pH 5.5). *N,N,N′,N′*-tetramethyl-1,4-phenylendiamine dihydrochloride (Sigma-Aldrich, St. Louis, MO, USA).

2.7 Measuring Affinity for Copper

1. Equipment: Isothermal titration calorimeter (e.g., a MicroCal VP-ITC, Malvern, Malvern, England).

2. A purified LPMO, Milli-Q water, a buffer that does not chelate copper ions (e.g., Chelex-treated 20 mM MES buffer, pH 5.5), and ethylenediaminetetraacetic acid (EDTA) and $CuSO_4$ (both form for example Sigma-Aldrich, St. Louis, MO, USA).

3. A NAP-5 column, used for desalting (GE Healthcare; *see* **Note 4**).

2.8 Measuring Affinity for the Substrate

1. Equipment: Isothermal titration calorimeter (e.g., a MicroCal VP-ITC, Malvern, Malvern, England).

2. A purified copper-saturated LPMO, Milli-Q water and a buffer that does not chelate copper ions (e.g., Chelex-treated 20 mM MES buffer, pH 5.5), Avicel, xyloglucan, and/or 1,4-β-D-cellohexaose.

---

3 Methods

The protocols provided cover both common and less common methods for characterizing LPMOs. For more details, readers are directed to the original publications related to the method in question [6, 10, 11, 13, 26, 27, 33]. Where applicable, notes are appended in the Subheading 4. Additional methods, including alternative chromatographic methods, appear in another recent book in the Methods in Molecular Biology series ([27]; Subheadings 2.1, 2.2, 3.1 and 3.2, with notes adapted from this paper. For information on various electrochemical methods and EPR, the reader is referred to papers by for example the Ludwig (e.g., [16]) and Walton (e.g., [25]) groups.

3.1 MALDI-ToF MS with or Without Lithium Doping

1. To prepare samples for MALDI-ToF analysis, reactions should be run at low buffer concentrations (as a rule of thumb, less than 50 mM, preferably much less), and the use of MS-incompatible ions such as phosphate and nitrate should be avoided.

2. Centrifuge samples in an Eppendorf centrifuge at maximum speed at room temperature for 2 min.

3. Apply 2 μL saturated DHB solution to a MALDI plate.

4. Apply 1 μL sample, and mix with the DHB solution (on the MALDI plate).

5. Dry the spot under a stream of warm air.

6. Analyze the sample on a MALDI-ToF instrument (*see* **Note 3**). Alternative method, if the mass spectrum needs simplification (*see* **Note 3** and [27] for details):

   (a) Mix 1 μL sample with 9 μL LiCl solution and vortex for 5 s.

   (b) Apply 2 μL saturated DHB solution to a MALDI plate. DHB is the standard matrix used for all MALDI experiments, but other matrices may work equally well.

   (c) Add 1 μL of the lithium-doped sample (point 7) to the matrix droplet on the MALDI plate.

   (d) Dry the spot under a stream of warm air.

   (e) Analyze the sample on a MALDI-ToF instrument.

3.2 HPAEC-PAD	Use an instrumental setup as described in Subheading 2.2 or similar.

1. Centrifuge samples for 3 min in an Eppendorf centrifuge at maximum speed and transfer supernatants to HPLC vials; normally no further adjustments of the samples is needed (*see* **Note 4**). If samples already have been filtered (*see* Subheading 3.5), this centrifugation step is not necessary.

2. Set the column temperature 30 °C and use 0.25 mL/min flow rate.

3. Use mobile phases containing 0.1 M NaOH (A) and 0.1 M NaOH, 1 M sodium acetate (B) (*see* **Note 5**).

4. Inject 2–10 μL sample.

5. Use the following gradient: a 10 min linear gradient from 100% A (starting condition) to 10% B, a 15 min (this step may be extended if higher resolution is needed [34] or shortened to get higher throughput) linear gradient to 30% B, a 5 min exponential gradient (Dionex curve 6) to 100% B.

6. Recondition the column by running initial conditions (100% A) for 9 min.

If there is a need for higher throughput, and if only C1-oxidations occur, a 10 min method for separation and detection of aldonic acids may be used [13]. For other applications and mass spectrometry adaptations *see* ref. 27.

3.3   Product Identification by NMR

3.3.1   NMR Sample Preparation

1. C1-oxidized products: Dissolve 0.2–2 mg of C1-oxidized product in 150 or 600 μL (for 3 or 5 mm tubes, respectively) 99.9% or 99.96% D_2O. Transfer the sample into a 3- to 5-mm high-quality NMR tube.

2. "External" method for chemical shift referencing: Insert a 3-mm NMR tube (or coaxial insert tube) into a 5-mm NMR tube with TSP (0.1‰ (w/v)) in 99.9% D_2O.

3. C4-oxidized products: Add 500 μL cellopentaose stock solution (0.1% (v/w)) to a 5 mm NMR tube together with either 33 μL of cellobiose dehydrogenase (CDH, *Mt*AA3) to a final concentration of 0.9 μM) or 33 μL of hydroquinone solution (to a final concentration of 3 mM). After addition of 17 μL LPMO (e.g., *Nc*LPMO9C) to a final concentration of 2.9 μM, flush the head-space of the NMR with oxygen gas (100%) for ~10 s before sealing the tube with parafilm around the cap. After incubation of the samples at 25 °C for 24 h, the reaction products can be analyzed by NMR spectroscopy (*see* **Note 1**).

3.3.2   NMR Data Acquisition and Processing

1. Set the temperature to 25 °C on the NMR spectrometer. Insert the sample in the NMR spectrometer and let the sample temperature equilibrate for ~10 min. After equilibration, lock on

the deuterium signal and calibrate the tune, match and shims of the spectrometer according to standard operating procedures. Calibrate the ^{1}H and ^{13}C pulses and set spectral widths as well as other parameters in the subsequent spectra.

2. For structural elucidation of C1- and C4-oxidized products, the following spectra can be used (the recommended pulse programs and key parameter settings are listed in brackets): 1D proton [zg30; sw 14 ppm; TD: 32 k], 2D double quantum filtered correlation spectroscopy (DQF-COSY) [cosydfphpr; TD: 2k, 512; SW 11 ppm,11 ppm], 2D In-phase correlation spectroscopy (IP-COSY) [35] [ipcosygppr-tr; TD: 2k, 256; SW 11 ppm, 11 ppm; constant-time evolution in the indirect dimension], 2D total correlation spectroscopy (TOCSY) with 70 ms mixing time [mlevphpr; DS 128; TD: 2k, 512; SW 11 ppm,11 ppm; d1 2 s; d9 70 ms (spin-lock mixing time)], 2D ^{13}C heteronuclear single quantum coherence (HSQC) with multiplicity editing [hsqcetgpsisp2.3; TD: 2k, 256; SW 11 ppm, 60 ppm, o1p 80 ppm; Cnst2 135 Hz (^{1}J$_{CH}$)], 2D ^{13}C HSQC-[^{1}H,^{1}H]TOCSY with 70 ms mixing time on protons [hsqcdietgpsisp.2; DS 128; TD: 2k, 256; SW 11 ppm, 60 ppm, o1p 80 ppm; Cnst2 135 Hz (^{1}J$_{CH}$); d1 2 s (avoid RF-heating); d9 70 ms (Spin-lock mixing time)], and 2D heteronuclear multiple bond correlation (HMBC) with BIRD filter to suppress first order correlations [hsqcetgpml; TD: 2k, 256; SW 11 ppm, 160 ppm, o1p 120 ppm; Cnst2 135 Hz (^{1}J$_{CH}$), Cnst6 125 Hz (lower filter to remove ^{1}J$_{CH}$), Cnst7 150 Hz (higher filter to remove ^{1}J$_{CH}$), Cnst13 8 Hz (coupling constant for long range coupling)].

3. Process and analyze the spectra using a standard NMR processing software, such as TopSpin, Mestrelab, or ACD/NMR Processor.

3.3.3 Assignment of Chemical Shifts

1. The individual monosaccharide residues are assigned by starting at the anomeric signal and/or at the primary alcohol group at C6 and then following the proton–proton connectivity using TOCSY, DQF-COSY/IP-COSY, and ^{13}C HSQC-[^{1}H,^{1}H] TOCSY spectra. ^{13}C-HSQC is used for assigning the carbon chemical shifts. The ^{13}C HMBC spectrum provides information on connectivity between the individual monosaccharide residues as well as the chemical shifts for the oxidized carbons C1 or C4 via two-bond couplings (^{2}J$_{CH}$) (*see* **Note 6**).

3.4 Copper Saturation

1. Add a threefold molar excess of the CuSO$_4$ to the protein solution and incubate at room temperature for 30 min (*see* **Note 7**)

2. Equilibrate the PD midiTrap G-25 (3.5 mL) column (GE Healthcare) with 15 mL of the same buffer that is used in the activity assay (*see* **Notes 2** and **8**)

3. Add a maximum of 1.0 mL sample to the column and let it enter (*see* **Note 9**).

4. Elute the protein with 1.0 mL of buffer (*see* **Note 10**).

3.5 *Quantitative Activity Assays*

1. Mix the desired amount of substrate with water, buffer, and the LPMO (usually 1.0 µM). The final reaction volume should not exceed 1.5 mL.

2. Preincubate the sample at the reaction temperature for 15 min (*see* **Note 11**).

3. Start the reaction by adding 1.0–5.0 mM reducing agent (*see* **Note 12**).

4. Incubate the sample at 40 °C, shaking at ~800 rpm.

5. Collect samples in regular intervals to follow the progress of the reaction.

6. If solubilized products only are to be quantified, stop the reactions by filtration (*see* **Note 13**) or by boiling (note that some enzymes have are remarkably stable when substrate is present, so make sure to include proper negative controls to check for possible, however unlikely, residual activity). Filtration is preferred since this is required before analysis by HPLC to prevent particles from being injected into the system.

   (a) If the amount of the individual oxidized oligosaccharides is to be quantified, the sample can be analyzed directly with HPAEC (Subheading 3.2). If the total amount of soluble oxidized oligosaccharides is to be quantified, add a cellulase (e.g., the GH5 endoglucanase *Tf*Cel5A from *T. fusca* to a final concentration of 1 µM) to the filtrates and incubate at 40 °C overnight. This will reduce the complexity of the soluble products generated by the LPMO to a mixture of glucose, cellobiose, and oxidized products with a degree of polymerization of 2 and 3 (i.e., GlcGlc1A, Glc_2Glc1A, Glc4GemGlc, and $Glc4GemGlc_2$), allowing quantification by HPEAC (Subheading 3.2) using the appropriate standards (*see* **steps 8** and **9** below for the generation of standards).

   (b) If only aldonic acids are formed by the LPMO, it may also be convenient to use a β-glucosidase to reduce the complexity of the product mixture (e.g., [36] added one unit of β-glucosidase from *Thermotoga maritima* (Megazyme) to 40 µL aliquots containing soluble aldonic acid cello-oligosaccharides of various DP followed by incubation for

16 h at 37 °C to obtain ~100% conversion of the products to glucose and gluconic acid).

(c) If insoluble oxidized sites are to be quantified, stop the reaction by boiling for 15 min and degrade the sample using a LPMO-poor cellulase cocktail under conditions that do not promote LPMO activity prior to product analysis according to Subheading 3.2 (20 mM EDTA may be added to the reaction to abolish LPMO activity, but this may give problems with downstream analysis).

7. After cellulase treatment, filter or centrifuge the samples to remove substrate remnants, before analysis by HPAEC according to Subheading 3.2.

8. A C4-oxidized standard may be prepared as described by Müller et al. [15]: Dissolve 2.5 mg of the native cello-oligosaccharide Glc_5 in 500 μL 20 mM Tris–HCl pH 8.0. Add 0.5 g/L NcLPMO9C (or another LPMO active on soluble cello-oligosaccharides) and 2 mM ascorbic acid and add 20 mM Tris–HCl pH 8.0 to 1.0 mL. Incubate at an appropriate reaction temperature, e.g., 33 °C for NcLPMO9C, for 24 h. Note that there are stability issues here (*see* **Notes 14** and **15**).

9. C1 oxidized standards: Dissolve an appropriate amount of native cello-oligosaccharides (e.g., 2.5 mg) in 500 μL 25 mM Bis-Tris pH 6.0 and add cellobiose dehydrogenase to a final concentration of 2.0 μM. Incubate the sample for 48 h at 40 °C to ensure oxidation of the cellobiose (*see* **Notes 16** and **17**).

3.6 Redox Potential

1. Use an instrumental setup as described in Subheading 2.6, or similar.

2. Mix oxygen-free solutions: 50 μL 200 μM of N,N,N',N'-tetramethyl-1,4-phenylenediamine (TMP_{red}) in its reduced form and 50 μL, 70 μM Cu^{2+}-saturated LPMO in Chelex-treated 20 mM MES buffer pH 5.5, $t = 25$ °C in a cuvette. LPMO-Cu^{2+} solutions can be made oxygen free by sequential degassing and adding of N_2 over the solution in a stoppered vial with a rubber stopper using standard Schlenk techniques. The same approach can be used for the TMP_{red} solution. It is important that the LPMO solution is anaerobic prior to addition of TMP_{red}. An alternative method is to bubble N_2 (g) through the buffer for 1 h before the addition of TMP_{red}.

3. Determine the extent of the reaction by measuring the absorbance of the TMP radical cation (TMP_{ox}) at $\lambda = 610$ nm. Concentrations of TMP_{ox}, which equal concentrations of LPMO-Cu^{1+}, are calculated by using an extinction coefficient of 14.0/mM/cm [37].

4. This determination allows for calculations of the equilibrium concentrations for the electron transfer reaction (Eq. 1) and hence the equilibrium constant (Eq. 2).

$$TMP_{red} + LPMO\text{-}Cu^{2+} \rightleftarrows TMP_{ox} + LPMO\text{-}Cu^{1+} \qquad (1)$$

$$K = \frac{[TMP_{ox}]\,[LPMO\text{-}Cu^{1+}]}{[TMP_{red}]\,[LPMO\text{-}Cu^{2+}]} \qquad (2)$$

The relationship between the free energy change (ΔG_r°), the equilibrium constant (K), and the cell potential (E°) of the reaction is shown in Eq. 3:

$$\Delta G_r^\circ = -RT\ln K = -nFE^\circ \qquad (3)$$

where R is the gas constant, T is the temperature in Kelvin, n is the number of electrons transferred in the reaction, and F is the Faraday constant. Summation of the measured cell potential for the equilibrium reaction (Eq. 1) with the cell potential of the TMP_{ox}/TMP_{red} redox couple (273 mV vs. normal hydrogen electrode; [38]) yields the cell potential for the $LPMO\text{-}Cu^{2+}/LPMO\text{-}Cu^{1+}$ redox couple.

3.7 Measuring Affinity for Copper

1. Use an instrumental setup as described in Subheading 2.7 or similar.

2. Prepare an apo-LPMO solution by treating purified enzyme with 100 mM EDTA, followed by desalting using a NAP-5 column (GE Healthcare) equilibrated with Chelex-treated 20 mM MES buffer, pH 5.5.

3. Thoroughly degas the LPMO solution prior to experiments to avoid air bubbles in the calorimeter. The recommended concentration of apo-LPMO is 5 μM and the suggested temperature of the reaction is $t = 10\ °C$ (*see* **Note 18**). The volume of the LPMO solution added to the calorimeter should exceed the volume of the reaction cell. In this example, for a microcal VP-ITC with a reaction cell volume of 1.42 mL, it is beneficial to (over)fill the reaction cell with 1.6 mL of LPMO solution.

4. Place 300 μL of 150 μM $CuSO_4$ in Chelex treated 20 mM MES buffer, pH 5.5 in the syringe. Inject 40 aliquots of 4 μL of the copper solution into the reaction cell, at 180 s intervals, with a stirring speed of 260 rpm. The software accompanying the calorimetric system collects ITC data automatically.

5. Prior to further analysis, correct for the heat of dilution by subtracting the heat produced by the injections of ligand into the reaction cell after completion of the binding reaction.

A complementary approach is to subtract the heats produced by titrating the ligand into buffer alone in a parallel experiment. Ideally, these heats should have the same magnitude.

6. Fit data using a nonlinear least squares algorithm using a single-site binding model employed by the software that accompanies the calorimetric system to yield the stoichiometry (n), the equilibrium binding association constant (K_a), and the enthalpy change (ΔH_r°) of the reaction. The changes in reaction free energy (ΔG_r°) and entropy (ΔS_r°) as well as the dissociation constant (K_d) are calculated using the relationship in Eq. 4:

$$\Delta G_r^\circ = -RT \ln K_a = RT \ln K_d = \Delta H_r^\circ - T\Delta S_r^\circ \qquad (4)$$

Errors in ΔH_r°, K_d, and ΔG_r° are obtained as standard deviations of at least three experiments. Errors in ΔS_r° are obtained through propagation of errors.

7. It is worth noting that by combining the K_d for Cu^{2+} resulting from this experiment and the redox potential resulting from the method described in Subheading 3.6, one may obtain the K_d for Cu^+, as described in detail in Fig. S2 of Aachmann et al. [26].

3.8 Measuring Affinity for the Substrate

1. Use an instrumental setup as described in Subheading 2.8 or similar.

2. For binding of small, soluble oligomers (i.e., 1,4-β-D-cellohexaose or hexa-*N*-acetyl chitohexaose), prepare a 30 μM copper-saturated LPMO solution in Chelex-treated 20 mM MES buffer, pH 5.5 (*see* **Note 19**).

3. Thoroughly degas the LPMO solution prior to experiments to avoid air bubbles in the calorimeter.

4. The volume of the LPMO solution added to the calorimeter should exceed the volume of the reaction cell. In this example, for a microcal VP-ITC with a reaction cell volume of 1.42 mL, it is beneficial to (over)fill the reaction cell with 1.6 mL of LPMO solution. The suggested temperature of the reaction is $t = 25\ ^\circ C$.

5. Place 300 μL of 11 mM of the ligand in Chelex treated 20 mM MES buffer, pH 5.5 in the syringe. Inject 40 aliquots of 8 μL of the ligand solution into the reaction cell, at 180 s intervals, with a stirring speed of 260 rpm. The software accompanying the calorimetric system collects ITC data automatically.

6. Prior to further analysis, correct for the heat of dilution by subtracting the heat produced by titrating the ligand into buffer alone in a parallel experiment.

7. Fit data using a nonlinear least squares algorithm using a single-site binding model employed by the software that accompanies the calorimetric system where the stoichiometry (n) is set to be 1 (*see* **Note 19**). The fitting yields the equilibrium binding association constant (K_a), and the enthalpy change (ΔH_r°) of the reaction. The changes in reaction free energy (ΔG_r°) and entropy (ΔS_r°) as well as the dissociation constant (K_d) are calculated using the relationship in Eq. 4. Errors in ΔH_r°, K_d, and ΔG_r° are obtained as standard deviations of at least three experiments. Errors in ΔS_r° are obtained through propagation of errors

8. For binding to large, soluble oligomers or insoluble polymers (i.e., xyloglucan, cellulose, chitin, etc.), place the LPMO in the syringe and the substrate in the reaction cell (*see* **Note 20**).

9. Prepare cellulose (phosphoric acid-swollen cellulose, PASC, 0.15 mg/mL) or xyloglucan (22 kDa, from tamarind seeds, 0.9 μM) in Chelex-treated 20 mM MES buffer, pH 5.5 (*see* **Note 20**).

10. Degas and add PASC or xyloglucan in the reaction cell as described in **steps 3** and **4**. Suggested temperature of the reaction is $t = 25$ and 10 °C, for PASC and xyloglucan, respectively.

11. Place 300 μL of 500 μM of the LPMO in Chelex treated 20 mM MES buffer, pH 5.5 in the syringe. Inject 50 aliquots of 6 μL at 180 s intervals of the ligand solution into the reaction cell with a stirring speed of 260 rpm.

12. Prior to further analysis, correct for heat of dilution by subtracting the heat produced by the injections of ligand into the reaction cell after completion of the binding reaction. A complementary approach is to subtract the heats produced by titrating the ligand into buffer alone in a parallel experiment. Ideally, these heats should have the same magnitude.

13. Fit data as described in **step 7** *without* fixing the stoichiometry (n).

---

4   Notes

1. The stoichiometric ratio between the LPMO and the CDH may need to be optimized for each LPMO and may also need adjustment if another CDH is used.

2. Desalting can be achieved using several standard columns and two of these appear in this paper: PD midiTrap G-25 & NAP-5, both from GE Healthcare. Note that certain carbohydrate-binding modules, which are present on some LPMOs, may bind to some of the column materials used.

3. Under standard conditions, several adducts tend to be observed during MALDI analysis (Fig. 1), the most common being sodium and potassium adducts. The fact that the m/z difference between a sodium and a potassium adduct equals the m/z difference connected to an oxygen atom complicates interpretation of mass spectra, although such interpretation is not impossible (Fig. 1). One simple way of overcoming this multiplicity of signals, which may hamper product identification, is ion doping to force the adduct composition to a defined adduct type, as described in detail in [27].

4. Samples may contain most buffers used in biochemistry labs, but avoid organic solvents (e.g., acetonitrile, methanol) in the sample matrix since several of these affect the PA-detection. Injection volumes between 2 and 10 μL may be used. Five microliter injection has proven a suitable compromise between resolution and sensitivity. Two microliters will result in slightly improved resolution due to less sample diluent effects in HPLC. Increasing the injection volume to 10 μL may be considered, but be aware that the sample diluent effect will affect resolution to some extent unless your sample diluent is equivalent to the eluent.

5. When eluents are prepared, note that the 50% NaOH solution has limited use due to carbonate contamination; we usually discard these solutions when approximately half of it has been used. It is critical to follow this and other procedures for mobile phase preparation or to follow equivalent recommendations by instrument vendors, in order to achieve satisfactory results. The most important things to pay attention to are (1) the quality of water and chemicals, (2) sufficient degassing for removal of dissolved carbon dioxide, (3) storage in an atmospheres with reduced content of carbon dioxide (N_2 or He-saturated headspace), (4) regular exchange of mobile phases (2–3 days shelf life) and (5) to avoid all kinds of detergents in mobile phases (do not use detergent washing of mobile phase bottles between eluent preparations, but restrict cleaning to rinsing with Milli-Q water, Type I, 18.2 MΩ cm). Extensive exchange of mobile phases on the column and careful column regeneration after each change of eluent are also important in order to remove accumulation of carbonate contaminations on the column.

6. ^{1}H-^{13}C HMBC spectra provide further insight into the nature of the products. In the HMBC spectra for the C4-oxidized products, correlations can be observed from H/C-5 and H/C-3 to a carbon chemical shift of 95.9 ppm and 175.2 ppm. These ^{13}C shifts correspond to a keto group (175.2 ppm) and its hydrated geminal diol form (95.9 ppm), and the intensity ratio between these forms depends on the pH of the sample.

For the C1-oxidized products, the hydrated form (an aldonic acid) dominates; this is observed as a correlation from H/C-2 to a carbon chemical shift of 181.1 ppm.

7. The protein concentration should not exceed ~15–20 g/L. Some proteins tend to be unstable at higher concentrations when the excess of copper is added. Consider also the type of buffer that is used, since some buffers are good copper chelators.

8. Use the same buffer that is used in the activity assays. Some buffers can interfere with analytical methods. For example, MES can be detected in MALDI-TOF and HPAEC analyses.

9. It is advantageous to load lower sample volumes (~300–500 µL) with a high concentration in order to limit dilution during desalting. Let the sample enter and add buffer to fill up the sample volume to 1.0 mL. This implies for example that one loads 400 µL of sample followed by 600 µL of buffer, instead of loading 1 mL of sample.

10. In order to avoid carryover of excess copper, the elution volume is reduced to 1.0 mL instead of the 1.5 mL described in the manufacturer's protocol. A second elution step using 500 µL of buffer may be performed to collect the residual protein from the column. Note that some of the excess copper will elute in this second elution step.

11. Preincubation allows the LPMO to bind to the substrate before the reaction is started. By adding the reductant as the last component in the reaction, enzyme inactivation due to self-oxidation is prevented in the initial stage of the reaction. Enzyme inactivation is often observed, which is why it is important to always follow the progress of the reaction and to report progress curves.

12. It is important that the amount of reductant is suitable for the reaction. A good starting point is 1.0 mM of chemical reductant or 1.0 µM cellobiose dehydrogenase and 5.0 mM lactose. It has recently been described that it is also possible to drive LPMO reactions by adding H_2O_2 [6]. In that case, H_2O_2 need to be kept low and the amount of reductant added may be substoichiometric. Regular addition of fresh H_2O_2 (and, possibly, reductant) is then necessary, since these compounds will be consumed in the course of the reaction. Too high amounts of H_2O_2 should be avoided since this may harm the enzyme. As a starting point, one may try initiating reactions by adding 10 µM ascorbic acid and 100 µM H_2O_2.

13. Some LPMOs are active on soluble cello-oligosaccharides; in these cases, filtration will not stop the reaction. One may consider using ultrafiltration spin filters.

14. When using *Nc*LPMO9C under these conditions, approximately 70% of the native Glc_5 will be converted to Glc4gemGlc and Glc_3 (in equimolar amounts), whereas the rest will be un-cleaved or converted to small amounts of $Glc4gemGlc_2$ and Glc_2 (in equimolar amounts). By quantifying the native products, the amount of C4-oxidized product in the standard sample can be determined. It is of major importance to realize that C4-oxidized sugars are unstable at high pH, which implies, among other things, that on-column degradation will occur when carrying out the chromatographic analysis described in Subheading 3.2 [10]. Thus, quantitative analysis of C4-oxidized products according to the methods described here is not that accurate. It is obviously important to treat all samples (standard and samples from for example a progress curve) in exactly the same manner. Better quantification of C4-oxidized products may be obtained by carrying our chemical modifications prior to chromatographic analysis by HPAEC-PAD, and several methods are currently being considered and under development [28, 29]. It is also possible, albeit not straightforward, to use other chromatographic techniques, in particular PGC (Porous Graphitic Carbon chromatography) [10].

15. Note that the procedure for producing a C4-standard described above only yields an oxidized dimer, which, consequently, is the only species that can be quantified. This is still useful since the by far dominant of C4-oxidized products obtained after incubating a cellulosic substrate with an LPMO-containing cellulase cocktail such as Cellic CTec2 is the dimeric species [15]. When using the correct chromatographic setup, the C4-oxidized dimer yields a diagnostic peak in the HPAEC chromatograms that can easily be quantified.

16. The amount of oxidized products can be determined by quantifying the nonconverted native oligomer(s) and subtracting the amounts of nonconverted oligomers from the total amount of oligomers used in the reaction. Note that CDH is active on cello-oligosaccharides of varying length and that this approach thus can be used to quantify oxidized oligomers of varying lengths.

17. The dominant C1-oxidized products obtained after incubating a cellulosic substrate with an LPMO-containing cellulase cocktail such as Cellic CTec2 are gluconic acid and cellobionic acid [14].

18. It has been observed for several LPMOs that the stoichiometry of copper ions versus LPMO is higher than 1 (i.e., 2–3) at concentrations higher than 5 µM of the protein and at reaction

temperatures higher than $t = 10\ °C$. The origin of this behavior has not been investigated. A possible explanation is the occurrence of protein aggregation, which is more likely to take place at temperatures and protein concentrations that are higher than the concentrations that yield a 1:1 stoichiometry.

19. The shape of the ITC binding curve is described by the so-called Wiseman c value [39]. which can be expressed as follows: $c = nK_a[M]_t$, where n is the stoichiometry of the reaction, K_a is the equilibrium binding association constant, and $[M]_t$ is the protein concentration. It is well established that c values within the range of $10 < c < 1000$ are a prerequisite for meaningful calculations of K_a. It has been shown, however, that binding thermodynamics can be obtained even if c is in the range of $0.01–10$ if a sufficient portion of the binding isotherm is used for analysis [40]. This is achieved by ensuring a high molar ratio of ligand versus protein at the end of the titration, accurate knowledge of the concentrations of both ligand and protein, an adequate level of signal to noise in the data, and known stoichiometry. The latter implies that the value of n (i.e., 1 for a 1 to 1 binding system) needs to be fixed in the nonlinear fitting of experimental data to the theoretical model. In our experience, binding of small, soluble oligomers yields a c value below 10 [28, 39]. Figure 5a, b illustrate this concept, showing data for binding of Cu^{2+} to apo NcLPMO9C with a K_a of $3.0 \times 10^7\ M^{-1}$ ($K_d = 33$ nM) with a c value of 150 and 1,4-β-D-cellohexaose to NcLPMO9C-Cu^{2+} K_a of $1.2 \times 10^3\ M^{-1}$ ($K_d = 0.81$ mM) with a c value of 0.04. The binding isotherm changes from being sigmoidal (high c value) to hyperbolic (low c value). In the first case (Fig. 5a), the fitting can be used to determine the stoichiometry, n; in the second case (Fig. 5b), the stoichiometry needs to be set.

20. It is in practice impossible to place insoluble polymeric substrates in the syringe in an ITC experiment, but doing so with large soluble substrates, such as xyloglucan from tamarind seeds, would in principle be possible. Still, placing PASC or the xyloglucan in the reaction cell and the LPMO in the syringe carries an advantage, since this approach allows estimation of the number of LPMO molecules binding to the substrate. In the case of NcLPMO9C binding to xyloglucan from tamarind seeds, an average of 30 LPMO units bind to the xyloglucan with an estimated degree of polymerization of the main chain of 594 (calculated from the sugar composition) with a K_d of 2.3 μM (Fig. 5) [41]. This suggests binding of one LPMO per 20 sugar residues in the main chain. If the experimental setup was reversed, with the xyloglucan in the syringe and the LPMO in the reaction cell (as is the case for $(Glc)_6$ binding to NcLPMO9C), the same K_d would be observed with the

reciprocal (i.e., 0.033) stoichiometry. With respect to for example PASC, the concentration of the polymer needs to be set based on an estimation of an average chain length of the polymer. As an example, Avicel-derived PASC prepared according to Zhang and Lynd is estimated to have an average chain length of 200 glucose units [42]. This Avicel-derived PASC was used in the work of Borisova et al. [41]. Here, a concentration of 4.5 μM PASC, yielding 6.4 nmol polymer chains assuming 200 glucose units per chain, was placed in the reaction cell. A total of 300 μL of a 500 μM solution of NcLPMP9C (150 nmol) was needed to complete LPMO binding to the PASC, and the fitting of theoretical data to the experimental data suggested that ~3 and ~11 NcLPMO9C bind with a K_d of 0.013 μM and 0.64 μM, respectively, per polymer chain of PASC (Fig. 5).

Acknowledgments

This work was supported by the Norwegian Research Council through grants 214613, 216162, 214138, 226244, 221576, 226247, and 244259.

References

1. Vaaje-Kolstad G, Westereng B, Horn SJ et al (2010) An oxidative enzyme boosting the enzymatic conversion of recalcitrant polysaccharides. Science 330:219–222. https://doi.org/10.1126/science.1192231

2. Harris PV, Xu F, Kreel NE et al (2014) New enzyme insights drive advances in commercial ethanol production. Curr Opin Chem Biol 19:162–170. https://doi.org/10.1016/j.cbpa.2014.02.015

3. Eibinger M, Ganner T, Bubner P et al (2014) Cellulose surface degradation by a lytic polysaccharide monooxygenase and its effect on cellulase hydrolytic efficiency. J Biol Chem 289:35929–35938. https://doi.org/10.1074/jbc.M114.602227

4. Beeson WT, Vu VV, Span EA et al (2015) Cellulose degradation by polysaccharide monooxygenases. Annu Rev Biochem 84:923–946. https://doi.org/10.1146/annurev-biochem-060614-034439

5. Walton PH, Davies GJ (2016) On the catalytic mechanisms of lytic polysaccharide monooxygenases. Curr Opin Chem Biol 31:195–207. https://doi.org/10.1016/j.cbpa.2016.04.001

6. Bissaro B, Rohr AK, Muller G et al (2017) Oxidative cleavage of polysaccharides by monocopper enzymes depends on H_2O_2. Nat Chem Biol Adv 13(10):1123–1128. https://doi.org/10.1038/nchembio.2470. http://www.nature.com/nchembio/journal/vaop/ncurrent/abs/nchembio.2470.html#supplementary-information

7. Phillips CM, Beeson WT, Cate JH, Marletta MA (2011) Cellobiose dehydrogenase and a copper-dependent polysaccharide monooxygenase potentiate cellulose degradation by $Neurospora$ $crassa$. ACS Chem Biol 6:1399–1406. https://doi.org/10.1021/cb200351y

8. Coenen GJ, Bakx EJ, Verhoef RP et al (2007) Identification of the connecting linkage between homo- or xylogalacturonan and rhamnogalacturonan type I. Carbohydr Polym 70:224–235

9. Westereng B, Coenen GJ, Michaelsen TE et al (2009) Release and characterization of single side chains of white cabbage pectin and their complement-fixing activity. Mol Nutr Food Res 53:780–789. https://doi.org/10.1002/mnfr.200800199

10. Westereng B, Arntzen MØ, Aachmann FL et al (2016) Simultaneous analysis of C1 and C4 oxidized oligosaccharides, the products of lytic polysaccharide monooxygenases acting on cellulose. J Chromatogr A 1445:46–54. https://doi.org/10.1016/j.chroma.2016.03.064

11. Isaksen T, Westereng B, Aachmann FL et al (2014) A C4-oxidizing lytic polysaccharide monooxygenase cleaving both cellulose and cello-oligosaccharides. J Biol Chem 289:2632–2642. https://doi.org/10.1074/jbc.M113.530196

12. Vu VV, Beeson WT, Phillips CM et al (2014) Determinants of regioselective hydroxylation in the fungal polysaccharide monooxygenases. J Am Chem Soc 136:562–565. https://doi.org/10.1021/ja409384b

13. Westereng B, Agger JW, Horn SJ et al (2013) Efficient separation of oxidized cello-oligosaccharides generated by cellulose degrading lytic polysaccharide monooxygenases. J Chromatogr A 1271:144–152. https://doi.org/10.1016/j.chroma.2012.11.048

14. Cannella D, Hsieh CW, Felby C, Jorgensen H (2012) Production and effect of aldonic acids during enzymatic hydrolysis of lignocellulose at high dry matter content. Biotechnol Biofuels 5:26. https://doi.org/10.1186/1754-6834-5-26

15. Muller G, Varnai A, Johansen KS et al (2015) Harnessing the potential of LPMO-containing cellulase cocktails poses new demands on processing conditions. Biotechnol Biofuels 8:187. https://doi.org/10.1186/S13068-015-0376-Y

16. Kracher D, Scheiblbrandner S, Felice AKG et al (2016) Extracellular electron transfer systems fuel cellulose oxidative degradation. Science 352:1098–1101. https://doi.org/10.1126/science.aaf3165

17. Quinlan RJ, Sweeney MD, Lo Leggio L et al (2011) Insights into the oxidative degradation of cellulose by a copper metalloenzyme that exploits biomass components. Proc Natl Acad Sci USA 108(37):15079–15084. https://doi.org/10.1073/pnas.1105776108

18. Westereng B, Cannella D, Agger JW et al (2015) Enzymatic cellulose oxidation is linked to lignin by long-range electron transfer. Sci Rep 5:18561. https://doi.org/10.1038/srep18561

19. Frommhagen M, Sforza S, Westphal AH et al (2015) Discovery of the combined oxidative cleavage of plant xylan and cellulose by a new fungal polysaccharide monooxygenase. Biotechnol Biofuels 8:12. https://doi.org/10.1186/s13068-015-0284-1

20. Langston JA, Shaghasi T, Abbate E et al (2011) Oxidoreductive cellulose depolymerization by the enzymes cellobiose dehydrogenase and glycoside hydrolase 61. Appl Environ Microb 77:7007–7015. https://doi.org/10.1128/Aem.05815-11

21. Loose JS, Forsberg Z, Kracher D et al (2016) Activation of bacterial lytic polysaccharide monooxygenases with cellobiose dehydrogenase. Protein Sci 25:2175–2186. https://doi.org/10.1002/pro.3043

22. Garajova S, Mathieu Y, Beccia MR et al (2016) Single-domain flavoenzymes trigger lytic polysaccharide monooxygenases for oxidative degradation of cellulose. Sci Rep 6:28276. https://doi.org/10.1038/srep28276

23. Nekiunaite L, Petrovic DM, Westereng B et al (2016) FgLPMO9A from *Fusarium graminearum* cleaves xyloglucan independently of the backbone substitution pattern. FEBS Lett 590:3346–3356. https://doi.org/10.1002/1873-3468.12385

24. Kim S, Ståhlberg J, Sandgren M et al (2014) Quantum mechanical calculations suggest that lytic polysaccharide monooxygenases use a copper-oxyl, oxygen-rebound mechanism. Proc Nat Acad Sci 111:149–154. https://doi.org/10.1073/pnas.1316609111

25. Frandsen KEH, Simmons TJ, Dupree P et al (2016) The molecular basis of polysaccharide cleavage by lytic polysaccharide monooxygenases. Nat Chem Biol 12:298. https://doi.org/10.1038/Nchembio.2029

26. Aachmann FL, Sorlie M, Skjak-Braek G et al (2012) NMR structure of a lytic polysaccharide monooxygenase provides insight into copper binding, protein dynamics, and substrate interactions. Proc Natl Acad Sci USA 109:18779–18784. https://doi.org/10.1073/pnas.1208822109

27. Westereng B, Arntzen MO, Agger JW et al (2017) Analyzing activities of lytic polysaccharide monooxygenases by liquid chromatography and mass spectrometry. Methods Mol Biol 1588:71–92. https://doi.org/10.1007/978-1-4939-6899-2_7

28. Beeson WT, Phillips CM, Cate JHD, Marletta MA (2012) Oxidative cleavage of cellulose by fungal copper-dependent polysaccharide monooxygenases. J Am Chem Soc 134:890–892. https://doi.org/10.1021/Ja210657t

29. Frommhagen M, van Erven G, Sanders M et al (2017) RP-UHPLC-UV-ESI-MS/MS analysis of LPMO generated C4-oxidized gluco-oligosaccharides after non-reductive labeling with 2-aminobenzamide. Carbohydr Res

448:191–199. https://doi.org/10.1016/j.carres.2017.03.006

30. Potthast A, Radosta S, Saake B et al (2015) Comparison testing of methods for gel permeation chromatography of cellulose: coming closer to a standard protocol. Cellulose 22:1591–1613. https://doi.org/10.1007/s10570-015-0586-2

31. Vuong TV, Liu B, Sandgren M, Master ER (2017) Microplate-based detection of lytic polysaccharide monooxygenase activity by fluorescence-labeling of insoluble oxidized products. Biomacromolecules. https://doi.org/10.1021/acs.biomac.6b01790

32. Flitsch A, Prasetyo EN, Sygmund C et al (2013) Cellulose oxidation and bleaching processes based on recombinant Myriococcum thermophilum cellobiose dehydrogenase. Enzyme Microb Tech 52:60–67. https://doi.org/10.1016/j.enzmictec.2012.10.007

33. Courtade G, Wimmer R, Rohr AK et al (2016) Interactions of a fungal lytic polysaccharide monooxygenase with beta-glucan substrates and cellobiose dehydrogenase. Proc Natl Acad Sci USA 113:5922–5927. https://doi.org/10.1073/pnas.1602566113

34. Forsberg Z, Vaaje-Kolstad G, Westereng B et al (2011) Cleavage of cellulose by a CBM33 protein. Protein Sci 20:1479–1483. https://doi.org/10.1002/Pro.689

35. Xia YL, Legge G, Jun KY et al (2005) IP-COSY, a totally in-phase and sensitive COSY experiment. Magn Reson Chem 43:372–379. https://doi.org/10.1002/mrc.1558

36. Gardner JG, Crouch L, Labourel A et al (2014) Systems biology defines the biological significance of redox-active proteins during cellulose degradation in an aerobic bacterium. Mol Microbiol 94:1121–1133. https://doi.org/10.1111/mmi.12821

37. Sørlie M, Seefeldt LC, Parker VD (2000) Use of stopped-flow spectrophotometry to establish midpoint potentials for redox proteins. Anal Biochem 287:118–125. https://doi.org/10.1006/abio.2000.4826

38. Liu Y, Seefeldt LC, Parker VD (1997) Entropies of redox reactions between proteins and mediators: the temperature dependence of reversible electrode potentials in aqueous buffers. Anal Biochem 250:196–202. https://doi.org/10.1006/abio.1997.2222

39. Wiseman T, Williston S, Brandts JF, Lin LN (1989) Rapid measurement of binding constants and heats of binding using a new titration calorimeter. Anal Biochem 179:131–137. https://doi.org/10.1016/0003-2697(89)90213-3

40. Turnbull WB, Daranas AH (2003) On the value of c: can low affinity systems be studied by isothermal titration calorimetry? J Am Chem Soc 125:14859–14866. https://doi.org/10.1021/ja036166s

41. Borisova AS, Isaksen T, Dimarogona M et al (2015) Structural and functional characterization of a lytic polysaccharide monooxygenase with broad substrate specificity. J Biol Chem 290:22955–22969. https://doi.org/10.1074/jbc.M115.660183

42. Zhang YHP, Lynd LR (2005) Determination of the number-average degree of polymerization of cellodextrins and cellulose with application to enzymatic hydrolysis. Biomacromolecules 6:1510–1515. https://doi.org/10.1021/bm049235j

43. Forsberg Z, Mackenzie AK, Sorlie M et al (2014) Structural and functional characterization of a conserved pair of bacterial cellulose-oxidizing lytic polysaccharide monooxygenases. Proc Natl Acad Sci USA 111:8446–8451. https://doi.org/10.1073/pnas.1402771111

44. Kojima Y, Varnai A, Ishida T et al (2016) Characterization of an LPMO from the brown-rot fungus *Gloeophyllum trabeum* with broad xyloglucan specificity, and its action on cellulose-xyloglucan complexes. Appl Environ Microbiol 82:6557–6572. https://doi.org/10.1128/AEM.01768-16

Light-Induced Electron Transfer Protocol for Enzymatic Oxidation of Polysaccharides

David Cannella

Abstract

Lytic polysaccharide monooxygenases (LPMOs) are redox enzymes that oxidize the most recalcitrant polysaccharides and require extracellular electron donors. The role of electron donation to redox enzymes is pivotal since a nonefficient electron transfer might result in partial activity or reduced kinetics. In this protocol we show the effect of using excited photosynthetic pigments combined with reducing agents as efficient electron donors for monooxygenases. The light-induced electron transfer can enhance the oxidation ability of LPMOs up to ten times.

Key words LPMO, Chlorophyllin, PASC, Light-induced electron transfer, Cellulose oxidation

1   Introduction

Extracellular copper-based lytic polysaccharide monooxygenases (LPMOs) are capable of cleaving many recalcitrant polysaccharides through oxidation [1]. Even though we are still struggling in deciphering their exact mechanism of action [2], their physiological meaning and their role in pathogenic and nonpathogenic degradation of the plant cell wall, we have clarified that LPMOs aid in breaking down the recalcitrant stiff cellulose structure [3], thus softening the framework for the following hydrolases with which they share a strong synergism as regards biomass degradation [4]. Being an oxidoreductase, LPMO needs electrons to perform the oxidation of its substrate, plus a dioxygen molecule. With regard to substrates it has been shown that certain LPMOs can oxidize only one specific polysaccharide, like cellulose or starch or chitin; however one example of substrate promiscuity has been reported allowing for cellulose and hemicellulose oxidation with the same LPMO (e.g., TtLPMO9E) [5]. If the substrate recognition of this monooxygenase is very specific on the one hand, its ability as an electron sink is highly promiscuous on the other hand, allowing the LPMO to receive electrons from a huge variety of

Mette Lübeck (ed.), *Cellulases: Methods and Protocols*, Methods in Molecular Biology, vol. 1796,
https://doi.org/10.1007/978-1-4939-7877-9_17, © Springer Science+Business Media, LLC, part of Springer Nature 2018

reducing agents. General reducing agents like ascorbate or gallate, but also lignin [6, 7], phenols [8], enzymes (e.g., CDH) [9], and recently pigments [10] are all electron donors for LPMOs. This ambiguity led some researchers to question the exact definition of the enzymes' main activity, or substrate and cofactor role [8]. Obviously, the different sources of electron reflect also a different level of activity of the LPMOs in terms of speed of reaction and number of substrate oxidations performed. Using light-excited pigments a tenfold increase in reactivity was obtained when compared to ascorbate. This has been demonstrated on a variety of LPMOs using three different pigments [10]. We are now getting more evidence that the electron transfer to LPMOs is the crucial step for a deeper substrate oxidation and the photo-excited pigments can be an efficient tool for highlighting oxidative reactions that might too slow to be detected. This protocol provides an assay to highlight monooxygenase activity, specifically for the class of LPMOs, but might also be used to assess the supply of electrons to other redox enzymes.

To be noted is that dosage of the enzyme, pigments, and reducing agents, time of incubation, and temperatures were optimized for the specific enzymes tested, and if it has to be applied to other enzymes, these parameters should be adjusted accordingly to the optimum range of activity of the enzymes under study. LIET stands for light-induced electron transfer reaction where the polysaccharide substrate is oxidized by the LPMO and the electrons are transferred to the enzymes by the excited chlorophyllin and restored by the reducing agent ascorbate. Ascorbate as a reducing agent, a paramount discovery of Alexander Krasnovsky in the early 1950s [11], has been crucial in inspiring the LIET process.

2   Materials

2.1   Enzymes

The enzymes must be free of any residual cellulase or hemicellulase activities prior to the test.

1. *Thielavia terrestris* LPMO (*Tt*LPMO9E, previously *Tt*GH61E) (Novozymes A/S, Denmark).

2. *Thermoascus auranticus* (*Ta*LPMO9A) (Novozymes A/S, Denmark).

3. *Thermobifida fusca* AA10 (TfLPMO10A) (Nzytech Ltd., Portugal).

2.2   Chemicals and Buffers

1. 100 mM ascorbic acid stock solution in water. Store at $-20\ ^\circ$C in darkness.

2. 100 mM citrate-phosphate buffer pH 6.3. Store at $+4\ ^\circ$C.

3. 100 mM phosphate buffer pH 7.8.

2.3 Photopigments

1. Chlorophyllin produced by extraction of Festuca arundinacea (Chr. Hansen, Hørsholm, Denmark). 12% w/v 166 mM stock solution, in 100 mM citrate-phosphate buffer pH 6.3. Store at +4 °C in darkness and freshly prepare 2 h before use (*see* **Note 1**).

2. Plant thylakoid membranes are extracted from *Arabidopsis thaliana* (L.) [10]. The final concentration of the thylakoid membranes is between 3 and 4 mg Chl mL^{-1} (Chl a/b ratio = 3.0). Freshly prepare the needed aliquot and store at −20 °C in darkness. Dilute at 1:20 prior to use (*see* **Note 1**).

2.4 PASC Cellulose

1. Microcrystalline cellulose (Sigma-Aldrich, St. Louis, USA). Store at +4 °C (*see* **Note 2**).

2. 85% w/v phosphoric acid H_3PO_4, stored at +4 °C.

3. Ice-cold MilliQ water, stored at +4 °C (*see* **Note 3**).

4. 1% w/v $NaHCO_3$ solution, stored at +4 °C.

5. 100 mM citrate-phosphate buffer pH 6.3, stored at +4 °C.

6. 2-L glass beaker.

7. Coffee blender.

2.5 LIET Light-Induced Electron Transfer

1. Dark room with dim green light at 0 µmol of photons m^{-2} s^{-1} (Phillips TL-D (Ph W color green Lumen 3600, 540 nm) equipped with a centrifuge and desirably a UV-spectrophotometer for pigment dosage.

2. LED lamps for blue light (440 nm) and red light (625 nm).

3. Light probe (Spherical Micro Quantum Sensor US-SQS/ Lund).

4. Thermomixer.

5. Transparent centrifuge rack.

6. 2-mL Eppendorf polypropylene microfuge tubes.

7. Black cloth.

3 Methods

3.1 PASC Preparation

All the solutions used during PASC preparation should be stored at +4 °C until use [12]. A consistent temperature during cellulose suspension in concentrated phosphoric acid and washing steps is imperative for ensuring good quality of the resulting PASC.

Wear protective eye glasses and gloves; use a weighing microspatula to dissolve the Avicel into concentrated acid. Prepare the PASC solution at least 2 days prior to the LIET enzymatic reactions.

1. Put an empty glass beaker of 2 L volume inside a bigger container filled of ice and water. Then put both on top of a magnetic stirrer (*see* **Note 4**).

2. Suspend 4 g Avicel in 100 mL of H_3PO_4 (85% w/v) at +4 °C, and stir for 1 h (*see* **Note 5**).

3. Add 1900 mL of ice-cold MilliQ water and stir for 1 h.

4. Wash thoroughly the entire swollen cellulose suspension by replacing the 2 L volume 5× with ice-cold water (*see* **Note 6**).

5. After washing with water, wash again 2× with 2 L of 1% w/v $NaHCO_3$ solution each time. Now the pH should reach ~6.

6. After the pH is adjusted to ~6 wash additional 5× with 2 L of ice-cold water to remove all the carbonate.

7. Replace the last washing step with the desired buffer solution instead of water.

8. Filter the final mixture into a vacuum filtration apparatus using a nylon filter paper (*see* **Note 7**).

9. A thick gelatinous suspension should be obtained at this stage. Homogenization of the swollen cellulose suspension with a coffee blender might be necessary to remove big lumps.

10. Use again water or the desired buffer solution to adjust to the desired final concentration (suggested between 1 and 2% w/v).

11. Keep always the phosphoric acid swollen cellulose (PASC) at +4 °C in a glass bottle. It might last up to 1 year without contaminations if kept in water. The resulting degree of polymerization of PASC prepared with this protocol should be 100. If prepared at higher temperatures the degree of polymerization should drop, usually around 50 if prepared at room temperature.

3.2 Setup of the LED Lights and Thermomixer

1. The thermomixer should have a transparent tube holder for permitting the light to penetrate also from the side, especially in the case of sunlight exposure (*see* **Note 8**).

2. Preheat the modified thermomixer long enough prior to the enzymatic reaction (*see* **Note 9**).

3. Place the thermomixer below the LED lights such that light hits the thermomixer perpendicularly from the top.

4. Accurately adjust the distance between the light source and the bottom of the tubes where the reaction takes place (*see* **Note 10**).

5. Wrap the whole setup (LED light source plus thermomixer) with a black cloth to prevent any eye damage that could be caused by LED lights.

3.3 Standard Conditions for Enzymatic Reactions and LIET.

Carry out all the procedure in dim green light at 0 µmol of photons $m^{-2} s^{-1}$ (Phillips TL-D (Ph W color green Lumen 3600, 540 nm). Set the LED light to provide 150 µmol of photons $m^{-2} s^{-1}$ for matching the enzymatic reaction set as follows [cannella].

1. First add excess of PASC substrate into the 2 mL Eppendorf tubes. In our protocol we dose the PASC at a final concentration of 0.75% w/v diluting two times a concentrated PASC solution 1.5% w/v. So 100 µL of PASC is pipetted inside the tubes.

2. Add 4 µL of ascorbic acid solution (final concentration 2 mM).

3. Add 2 µL of chlorophyllin solution (final 1.6 mM) or 10 µL of thylakoid membrane suspension.

4. Add 93 µL of 100 mM of citrate-phosphate buffer pH 6.3 (*see* **Note 11**).

5. Close the tubes and let equilibrate and mix inside the thermo-mixer at the enzymatic reaction temperature for few minutes with the LED light still off.

6. Add the LPMO enzymes at a final dosage of 2 nmol. In the case of *Tt*LPMO9E which has a molecular weight of 22.5 kDa, and since the purified solution has a protein concentration of 10 mg/mL, this corresponds to 1 µL of the enzyme.

7. After the enzymes are added (and thus having a final volume of 200 µL), the tubes are closed, and the LED light and the mixer are switched on.

8. After the desired time period for the enzymatic reaction, the tubes are transferred directly into the centrifuge inside the dark room. Centrifuge at $14,000 \times g$ for 3 min and separate the supernatant (containing the water-soluble reaction products) from the pellet containing the insoluble substrate (*see* **Note 12**).

9. In cases where the samples have to leave the dark room for further analyses (HPLC, FTIR, NMR, etc.) it is highly suggested to prepare the samples inside the dark room and keep them isolated from light as much as possible during the transfer to the final analysis. For example in the case of HPLC analysis the vials containing the supernatants are prepared inside the dark room and wrapped in aluminum foil prior to leaving the darkroom. Instead in the case of UV spectroscopy for the pigments, it is highly suggested to place the UV-spec directly inside the dark room.

4 Notes

1. Keep the powder of chlorophyllin and other pigment stock solutions always in darkness; it is a good practice to let stand for 2–3 h in darkness any pigment solution prior to the experiment. Check regularly the UV absorbance of the pigment stock solutions after an appropriate dilution to match the

UV-spectrophotometer sensibility, and account for any loss of UV absorbance, then dose the pigment to the desired UV-absorbance capability.

pH also plays a crucial role; below pH 6.0 the chlorophyllin starts losing its stability [13].

2. Wash the microcrystalline cellulose and keep at +4 °C.

3. Prepare 5 L for each gram of Avicel to be swollen.

4. The glass beaker should be flat at bottom and handy in volume; if 2 L is too big to handle reduce the amount of cellulose and accordingly the volume of the beaker. The magnet stirrer must be brand new with no scratches and triangular shape to have maximum surface contact with the beaker. Do not use a cylindrical magnetic stirrer, as it will reduce the efficiency of the mixing at a later stage.

5. Avicel is added to phosphoric acid, not the other way around. Put first the phosphoric acid, it should cover the entire bottom surface of the beaker. Then add the triangular magnet stirrer and switch on the stirrer set at a slow pace. Add the Avicel very slowly. This step is of paramount importance. Four grams of Avicel should be added in one or few milligrams at a time. This is necessary to avoid formation of lumps, which in case appearing at this stage will be difficult to remove and will cause a loss of homogeneity. The density of the suspension will increase when adding the cellulose, so the stirring speed must be adjusted such that it slowly increases at any time.

6. The washing should be done by sedimentation, which usually takes 1 h for 2 L of pulp; alternatively a centrifuge can be used, with no more than few hundreds g for 3 min.

7. Manual stirring while filtering will speed up the vacuum filtration, since often the nylon cloth will be fouled.

8. In case that the transparent tubes holder is not provided, it can be set by replacing the standard tube holder of the thermomixer with the 96-well plate holder, and stacking on top of this a transparent (plastic, glass, or quartz) tube rack. In case of these modifications of the stirring plate, place the added tube rack at the very center to balance the weight.

9. Check the temperature difference between the display and inside the tubes since contact of the tubes with the heating surface is not optimal. A difference between 5 and 10 °C can be expected.

10. The desired light intensity or Lumen should be measured with the light probe on the bottom of a closed Eppendorf tube placed inside the tube holder.

11. Use the right buffer in relation to the enzyme specification.

12. The reaction can be stopped based on the further use of the samples, and thermal treatment should be the last choice; dropping the pH to more acidic values or adding a specific enzymatic inhibitor should always be preferred. Note that from the point of view of water-soluble oligosaccharides, the products are often found stable even few days after separation by centrifugation, meaning that this separation stops the reaction by removing the substrate and part of the enzymes.

References

1. Frandsen KEH, Lo Leggio L (2016) Lytic polysaccharide monooxygenases: a crystallographer's view on a new class of biomass-degrading enzymes. IUCrJ 3 (Pt 6):448–467

2. Möllers KB, Mikkelsen H, Simonsen TI et al (2017) On the formation and role of reactive oxygen species in light-driven LPMO oxidation of phosphoric acid swollen cellulose. Carbohydr Res 448:182–186. https://doi.org/10.1016/j.carres.2017.03.013

3. Villares A, Moreau C, Bennati-Granier C et al (2017) Lytic polysaccharide monooxygenases disrupt the cellulose fibers structure. Sci Rep 7:40262

4. Vaaje-Kolstad G, Westereng B, Horn SJ et al (2010) An oxidative enzyme boosting the enzymatic conversion of recalcitrant polysaccharides. Science 330:219–222

5. Bennati-Granier C, Garajova S, Champion C et al (2015) Substrate specificity and regioselectivity of fungal AA9 lytic polysaccharide monooxygenases secreted by *Podospora anserina*. Biotechnol Biofuels 8:90

6. Westereng B, Cannella D, Agger JW et al (2015) Enzymatic cellulose oxidation is linked to lignin by long-range electron transfer. Sci Rep 5:18561

7. Rodrigues-Zuniga UF, Cannella D, Giordano RC et al (2015) Lignocellulose pretreatment technologies affect the level of enzymatic cellulose oxidation by LPMO. Green Chem 17:2896–2903

8. Frommhagen M, Koetsier MJ, Westphal AH et al (2016) Lytic polysaccharide monooxygenases from *Myceliophthora thermophila* C1 differ in substrate preference and reducing agent specificity. Biotechnol Biofuels 9:186

9. Kracher D, Scheiblbrandner S, Felice AK et al (2016) Extracellular electron transfer systems fuel cellulose oxidative degradation. Science 352(6289):1098–1101

10. Cannella D, Möllers KB, Frigaard NU et al (2016) Light-driven oxidation of polysaccharides by photosynthetic pigments and a metalloenzyme. Nat Commun 7:11134

11. Krasnovsky AA (1948) Reversible photochemical reduction of chlorophyll by ascorbic acid. Dokl AN SSSR 60:421–424

12. Wood TM (1988) Preparation of crystalline, amorphous, and dyed Cellulase substrates. Methods Enzymol 160:19–25

13. Koca N, Karadeniz F, Burdurlu HS (2007) Effect of pH on chlorophyll degradation and colour loss in blanched green peas. Food Chem 100(2):609–615

Chapter 18

Purification and Characterization of the Total Cellulase Activities (TCA) of Cellulolytic Microorganisms

Ayyappa Kumar Sista Kameshwar and Wensheng Qin

Abstract

Cellulose is the earth's most abundant plant polysaccharide containing a large array of glucose units linked through β (1 → 4) linkages by existing in both crystalline and amorphous forms. Cellulose is widely distributed in plants, constituting up to 40–50% overall dry weight of the plant biomass. Majorly, microorganisms secrete three types of enzymes such as endoglucanases, exoglucanases, and beta-glucosidase for the hydrolysis of cellulose, contributing to the total cellulase activity. Industrially, the cellulolytic microorganisms are assessed based on their total cellulolytic activities. Similarly, total cellulase activity can also be used for the isolation and characterization of the cellulolytic microorganisms. In this chapter, we have specifically discussed about the methods used for the purification and characterization of the total cellulase activities of the microorganisms such as filter paper assay and cellulase zymogram assay. Our present chapter can be used as primer for characterizing cellulolytic abilities of cellulose-degrading microorganisms.

Key words Cellulose, Total cellulolytic ability, Endoglucanase, Exoglucanases and β-glucosidases

1 Introduction

Over the past few decades, studies on the production of plant biomass-based biofuels and biorefineries have significantly increased. Especially much attention was given to cellulose-based production of ethanol. Microbial degradation of cellulose (and other plant biomass units) is considered an environmentally friendly, cost-effective, and potentially efficient method. Majorly cellulose degrading enzymes secreted by microorganisms can be classified as endoglucanases or 1,4-β-D-glucan-4-glucanohydrolases (EC 3.2.1.4) which are involved in random fragmentation of amorphous cellulose units resulting in oligosaccharides units of varying lengths. Exoglucanases which includes cellodextrinases or 1,4-β-D-glucan glucanohydrolases (EC 3.2.1.74) and cellobiohydrolases or 1,4-β-D-glucan cellobiohydrolases (EC 3.2.1.91) these set of enzymes act on reducing and nonreducing ends of cellulose or microcrystalline cellulose and β-glucosidases or β-glucoside

Mette Lübeck (ed.), *Cellulases: Methods and Protocols*, Methods in Molecular Biology, vol. 1796,
https://doi.org/10.1007/978-1-4939-7877-9_18, © Springer Science+Business Media, LLC, part of Springer Nature 2018

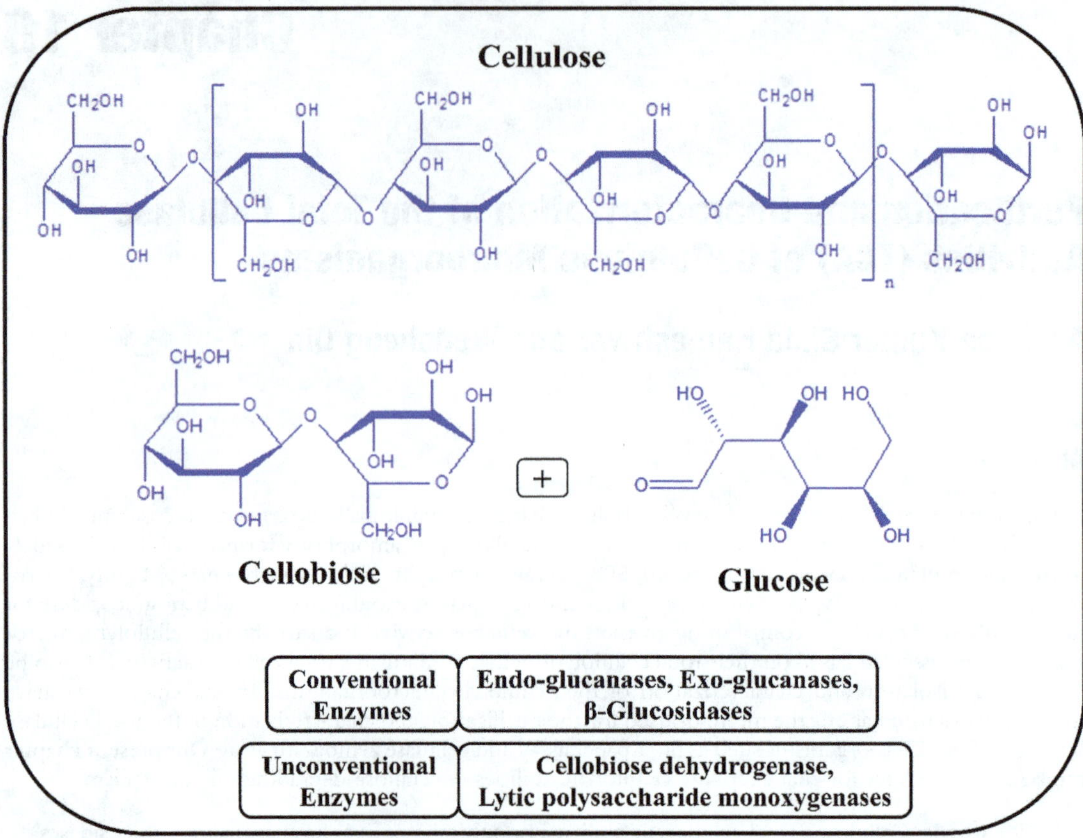

Fig. 1 Schematic representation of enzymatic breakdown of cellulose by the microbial enzyme systems

glucohydrolases (EC 3.2.1.21) are required for solubilizing cellodextrins and cellobiose to simple glucose residues [1]. Several extensive and informative reviews are already available on microbial degradation of cellulose units [1–7] (Fig. 1).

Thus, it is important to understand different methods and techniques involved in the isolation and characterization of cellulose degrading microorganisms, various studies have already developed various easy and significant methods for microbe isolation. In this chapter, we have clearly explained about various quantitative and qualitative methods used for the measurement of cellulases secreted by the microorganisms.

2   Materials

2.1   Filter Paper Assay

1. DNS Reagent: 3,5-dinitro salicylic acid, sodium hydroxide (19.3 N), deionized water, sodium potassium tartrate 3H₂O, double distilled H₂O.

2. Citrate buffer: citric acid monohydrate 210 g, deionized water, NaOH.

3. Filter paper: cut approximately 1.0×6.0 cm of Whatman No.1 filter paper, weight must be 50 mg (*see* **Note 1**).

4. Glucose standard: glucose 0.4 g dissolved in 200 ml of ddH$_2$O, citrate buffer (pH 4.8) solutions.

2.2 Purification of Cellulase

2.2.1 Ammonium Sulfate Precipitation and Dialysis

1. Fresh and desiccated ammonium sulfate [(NH$_4$)$_2$SO$_4$].

2. Magnetic stirrer, 1 L graduated cylinder, hot air oven, and centrifuge.

3. Dialysis tubing and the elution buffer (100 mM sodium phosphate buffer pH 7.0).

2.2.2 Ion Exchange Chromatography Using DEAE-Sephadex A-50

1. Take a glass column with the dimensions 1.5×40 cm.

2. DEAE (diethyl-aminoethyl) Sephadex A-50.

3. 100 mM sodium phosphate buffer with pH 7.0.

4. Elution buffer: 10 ml each of sodium chloride in increasing concentrations from 0.1 to 0.5 M.

2.2.3 Gel Filtration Chromatography Using Sephadex G-75

1. Take a glass column with the dimensions of 1.0×60.0 cm.

2. Sephadex G-75 commercial grade (Sigma-Aldrich Pvt. Ltd., USA, 1.5×40 cm).

3. 100 mM sodium phosphate buffer with pH 7.0.

2.3 Lowry's Method

1. Stock solution A: 2% (w/v) Na$_2$CO$_3$ (sodium carbonate) in distilled water.

2. Stock solution B: 1% (w/v) CuSO$_4$·5H$_2$O (Copper (II) sulfate pentahydrate) in distilled water.

3. Stock solution C: 2% (w/v) KNaC$_4$H$_4$O$_6$·4H$_2$O (sodium potassium tartarate) in distilled water.

4. 2 N NaOH (sodium hydroxide).

5. Folin's reagent (commercially available): Use at 1 N concentration.

6. Bovine serum albumin or any other standard protein is used for standard protein stock solution.

7. Stock solutions of standard protein (2 mg/ml) is dissolved in distilled water and stored at -20 °C.

8. Standard solutions of different concentrations can be prepared by diluting the stock solution with distilled water. Increasing concentrations of standard protein can be achieved by mixing different proportions of contents as shown below (Table 1) [8].

2.4 SDS-PAGE

1. Analytical grade gel preparation reagents: Sodium dodecyl sulfate (SDS), TEMED (4-4-tetraethylenemethylene diamine), ammonium per sulfate, acrylamide, bis-acrylamide,

Table 1
Increasing dilutions of the standard protein samples

Stock solution (μl)	0	2.5	5	12.5	25	50	125	250	500
Water (μl)		500	498	495	488	475	450	375	250	0
Protein concentration (μg/ml)	0	10	20	50	100	200	500	1000	2000

Bromophenol Blue, Tris base, Glycine, EDTA, Glycerol, Isopropanol, Tris–HCl (pH 6.8), β-mercaptoethanol, and Coomassie Blue.

2. Protein molecular weight marker (Bio-Rad Laboratories, USA).

3. Prepare 30% acrylamide stock solution by adding acrylamide–bis-acrylamide at a ratio of 37.5:1.

4. Prepare (10×) electrophoresis running buffer by adding 30.3 g of Tris base, 144.0 g D-glycine, 10.0 g SDS. Dissolve the above ingredients in 880 ml of double distilled H_2O after attaining a clear solution make it up to 1000 ml using ddH_2O. The pH of the electrophoresis running buffer is maintained at 8.3 and no pH adjustment is required further [9].

5. Prepare (2×) protein sample buffer solution by adding 1.25 ml 1 M Tris–HCl (pH 6.8), 4.0 ml 10% (w/v) SDS, 2.0 ml glycerol, 0.5 ml 0.5 M EDTA, 4 mg Bromophenol Blue, 0.2 ml β-mercaptoethanol (14.3 M). Dissolve the above ingredients in 10 ml ddH_2O [9].

2.5 Cellulase Zymogram Assay

1. Prepare CMC zymogram by adding 0.1% of carboxymethyl cellulose (CMC) to 12% separating gel (SDS-PAGE gel can be prepared as described in Subheading 2.4).

2. Prepare wash buffer containing sodium citrate buffer (50 mmol/L) pH 5.5, 1% Triton X-100.

3. 0.1% Congo Red.

4. 1 mol/L NaCl.

3 Methods

3.1 Filter Paper Assay (FPA)

It is one of the highly studied and used enzyme assay recommended by International Union of Pure and Applied Chemistry (IUPAC) for the determination of cellulase enzyme. FPA is based on the conversion of defined amounts of substrates (2 mg of glucose released from 50 mg of filter paper measured using DNS assay at given time of 60 min). The wide availability and reasonable susceptibility of the filter paper towards cellulase activities are major advantages of FPA method. The detailed protocol of the filter paper assay was clearly and extensively explained earlier [10]. In this chapter, we have discussed about the microplate method of the filter paper assay explained by Xiao et al. [11]. FPA is highly used for the determination of cellulases secreted by the aerobic microorganisms, whereas for the determination of cellulases and cellulosomal enzyme activities Avicel assay is highly preferred. The avicel assay protocol for anaerobic cellulases was clearly and extensively explained earlier [10].

DNS Reagent

1. Mix distilled water 400 ml with 3,5-dinitrosalicylic acid 10.0 g and 19.3 N (50% w/w) sodium hydroxide 20.75 ml; slightly heat to help dissolving.

2. Add Rochelle salts (sodium potassium tartrate tetrahydrate) 300 g; make sure that all the salts are dissolved; add deionized water up to 1 L.

3. Store the above prepared reagents in dark conditions (in - amber-colored bottles or by covering the bottles using aluminum foil) (*see* **Note 2**).

Citrate Buffer

1. Mix citric acid monohydrate 210 g with DI water 750 ml.

2. Dilute to 1 L and check pH.

3. Adjust by adding NaOH until pH equals 4.5; the solution above should be 1 M citrate buffer.

4. Dilute 1 M buffer to 0.05 M by adding 1 : 19 folders DI water.

Glucose Standard

1. Prepare working stock solution of anhydrous glucose (2 mg/ml) by dissolving 0.4 g glucose up to 200 ml pH 4.8 citrate buffer, use immediately or tightly seal and store frozen (store at 4 °C will not last long) (*see* **Note 3**).

2. Vortex the solution before use to ensure adequate mixing.

3. Dilute the working stock solution to different concentration by the following manner: 0.5 ml + 0 ml buffer = 1.0 mg/0.5 ml, 400 μl + 100 μl buffer = 0.8 mg/0.5 ml, 300 μl + 200 μl buffer = 0.6 mg/0.5 ml, 200 μl + 300 μl buffer = 0.4 mg/0.5 ml, 100 μl + 400 μl buffer = 0.2 mg/0.5 ml.

4. Add 0.5 ml of each of the above glucose dilutions to 1.0 ml citrate buffer in test tube with cap, incubate at 50 °C for exactly 60 min.

5. Remove each assay tube from the 50 °C bath; add 3.0 ml DNS reagent to each tubes and mixing.

3.2 Filter Paper Assay Using Microplate

1. Whatman No.1 filter paper (using a paper cutter cut into 1.0 × 6.0 cm strips for 1.5 ml filter paper assay and for 96-well plate assay cut the paper into 7.0 mm using a punch machine) (*see* **Notes 1** and **3**) [11].

2. Prepare 50 mM NaAc (Sodium acetate buffer) with pH 4.8.

3. DNS solution preparation is same as described above Subheading 2.2.1.

4. The 7 mm diameter paper disk with an area of 38.5 mm^2 is used in the 96-well plate assay. Where as in the standard FPA assay one filter paper strip of 600 mm^2 (1 × 6 cm) is used as the substrate.

5. The final reaction volume contains 32 μl aliquot of the diluted enzyme extract, 64 μl of 50 mM NaAc buffer (pH 4.8).

6. Incubate the 96 μl-well plates at 50 °C in a temperature cycler after 60 min of incubation, transfer 50 μl of the above reaction volume is transferred to the corresponding well of second 96-well plate, containing 100 μl of DNS.

7. Incubate the 96-well plate at 95 °C for 5 min in a temperature cycler.

8. Once the color is developed, transfer 36 μl of sample solution to the flat bottomed 96-well plate containing 160 μl of H_2O.

9. Monitor the reaction using a spectrophotometer or a plate reader, at an absorbance of 540 nm.

10. According to Xiao et al., in relation to the amounts of glucose equivalents expected to be produced from standard FPA assay, amount of enzyme used in 96-well plate should release around 128 μg of glucose equivalents in each well [11].

11. Xiao et al. also proposed a 60 μl format FPA assay, where the reaction mixture contains 20 μl aliquot of diluted enzyme, 40 μl of 50 mM NaAc buffer and a filter paper disk of 7 mm diameter [11].

12. After adding all the contents of reaction mixture mentioned above, 96-well plate is incubated in a temperature cycler at 50 °C for 60 min.

13. After the incubation, 120 μl of DNS solution is added to each well, and further incubated at 95 °C for 5 min in a temperature cycler.

14. Finally, transfer 36 μl of aliquot of each sample to a flat bottomed 96-well plate containing 160 μl of H_2O.

15. Monitor the reaction using a spectrophotometer or a plate reader, at an absorbance of 540 nm (*see* **Note 4**).

16. Calculation of filter paper units (FPU), one filter paper unit is defined as average of 1 μmol of glucose equivalents released per minute in the assay reaction (*see* **Note 5**).

$$\text{FPU/ml} = \frac{(\text{Sample } A_{540})}{(\text{Glucose solution } A_{540}/\text{mg})} \quad (5.55 \text{ μmole/mg})$$

$$\times \left(\frac{1}{60 \text{ min}}\right)\left(\frac{1}{0.02 \text{ ml}}\right)$$

3.3 Purification of Cellulase

3.3.1 Ammonium Sulfate Precipitation

1. Measure the culture filtrate obtained from the CMC broth cultures using a graduated cylinder.

2. Transfer the filtrate to a beaker with twice the capacity of the measured filtrate solution.

3. Place the beaker in a bed of ice pack.

4. Use fresh, desiccated ammonium sulfate and dry it overnight in a hot air oven at 120 °C in a large beaker.

5. Transfer dried ammonium sulfate to a clean mortar and grind the ammonium sulfate to fine powder using the pestle (*see* **Note 5**).

6. To achieve 80% saturation, use 53.2 g of ammonium sulfate is added for 100 ml of filtrate.

7. Add small amounts of ammonium sulfate powder to the filtrate solution and constantly mix with the help of a stirrer (or a magnetic stirrer).

8. Do not stir vigorously as it might result in denaturation and avoid the foam formation.

9. Continue stirring for some time even after the addition of ammonium sulfate to ensure complete precipitation (*see* **Note 6**).

10. Transfer the filtrate protein solution to a polycarbonate centrifuge tubes by using a balancing tube filled with water centrifuge at 10,000 × *g* for 15 min at 4 °C (*see* **Notes 7** and **8**).

11. Collect the supernatant from the centrifuge tube, and resuspend the protein pellet in the minimum volume of (1–2 pellets volume) 0.02 M sodium phosphate buffer (pH 7.0).

12. Repeat the **step 11** for 2–3 times to ensure maximum protein collection [12, 13] (*see* **Notes 8** and **9**).

3.3.2 Ion Exchange Chromatography Using DEAE-Sephadex A-50

1. Take column ranging 1.5 × 40 cm, thoroughly clean and dry the column.

2. Use fresh and hot air oven dried DEAE-Sephadex A-50 to pack the vertically mounted column.

3. The column is equilibrated using 100 mM sodium phosphate buffer of pH 7.0 and is allowed to pass through the column at 30 ml/h flow rate.

4. The enzyme filtrate obtained after ammonium sulfate precipitation is subjected to dialysis by transferring the enzyme filtrate to a dialysis tubing.

5. Dialysis is performed against 100 mM sodium phosphate buffer (pH 7.0) overnight kept at a constant stirring using a magnetic stirrer.

6. The buffer used for the dialysis is replaced twice at constant time interval with the fresh buffer.

7. After dialysis, the enzyme filtrate is run through on the vertical column by constantly adding 100 mM sodium phosphate buffer (pH 7.0).

8. The unbound fractions are constantly collected from the column and further analyzed for cellulase activity and total protein content.

9. Later the bound enzymes are eluted using the elution buffer containing 100 mM sodium phosphate buffer (pH 7.0) with increasing concentration of 0–0.5 M NaCl.

10. The bound filtrates are continuously collected using test tubes and tested for the total cellulase activity and total protein activity [12, 13].

3.3.3 Gel Filtration Chromatography Using Sephadex G-100

1. Take column ranging 1.0×60.0 cm, thoroughly clean and dry the column (*see* **Notes 10** and **11**).

2. Use fresh and hot air oven dried Sephadex G-100 slurry is packed into the vertically mounted column.

3. The column is equilibrated using 100 mM sodium phosphate buffer of pH 7.0 and is allowed to pass through the column at 10 ml/h flow rate.

4. The enzyme filtrates exhibiting highest cellulase activities from the ion exchange chromatography are applied on to the Sephadex G-100 column.

5. The 3 ml of the eluted fractions are constantly collected at a flow rate of 10 ml/h in clean and dry test tubes.

6. These collected enzyme filtrates are tested for the total protein concentration using spectrophotometer at 280 nm wavelength.

7. Eluted fractions with higher protein activity are pooled and collected together in test tubes.

8. The enzyme filtrates are also tested for the total cellulase activities [12, 13].

3.4 Protein Estimation Using Lowry's Method

1. Prepare 1 mg/ml stock solution of bovine serum albumin.

2. Prepare solution A containing 2% (w/v) sodium carbonate (Na_2CO_3) in distilled water.

3. Prepare solution B containing 1% (w/v) copper sulfate ($CuSO_4 \cdot 5H_2O$) in distilled water.

4. Prepare solution C containing 2% (w/v) sodium potassium tartarate in distilled water.

5. Prepare alkaline copper sulfate reagent by mixing 100 ml of solution A and 1 ml of solution B and 1 ml of solution C.

6. Take a clean and dry test tube add different dilutions of BSA stock solutions from (0.25–5 ml) to prepare the standard curve.

7. Take a clean and dry test tube add 0.1 ml of test sample and 0.1 ml of 2 N NaOH.

8. Place the test tube in a heating block or boiling water bath at 100 °C temperature for 10 min to hydrolyze the protein.

9. Allow the test tubes to attain room temperature.

10. Add 1 ml of alkaline copper sulfate reagent and allow the test tubes at room temperature for 10 min.

11. Add 0.1 ml of Folin–Ciocalteau reagent using a vortex mixer and incubate the test tubes for 30–60 min at room temperature under dark conditions.

12. Read the absorbance at 750 nm using a spectrophotometer or calorimeter.

13. Plot the standard curve based on the values obtained from the BSA stock solution, use these values for finding out the concentration of test protein sample [8].

3.5 SDS-PAGE for Cellulase Size Estimation

(A) SDS-PAGE Gel Preparation

1. Whole SDS-PAGE units containing glass plates, combs, and spacers are thoroughly clean and dried.

2. Based on the manufacturer instructions assemble the gel cassette appropriately.

3. Prepare 10% separating gel by adding the following reagents appropriately 2 ml double distilled H_2O, 1.67 ml 30% acrylamide–bis, 1.25 ml 1.5 M Tris (pH 8.8), 25 µl 20% SDS, 25 µl 10% ammonium per sulfate (freshly prepared and stored at 4 °C), 2.5 µl TEMED, total volume of the separating gel is 5 ml (*see* **Notes 12–15**).

4. Mix the above solution thoroughly and transfer it between the glass plates present in the casting chamber using 1 ml micropipette.

5. Fill the separating gel up to 0.7 cm below the bottom of the comb, by arranging the comb in its place.

6. Add few microliters of isopropanol to the top of the separating gel to aid in polymerization and straightening the level of the gel.

7. Once the separating gel is polymerized, using a filter paper remove the isopropanol on the top of the separating gel.

8. Prepare the 5% stacking gel by adding the following solutions 2.088 ml of double distilled H_2O, 0.506 ml of 30% acrylamide–bis, 0.375 ml 1 M Tris (pH 6.8), 15 µl 20% (w/v) SDS, 15 µl 10% ammonium per sulfate, and 1.5 µl of TEMED, total volume of the stacking gel is 3 ml (*see* **Notes 12–15**).

9. After adding TEMED, the stacking gel solution is immediately transferred to the casting chamber using 1 ml micropipette (*see* **Notes 16** and **17**).

10. Slowly insert the appropriate comb on to the casting chamber.

11. Allow the gel to solidify, once the gel is solidified carefully remove the comb.

12. The above prepared gels can be stored for 1–2 weeks at 4 °C, by plastic wrapping the whole casting chamber along with combs and placing it on a second chamber containing paper towel moistened with double distilled sterile water.

(B) Cellulase Sample Preparation

1. Prepare 0.1 ml of purified protein sample solution.

2. To the purified protein sample solution, add same amount of 2× protein sample buffer to each protein sample prepared.

3. Transfer the sample solutions to 1 ml Eppendorf tubes.

4. Mix the protein samples thoroughly and boil the samples at 95 °C in a heating block module for 10 min.

5. Using a table top centrifuge spin the sample tubes at maximum speed for 1 min.

6. Leave the sample tubes at room temperature until they are loaded to the gel.

7. The above sample tubes can be stored at −20 °C and can be loaded on to the gel by reheating the sample tubes at 95 °C for 5 min.

(C) Electrophoresis

1. Carefully, remove the gel cassette from the casting chamber and place the gel cassette in the electrode assembly by leaving short plate inside.

2. Press down the electrode assembly while clamping the frame to secure the electrode assembly, and place the clamping frame in to the electrophoresis tank.

3. Slowly, add 1× electrophoresis running buffer into the casting frame opening and between the gel cassettes.

4. Buffer is added until the wells of the gels are filled; also fill the region outside the frame with 1× running buffer.

5. Add 10 µl if the final concentration is 2 mg/ml (if the well dividers are in good shape add 20 µl), add same volume of the protein samples into each well.

6. Add 10 µl of protein molecular weight marker obtained from Bio-rad, USA, which includes β-galactosidase (120 kDa), Bovine Serum Albumin (91 kDa), Serum Albumin (66 kDa), Glutamic dehydrogenase (56 kDa), Ovalbumin (48 kDa) Carbonic anhydrase (4 kDa), Myoglobin (26 kDa), and Lysozyme (19 kDa).

7. After all the protein samples are added, connect the power supply.

(D) Detection of Protein

1. After the completion of run carefully dismantle the cassette and place the gel resolving gel in Coomassie Blue stain with a gentle agitation for 30 min.

2. After 30 mins pour of the Coomassie Blue stain and add destaining solution.

3. Repeat **step 2** for about 4–5 times.

4. To fasten the destaining process place a piece of paper towel or filter paper to absorb the stain.

5. Once the gel is destained, dry the gel by placing it on Whatman's no.1 filter paper (Fig. 2).

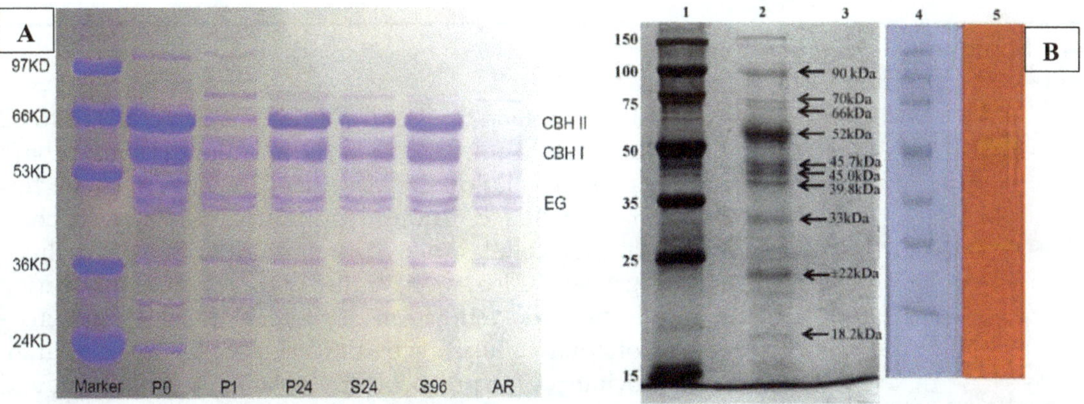

Fig. 2 Shows the separation of pure cellulase enzyme extracts (**a**) SDS-PAGE analysis of cellulase in supernatant during prehydrolysis for 0 (P0), 1 (P1), and 24 h (P24); SSF for 24 h (S24) and 96 h (S96); and after readsorption (AR). (Reprinted with permission from Ruoyu et al. [14].) (**b**) Lane 1—150 kDa protein marker; Lane 2—crude enzyme; Lane 3—medium; Lane 4—150 kDa protein marker; Lane 5—CMCase (CMC-Zymogram). (Reprinted with permission from Ang et al. [15])

3.6 Cellulase Zymogram Assay

1. CMC-zymogram prepare 12% separating gel by adding the following ingredients 3.2 ml of ddH$_2$O, 4 ml of 30% acrylamide–0.8% of bis-acrylamide, 1.5 M Tris base (pH 8.8), 10% (w/v) SDS, 0.1% of carboxy methyl cellulose (CMC), 10% (w/v) ammonium per sulfate and 10 μl TEMED, total volume of the separating gel is 10 ml (*see* **Notes 13–16** and **19**).

2. Prepare the 5% stacking gel by adding the following solutions 2.088 ml of double distilled H$_2$O, 0.506 ml of 30% acrylamide-bis, 0.375 ml 1 M Tris (pH 6.8), 15 μl 20% (w/v) SDS, 15 μl 10% ammonium per sulfate and 1.5 μl of TEMED, total volume of the stacking gel is 3 ml (*see* **Notes 16–18**).

3. For the assembly of gel cassette, arrangement of casting chamber, preparing protein samples and process of electrophoresis, the protocol is same as Subheading 2.4.

4. After the completion of SDS-PAGE run, take out the gel dismantle the assembly of gel cassette and carefully separate the gel.

5. The gel is transferred into a clean container at room temperature and further rinsed with sodium citrate wash buffer (pH 5.5) for 1 h.

6. Pour off the buffer solution from the container and add fresh sodium citrate wash buffer to the container (pH 5.5) and incubated at 50 °C for 4-h.

7. Pour off the solution from the container and stain the gels with 0.1% Congo red for 30 min.

8. After 30 min, destain the gels with 1 M NaCl (for 1 h) or until the zones of clearance around the enzymes are observed [16–19] (Fig. 3).

---

4 Summary

Increase in global temperatures and continuous depletion of fossil fuels to meet up the human needs are the major reasons behind inflated research in cellulose-based ethanol production. In this chapter, we have discussed various quantitative assays for the characterization of total cellulase activities among the isolated cellulolytic microorganisms. Also, we have discussed about methods used for the purification and estimation of cellulase activities and characterization of total cellulase activities using cellulase zymogram assays. Employing robust protocols for isolation, purification, and characterization of cellulose-degrading microorganisms will significantly help the growing lignocellulose-based biofuel and biorefining industries.

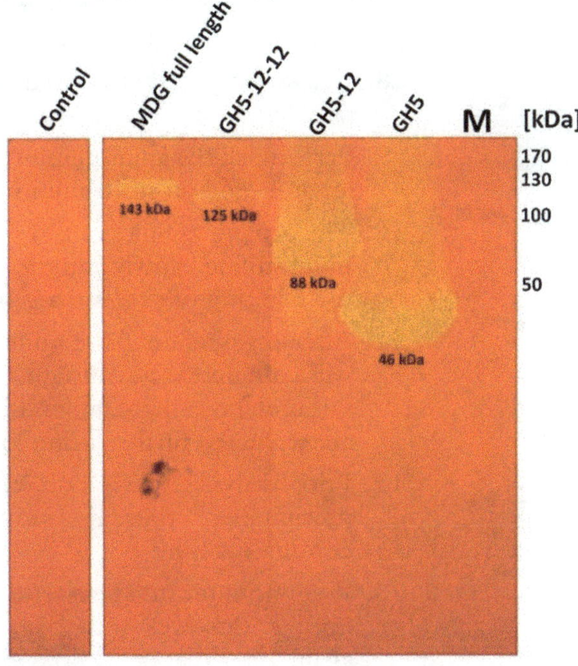

Fig. 3 Shows the CMC-Zymogram gels with multidomain glycosidase full length (MDG: Lane 1), glycoside hydrolase (GH5-12-12: Lane 2), (GH5-12: Lane 3) and GH5 on Lane 4. (Reprinted with permission from Gavrilov et al. [19])

5   Notes

1. It is important to make sure that weight of each paper strip do not vary more than 1 mg of weight. As filter paper assay (FPA) is subjected to the weight of filter paper.

2. The PBS and DNS (in darkness) reagents can be stored at 4 °C for at least 1 month. However, DNS reagents older than 1 month must be avoided for the quantitative assays as it could lose its reducing ability after long storage periods.

3. Handle the filter paper strips with forceps or gloved hands, do not use bear hands to transfer the paper strips.

4. The glucose standard solution is tightly sealed and stored frozen, the glucose solution is thawed and mixed well before usage.

5. International Unit (IU) is defined as 1 mmol/min, based on the initial hydrolysis rate, and is different from FPU assay, which is a fixed conversion assay.

6. For gel filtration chromatography pore size (matrix) plays a crucial role in separation of the protein mixtures.

7. The levels of β-D-glucosidase enzyme present in the cellulase mixture will significantly influence the FPA assay as the total

ratio of reducing ends (glucose, cellobiose and longer cellodextrins) will strongly impact on the DNS readings [10].

8. Use a face mask to avoid contact with ammonium sulfate powder.

9. Perform all the ammonium sulfate protein purification steps on ice at 4 °C, also store ammonium sulfate solution at 4 °C.

10. The protein samples must be kept at cold when adding ammonium sulfate. Slowly add the salt by constantly stirring to avoid the addition of excess than the desired concentration.

11. After resuspending the ammonium sulfate pellet, the sample will contain excess amounts of salt. Please remove the excess salt present in the samples using dialysis, if the protein samples are separated further using ion exchange chromatography.

12. Prior dialysis step is not required if the ammonium sulfate precipitated protein samples are subjected to gel chromatography.

13. If ammonium sulfate is settling at the bottom, stir the filtrate until it dissolves and then add ammonium sulfate powder again.

14. Wear a mask, lab coat and goggles when you weight acrylamide, as acrylamide is considered as a potential carcinogen and neurotoxic reagent (please refer MSDS sheet for further details).

15. Avoid exposure with acrylamide and cover the weighing boat containing acrylamide with another weighing boat and transport it to the fume hood.

16. Transfer the above weighed acrylamide to the cylinder inside the fume hood and mix with a glass rod/stirrer.

17. Avoid contact with unpolymerized acrylamide as it is a neurotoxin.

18. Gel solution should be quickly transferred to the casting chamber as to avoid the gel polymerization after the addition of TEMED.

19. Mix CMC thoroughly and ensure no lumps are formed.

References

1. Lynd LR, Weimer PJ, Van Zyl WH, Pretorius IS (2002) Microbial cellulose utilization: fundamentals and biotechnology. Microbiol Mol Biol Rev 66(3):506–577

2. Dashtban M, Maki M, Leung KT, Mao C, Qin W (2010) Cellulase activities in biomass conversion: measurement methods and comparison. Crit Rev Biotechnol 30(4):302–309

3. Béguin P, Aubert J-P (1994) The biological degradation of cellulose. FEMS Microbiol Rev 13(1):25–58

4. Horn SJ, Vaaje-Kolstad G, Westereng B, Eijsink V (2012) Novel enzymes for the degradation of cellulose. Biotechnol Biofuels 5(1):1

5. Baldrian P, Valášková V (2008) Degradation of cellulose by basidiomycetous fungi. FEMS Microbiol Rev 32(3):501–521

6. Bégum P, Lemaire M (1996) The cellulosome: an exocellular, multiprotein complex specialized in cellulose degradation. Crit Rev Biochem Mol Biol 31(3):201–236

7. Liao JC, Mi L, Pontrelli S, Luo S (2016) Fuelling the future: microbial engineering for the production of sustainable biofuels. Nat Rev Microbiol 14(5):288–304

8. Waterborg JH (2009) The Lowry method for protein quantitation. In: Walker JM (ed) The protein protocols handbook. Human Press, New York, pp 7–10

9. Laemmli UK (1970) Cleavage of structural proteins during the assembly of the head of bacteriophage T4. Nature 227:680–685

10. Zhang YP, Hong J, Ye X (2009) Cellulase assays. Methods Mol Biol 581:213–231

11. Xiao Z, Storms R, Tsang A (2004) Microplate-based filter paper assay to measure total cellulase activity. Biotechnol Bioeng 88(7):832–837

12. Bakare M, Adewale I, Ajayi A, Shonukan O (2005) Purification and characterization of cellulase from the wild-type and two improved mutants of Pseudomonas fluorescens. Afr J Biotechnol 4(9)

13. Gaur R, Tiwari S (2015) Isolation, production, purification and characterization of an organic-solvent-thermostable alkalophilic cellulase from Bacillus vallismortis RG-07. BMC Biotechnol 15(1):19

14. Du R, Su R, Zhang M, Qi W, He Z (2014) Cellulase recycling after high-solids simultaneous saccharification and fermentation of combined pretreated corncob. Front Energy Res 2:24

15. Ang SK, Shaza E, Adibah Y, Suraini A, Madihah M (2013) Production of cellulases and xylanase by Aspergillus fumigatus SK1 using untreated oil palm trunk through solid state fermentation. Process Biochem 48 (9):1293–1302

16. Geib SM, Tien M, Hoover K (2010) Identification of proteins involved in lignocellulose degradation using in gel zymogram analysis combined with mass spectroscopy-based peptide analysis of gut proteins from larval Asian longhorned beetles, Anoplophora glabripennis. Insect Sci 17(3):253–264

17. Thomson I (2015) Chamaiporn Champasri, Thongchai Champasri and Khanutsanan Woranam. Asian J Biochem 10(5):190–204

18. Anuradha Jabasingh S, Varma S, Garre P (2014) Production and purification of cellulase from Aspergillus nidulans AJSU04 under solid-state fermentation using coir pith. Chem Biochem Eng Q 28(1):143–151

19. Gavrilov SN, Stracke C, Jensen K, Menzel P, Kallnik V, Slesarev A, Sokolova T, Zayulina K, Bräsen C, Bonch-Osmolovskaya EA (2016) Isolation and characterization of the first xylanolytic hyperthermophilic euryarchaeon Thermococcus sp. strain 2319x1 and its unusual multidomain glycosidase. Front Microbiol 7:552

Part IV

Application of Cellulases for Lignocellulose Degradation

Chapter 19

On-Site Production of Cellulolytic Enzymes by the Sequential Cultivation Method

Cristiane S. Farinas, Camila Florencio, and Alberto C. Badino

Abstract

The conversion of renewable lignocellulosic biomass into fuels, chemicals, and high-value materials using the biochemical platform has been considered the most sustainable alternative for the implementation of future biorefineries. However, the high cost of the cellulolytic enzymatic cocktails used in the saccharification step significantly affects the economics of industrial large-scale conversion processes. The on-site production of enzymes, integrated to the biorefinery plant, is being considered as a potential strategy that could be used to reduce costs. In such approach, the microbial production of enzymes can be carried out using the same lignocellulosic biomass as feedstock for fungal development and biofuels production. Most of the microbial cultivation processes for the production of industrial enzymes have been developed using the conventional submerged fermentation. Recently, a sequential solid-state followed by submerged fermentation has been described as a potential alternative cultivation method for cellulolytic enzymes production. This chapter presents the detailed procedure of the sequential cultivation method, which could be employed for the on-site production of the cellulolytic enzymes required to convert lignocellulosic biomass into simple sugars.

Key words Enzymatic hydrolysis, biorefinery, enzymes, cellulases, fermentation, on-site production

1 Introduction

The use of renewable lignocellulosic biomass for the production of biofuels, chemicals, and high-value materials has been identified as one of the most promising routes to build new industries of the future, in accordance with the biorefinery concept [1–4]. The enzymatic conversion of the polysaccharides present in lignocellulosic biomass will certainly be a key technology in such biorefineries, since the biochemical pathway using biocatalysts is highly advantageous both technically and environmentally, being also compatible with the demands of the current bioeconomy [4–7]. However, one of the key technological challenges that are still holding back the implementation of industrial large-scale biorefineries is the high cost of the cellulolytic enzymatic cocktails

Mette Lübeck (ed.), *Cellulases: Methods and Protocols*, Methods in Molecular Biology, vol. 1796,
https://doi.org/10.1007/978-1-4939-7877-9_19, © Springer Science+Business Media, LLC, part of Springer Nature 2018

required in the saccharification step. Due to the complexity and recalcitrance of lignocellulosic materials, high enzyme loadings are needed in the conversion process and the cost of these enzymes affects significantly the economic feasibility of the whole process [8–10]. There is therefore a need to develop bioprocess engineering strategies to produce enzymatic cocktails with improved performance in lignocellulose hydrolysis.

In order to address this important issue, several studies have focused on increasing the production efficiency of cellulolytic enzymes by selection of microorganisms capable of secreting a high and diversified amount of enzymes [11–15] as well as by optimizing the composition of the cellulolytic cocktail [16–18]. Studies addressing bioprocess engineering strategies to improve cellulolytic enzymes production by manipulation of process variables, bioreactor type, and cultivation methods have also been reported [15, 19–24]. The on-site production of enzymes within the biorefinery is also being considered as a potential strategy that could be used to reduce costs [8, 10, 22, 25–28]. Furthermore, it has been suggested that use of the enzymes secreted from microorganisms grown on the same lignocellulosic material that will be converted to ethanol could be a possible means of better modulating the enzymatic complex [15, 25, 29].

Multiple enzymes are required for the complete hydrolysis of lignocellulosic materials, including cellulases, hemicellulases, pectinases, ligninases, and other accessory enzymes [30, 31]. The cellulolytic enzymes comprise a set of glycoside hydrolases whose action involves hydrolysis of the β-1,4-glycosidic bonds of cellulose [30]. The most widely accepted mechanism of action of cellulases involves three classes of enzymes: endoglucanases, exoglucanases, and β-glucosidases. Endoglucanases hydrolyze accessible intramolecular β-1,4-glycosidic bonds of the cellulose chains randomly, producing new chain ends; exoglucanases progressively cleave cellulose chains at the ends to release soluble cellobiose or glucose; and β-glucosidases hydrolyze cellobiose to glucose [32, 33]. Other important enzymes required for depolymerization of hemicellulose are the endo-1,4-β-xylanase (xylanase) enzymes, which cleave the β-1,4-glycosidic linkage between xylose residues in the backbone of xylans [34]. For instance, supplementation with feruloyl esterases and xylanase enzymes produced on-site improved the hydrolysis of sugarcane bagasse by up to 36% [17]. Moreover, the role of oxidative enzymes such as lytic polysaccharide monooxygenases (LPMO) and other accessory proteins in increasing the degradation of cellulose suggests that the action of the classical hydrolytic cellulases is also facilitated by the lytic action of the LPMO [31, 35, 36].

The microbial cultivation processes for enzyme production can be conducted in a solid medium, called solid-state fermentation (SSF), or in liquid medium, called submerged fermentation (SmF). Although most of the advances related to the microbial production of cellulases have been developed for SmF, the growth of filamentous fungi, the main producers of cellulolytic enzymes, occurs naturally under conditions similar to SSF [37]. Both processes have advantages as well as limitations, which should be considered according to the desired product and the selected microorganism [37–40]. A potential advantage of the SSF is that it enables the use of agroindustrial residues such as sugarcane bagasse as carbon source and inducer for microbial enzyme production [11, 22, 24, 37, 41–43].

Recently, a combination of the SSF and SmF cultivation techniques, defined as sequential fermentation (SeqF), has been effectively applied for the production of cellulolytic enzymes using sugarcane bagasse as carbon source and inducer [15, 23, 42, 44–46]. The sequential fermentation is characterized by a preculture preparation with initial stage of fungal growth under solid state, followed by a transition to submerged state (Fig. 1). The SeqF presented significant results in relation to the conventional submerged process of cellulase production, both in agitated flasks [15, 22, 42, 45, 46] and in conventional stirred-tank bioreactors as well as in airlift type bioreactors [23, 44]. Endoglucanase productivity was threefold higher in SeqF compared to conventional SmF, suggesting the potential of the technique as a promising alternative for the on-site production of cellulolytic enzymes [44]. A possible explanation for the increased endoglucanase activity could be differences in the morphology of the *A. niger* fungi, because in SeqF there was a predominance of filamentous dispersed mycelia, while in the conventional SmF process there was early formation of fungal pellets (Fig. 2). Another possible reason for the higher endoglucanase activity achieved under SeqF is that the solid medium used in the SSF preculture step could act as an inducer for cellulase production during the early stage of cell development, hence contributing to enzyme production [45].

The SeqF methodology was also validated for strains of the genus *Trichoderma*, resulting in an enzymatic profile with greater activities of xylanase, endoglucanase, β-glucosidase, avicellase, and FPase [45]. The secretome of the *T. reesei* and *A. niger* strains cultivated in SmF and SeqF also revealed significant differences between these cultivation methods [15]. The proteomic analysis of the *A. niger* strain showed that the SeqF presented a higher number of proteins identified and higher enzymatic activities as well in comparison to conventional SmF. In addition, the higher enzymatic activities and/or a better balance of the secretome composition from fungal cultivation under SeqF lead to a threefold increase in the saccharification of pretreated sugarcane bagasse [15]. In overall, these findings suggest that the integration of the on-site enzyme production process using lignocellulosic biomass as feedstock for the sequential fermentation is of potential interest for the implementation of future biorefineries.

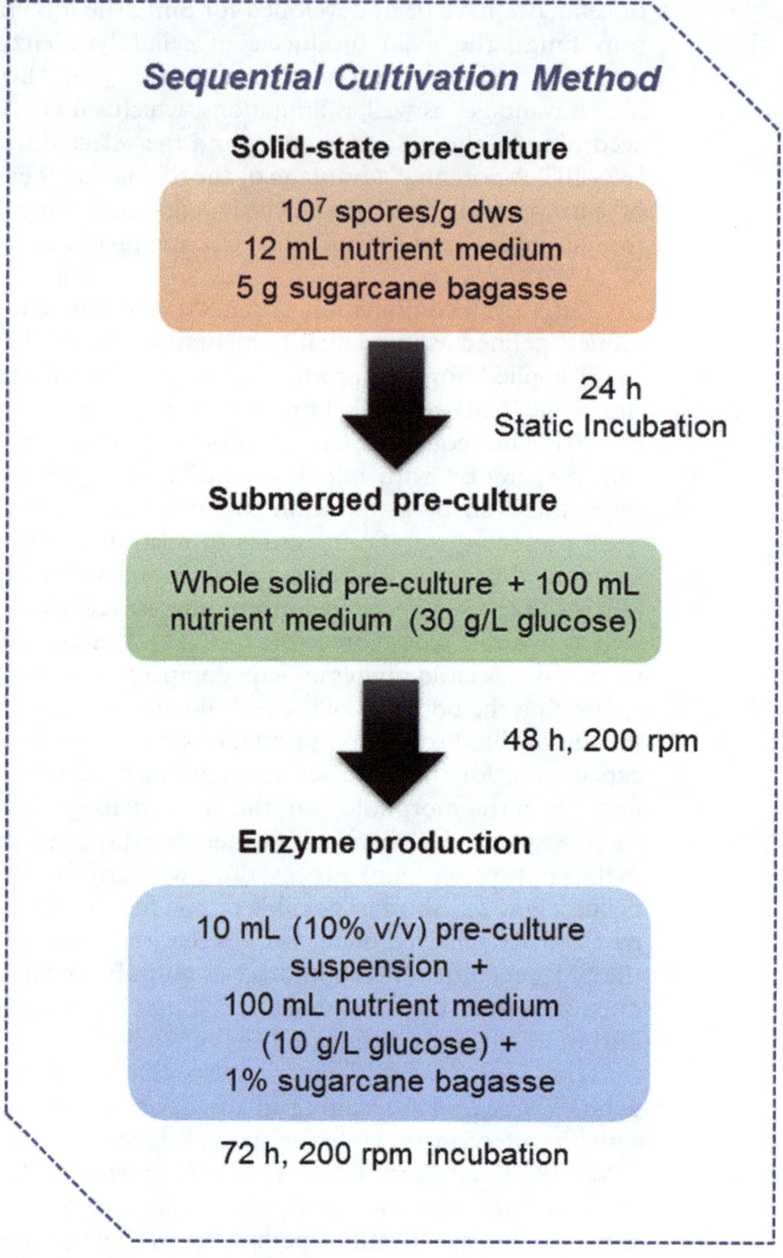

Fig. 1 Schematic of the experimental procedure for the sequential solid-state and submerged cultivation method used for cellulolytic enzyme production. Typical conditions used for the cultivation of *A. niger* using sugarcane bagasse as the solid substrate and inducer

This chapter presents the detailed procedure of the sequential cultivation method, which could be employed for the on-site production of the enzymatic cocktails required to convert lignocellulosic biomass into simple sugars.

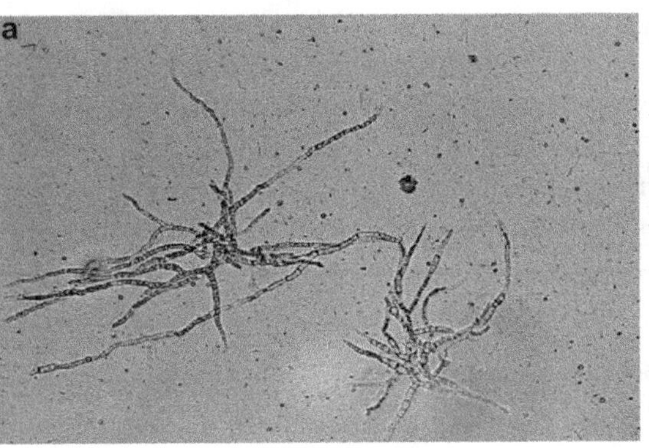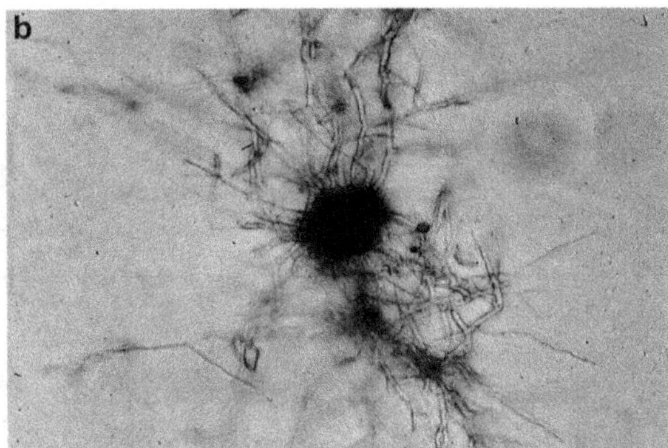

Fig. 2 Typical morphologies (at 200× magnification) of *A. niger* using the two different cultivation methods: (**a**) filamentous mycelium in the sequential fermentation (SeqF); (**b**) pellet in conventional submerged fermentation (SmF)

2   Materials

2.1   Microorganism

Filamentous fungi such as *Aspergillus niger* and *Trichoderma reesei* can be used as source of cellulolytic enzymes (*see* **Note 1**). Microorganism activation is usually carried out in potato dextrose agar (PDA) medium incubated for 7 days at 32 °C. Suspensions of spores is prepared by the addition of 10 mL of Tween 80 (0.3%, v/v), and the spore concentrations is determined using a Neubauer chamber and an optical microscope.

2.2   Inducer Substrate

Different sources of lignocellulosic biomass such as sugarcane bagasse can be used as inducer substrate for cellulolytic enzyme production, either pretreated or without pretreatment (*see* **Note 2**). The biomass should be ground and sieved to small particles (particle size between 1 and 2 mm) before use. If the substrate is wet, it has to be dried prior to milling. The drying step is done at either room temperature or at 45–50 °C to prevent and minimize collapsing of pores in the initial material.

2.3   Liquid Nutrient Medium

The liquid nutrient medium used in the preculture and during cellulolytic enzymes production is adapted from the medium described by [47], and contains (w/v): 0.14% $(NH_4)_2SO_4$, 0.20% KH_2PO_4, 0.03% $CaCl_2$, 0.02% $MgSO_4\cdot7H_2O$, 0.50% peptone, 0.20% yeast extract, 0.03% urea, 0.10% Tween 80, and 0.10% of salt solution (5 mg/L $FeSO_4\cdot7H_2O$, 1.6 mg/L, $MnSO_4\cdot H_2O$, 1.4 mg/L $ZnSO_4\cdot7H_2O$, and 2.0 mg/L $CoCl_2$). The medium should be sterilized in an autoclave for 20 min at 121 °C (*see* **Note 3**).

2.4 Bioreactor

In bioreactor cultivation, a stirred tank bioreactor such as the Bioflo IIC (New Brunswick Scientific, USA) with a 4-L working volume and equipped with two 6-blade Rushton turbine impellers, as well as dissolved oxygen and pH probes can be used to conduct laboratory-scale experiments.

3   Methods

3.1 Preculture

In the sequential cultivation method, the preculture is initiated as solid-state fermentation (SSF), using a lignocellulosic biomass such as sugarcane bagasse as the solid substrate. In the SSF cultivation step, the moisture content is adjusted to 70% (w/w) by the addition of 12 mL of the liquid nutrient medium into a 500-mL Erlenmeyer flask containing 5 g of dry solid substrate (*see* **Note 4**). A volume of spore suspension resulting in a concentration of 10^7 spores per gram of dry solid substrate is added, and cultivation is maintained as SSF under static conditions for 24 h at 32 °C (*see* **Note 5**). A volume of nutrient medium enriched with 30 g/L of glucose is added (40 parts of nutrient medium per gram of dry solid), and the cultivation is continued as SmF in an orbital shaker incubator for 48 h at 32 °C, with continuous agitation at 200 rpm.

3.1.1 Enzyme Production

A volume of preculture suspension corresponding to 10% (v/v) is transferred to cultivation medium to initiate enzymatic production in either shake flasks or in a stirred-tank bioreactor. The shake flask cultivations is performed in 500 mL Erlenmeyer flasks, containing 100 mL of nutrient medium enriched with 10 g/L of glucose and 10 g/L of lignocellulosic substrate, with incubation in a rotational incubator shaker at 200 rpm and 32 °C for 72 h (see **Note 6**). Samples are collected at 24-h intervals, centrifuged at 2500 × g for 10 min at 4 °C, and the crude enzymatic extract is used for quantification of enzymatic activity.

In bioreactor cultivation, a stirred tank bioreactor with a 4-L working volume (Bioflo IIC, New Brunswick Scientific, USA) can be used to carry out the cultivations. Typical conditions for cultivation of *A. niger* would be 30 h at 400 rpm, 0.5 vvm, and 32 °C (*see* **Note 7**). The pH is maintained at 5.0 by the addition of 1 mol/L HCl and 2 mol/L NaOH solutions. Samples are collected at 6-h intervals and centrifuged at 2500 × g for 10 min at 4 °C, and the crude enzymatic extract is used for quantification of enzymatic activity.

The reducing sugars concentration in the cultivation broth is determined by the DNS method [48] in order to monitor fungal growth according to the consumption of glucose and other reducing sugars. Activity of cellulolytic enzymes such as filter paper activity, endoglucanase, and β-glucosidase is determined by following the methodology proposed by Ghose [49].

4 Notes

1. The choice of filamentous fungi used for cellulolytic enzyme production could be based on preliminary screening studies, such as the selection using plate-assay with crystalline cellulose as the sole source of carbon [50]. Recognized cellulolytic strains such as *Aspergillus niger* and *Trichoderma reesei* can be obtained from different microbial culture collections, such as the Centre for Agricultural Bioscience International (CABI) culture collection in the UK. Either wild-type or genetically modified strains have been described as good producers of cellulolytic enzymes.

2. A variety of agroindustrial lignocellulosic wastes can be used as solid substrates for SSF, including sugarcane bagasse, cassava, soybean, and wheat bran, among others. The basic structures of these materials (consisting of cellulose, hemicellulose, lignin, starch, pectin, and fibers) determine the properties of the solid substrate and provide sources of carbon and energy for microorganism growth. In addition, physical characteristics of the substrate such as crystallinity, accessible area, surface area, porosity, and particle size have an important influence on the SSF process. Additional pretreatment process is usually employed in order to increase the accessibility to microbial degradation by assisting the adherence and penetration of the fungal hyphae [37].

3. Glucose solution will be also included in the medium, but it should be autoclaved separately from the nutrient medium and added the final concentration after cooling. The addition of glucose solution into the liquid nutrient medium should be carried out in a sterile place such as laminar flow cabinet.

4. An ideal humidity level must be used in SSF cultivation, since the water content is one of the most important operating parameters affecting the efficiency of the process. If the moisture content is too high, the void spaces of the solids are filled with water, resulting in restricted gaseous diffusion. At the other extreme, if the moisture content is low, the growth of the microorganism will be harmed [37, 41]. Consequently, identification of the optimal moisture content for each SSF substrate is key to promote favorable growth conditions and maximize the enzymatic production. The optimal moisture content value depends on both the solid substrate and the microorganism used, and it should be determined beforehand by carrying out preliminary SSF cultivations.

5. Temperature is an important process variables affecting SSF, because microbial growth under aerobic conditions results in the release of metabolic heat. The optimal temperature for

cellulolytic enzyme production by *Aspergillus* and *Trichoderma* is usually in the range of 28–35 °C, since these fungi are mesophilic organisms. The characterization of each particular microorganism in terms of the influence of temperature on the kinetics of growth and product formation is essential for SSF bioprocess development.

6. Agitation in SmF cultivations is directly related to the morphology of the filamentous fungi, which can vary from freely dispersed mycelium to highly dense clumps or pellets. The optimal morphology for enzyme production is usually related to each specific microorganism. However, changes in filamentous fungi morphology can impact the rheology of cultivation media, which consequently affects the flow patterns in the bioreactor and the overall performance of the cultivation. In terms of process requirements, a dispersed hyphal morphology may result in a medium with high viscosity, which will be more complicated to mix and aerate than a cultivation medium where pellets are the predominant morphology. It is not possible to draw firm conclusions concerning a relationship between fungal morphology and cellulolytic enzyme production, because of the existence of many other factors that might influence the cultivation process.

7. During SmF cultivations for cellulase production by filamentous fungi, the usual approach is to maintain the dissolved oxygen (DO) concentration above 20% air saturation by manipulating the agitation and aeration rates or by increasing the oxygen concentration in the sparge air. Therefore, finding an optimum condition for agitation and aeration is highly important to improve cellulolytic enzyme production.

Acknowledgments

The author would like to thank Embrapa, CNPq (Process 401182/2014-2), CAPES, and FAPESP (Process 2014/19000-3 and 2016/10636-8) (all from Brazil) for financial support, and the students and technicians from Embrapa Instrumentation for their invaluable contribution.

References

1. Moshkelani M, Marinova M, Perrier M, Paris J (2013) The forest biorefinery and its implementation in the pulp and paper industry: energy overview. Appl Therm Eng 50:1427–1436

2. Sheldon RA (2014) Green and sustainable manufacture of chemicals from biomass: state of the art. Green Chem 16:950–963

3. Kamm B, Kamm M (2004) Principles of biorefineries. Appl Microbiol Biotech 64:137–145

4. Mohan SV, Nikhil GN, Chiranjeevi P et al (2016) Waste biorefinery models towards sustainable circular bioeconomy: critical review and future perspectives. Bioresour Technol 215:2–12

5. Mu DY, Seager T, Rao PS, Zhao F (2010) Comparative life cycle assessment of lignocellulosic ethanol production: biochemical versus thermochemical conversion. Environ Manag 46:565–578

6. Lopes MSG (2015) Engineering biological systems toward a sustainable bioeconomy. J Ind Microbiol Biotech 42:813–838

7. Heux S, Meynial-Salles I, O'Donohue MJ, Dumon C (2015) White biotechnology: state of the art strategies for the development of biocatalysts for biorefining. Biotechn Adv 33:1653–1670

8. Johnson E (2016) Integrated enzyme production lowers the cost of cellulosic ethanol. Biofuels Bioproducts Biorefining Biofpr 10:164–174

9. Klein-Marcuschamer D, Oleskowicz-Popiel P, Simmons BA, Blanch HW (2012) The challenge of enzyme cost in the production of lignocellulosic biofuels. Biotechnol Bioeng 109:1083–1087

10. Liu G, Zhang J, Bao J (2016) Cost evaluation of cellulase enzyme for industrial-scale cellulosic ethanol production based on rigorous Aspen Plus modeling. Bioprocess Biosyst Eng 39:133–140

11. Delabona P, Pirota R, Codima C et al (2012) Using Amazon forest fungi and agricultural residues as a strategy to produce cellulolytic enzymes. Biomass Bioenergy 37:243–250

12. King BC, Waxman KD, Nenni NV et al (2011) Arsenal of plant cell wall degrading enzymes reflects host preference among plant pathogenic fungi. Biotechnol Biofuels 4:14

13. Guerriero G, Hausman JF, Strauss J et al (2015) Destructuring plant biomass: focus on fungal and extremophilic cell wall hydrolases. Plant Sci 234:180–193

14. Pirota R, Tonelotto M, Delabona PD et al (2015) Characterization of fungi isolated from the Amazon region for the potential of biomass-degrading enzymes production. Cienc Rural 45:1606–1612

15. Florencio C, Cunha FM, Badino AC et al (2016) Secretome analysis of *Trichoderma reesei* and *Aspergillus niger* cultivated by submerged and sequential fermentation processes: enzyme production for sugarcane bagasse hydrolysis. Enzym Microb Technol 90:53–60

16. Delabona PD, Cota J, Hoffmam ZB et al (2013) Understanding the cellulolytic system of *Trichoderma harzianum* P49P11 and enhancing saccharification of pretreated sugarcane bagasse by supplementation with pectinase and alpha-L-arabinofuranosidase. Bioresour Technol 131:500–507

17. Pinto Braga CM, Delabona PS, Lima DJS et al (2014) Addition of feruloyl esterase and xylanase produced on-site improves sugarcane bagasse hydrolysis. Bioresour Technol 170:316–324

18. Thomas L, Parameswaran B, Pandey A (2016) Hydrolysis of pretreated rice straw by an enzyme cocktail comprising acidic xylanase from *Aspergillus* sp for bioethanol production. Renew Energy 98:9–15

19. Pirota R, Tonelotto M, Delabona PS et al (2016) Bioprocess developments for cellulase production by *Aspergillus oryzae* cultivated under solid-state fermentation. Braz J Chem Eng 33:21–31

20. Delabona PD, Lima DJ, Robl D et al (2016) Enhanced cellulase production by *Trichoderma harzianum* by cultivation on glycerol followed by induction on cellulosic substrates. J Ind Microbiol Biotech 43:617–626

21. Pirota RDPB, Delabona PS, Farinas CS (2014) Simplification of the biomass to ethanol conversion process by using the whole medium of filamentous fungi cultivated under solid-state fermentation. Bioenergy Res 7:744–752

22. Vasconcellos VM, Tardioli PW, Giordano RLC, Farinas CS (2015) Production efficiency versus thermostability of (hemi)cellulolytic enzymatic cocktails from different cultivation systems. Process Biochem 50:1701–1709

23. Cunha FM, Esperanca MN, Florencio C et al (2015) Three-phasic fermentation systems for enzyme production with sugarcane bagasse in stirred tank bioreactors: effects of operational variables and cultivation method. Biochem Eng J 97:32–39

24. Rodriguez-Zuniga UF, Couri S, Neto VB et al (2013) Integrated strategies to enhance cellulolytic enzyme production using an instrumented bioreactor for solid-state fermentation of sugarcane bagasse. Bioenergy Res 6:142–152

25. Delabona P, Farinas C, da Silva M et al (2012) Use of a new *Trichoderma harzianum* strain isolated from the Amazon rainforest with pretreated sugar cane bagasse for on-site cellulase production. Bioresour Technol 107:517–521

26. Sorensen A, Teller PJ, Lubeck PS, Ahring BK (2011) Onsite enzyme production during bioethanol production from biomass: screening for suitable fungal strains. Appl Biochem Biotechnol 164:1058–1070

27. Kovacs K, Macrelli S, Szakacs G, Zacchi G (2009) Enzymatic hydrolysis of steam-pretreated lignocellulosic materials with *Trichoderma atroviride* enzymes produced in-house. Biotechnol Biofuels 2:11

28. Rana V, Eckard AD, Teller P, Ahring BK (2014) On-site enzymes produced from

Trichoderma reesei RUT-C30 and *Aspergillus saccharolyticus* for hydrolysis of wet exploded corn stover and loblolly pine. Bioresour Technol 154:282–289

29. van den Brink J, Maitan-Alfenas GP, Zou G et al (2014) Synergistic effect of *Aspergillus niger* and *Trichoderma reesei* enzyme sets on the saccharification of wheat straw and sugarcane bagasse. Biotechnol J 9:1329–1338

30. Payne CM, Knott BC, Mayes HB et al (2015) Fungal Cellulases. Chem Rev 115:1308–1448

31. Horn SJ, Vaaje-Kolstad G, Westereng B, Eijsink VGH (2012) Novel enzymes for the degradation of cellulose. Biotechnol Biofuels 5:45

32. Zhang Y, Himmel M, Mielenz J (2006) Outlook for cellulase improvement: screening and selection strategies. Biotechnol Adv 24:452–481

33. Lynd L, Weimer P, van Zyl W, Pretorius I (2002) Microbial cellulose utilization: fundamentals and biotechnology. Microbiol Mol Biol Rev 66:506–577

34. Saha BC (2003) Hemicellulose bioconversion. J Ind Microbiol Biotechnol 30:279–291

35. Cannella D, Mollers KB, Frigaard NU et al (2016) Light-driven oxidation of polysaccharides by photosynthetic pigments and a metalloenzyme. Nat Commun 7:8

36. Rodriguez-Zuniga UF, Cannella D, Giordano RD et al (2015) Lignocellulose pretreatment technologies affect the level of enzymatic cellulose oxidation by LPMO. Green Chem 17:2896–2903

37. Farinas CS (2015) Developments in solid-state fermentation for the production of biomass-degrading enzymes for the bioenergy sector. Renew Sust Energ Rev 52:179–188

38. Kuhad RC, Deswal D, Sharma S et al (2016) Revisiting cellulase production and redefining current strategies based on major challenges. Renew Sust Energ Rev 55:249–272

39. Thomas L, Larroche C, Pandey A (2013) Current developments in solid-state fermentation. Biochem Eng J 81:146–161

40. Singhania RR, Sukumaran RK, Patel AK et al (2010) Advancement and comparative profiles in the production technologies using solid-state and submerged fermentation for microbial cellulases. Enzym Microb Technol 46:541–549

41. Delabona PD, Pirota R, Codima CA et al (2013) Effect of initial moisture content on two Amazon rainforest *Aspergillus* strains cultivated on agro-industrial residues: biomass-degrading enzymes production and characterization. Ind Crop Prod 42:236–242

42. Cunha FM, Vasconcellos VM, Florencio C et al (2017) On-site production of enzymatic cocktails using a non-conventional fermentation method with agro-industrial residues as renewable feedstocks. Waste Biomass Valorization 8:517–526

43. Cunha FM, Kreke T, Badino AC et al (2014) Liquefaction of sugarcane bagasse for enzyme production. Bioresour Technol 172:249–252

44. Cunha FM, Esperanca MN, Zangirolami TC et al (2012) Sequential solid-state and submerged cultivation of *Aspergillus niger* on sugarcane bagasse for the production of cellulase. Bioresour Technol 112:270–274

45. Florencio C, Cunha FM, Badino AC, Farinas CS (2015) Validation of a novel sequential cultivation method for the production of enzymatic cocktails from *Trichoderma* strains. Appl Biochem Biotechnol 175:1389–1402

46. Cunha FM, Badino AC, Farinas CS (2017) Effect of a novel method for in-house cellulase production on 2G ethanol yields. Biocatal Agric Biotechnol 9:224–229

47. Mandels M, Sternberg D (1976) Recent advances in cellulase technology. J Ferment Technol 54:267–286

48. Miller G (1959) Use of dinitrosalicylic acid reagent for determination of reducing sugar. Anal Chem 31:426–428

49. Ghose T (1987) Measurement of cellulase activities. Pure Appl Chem 59:257–268

50. Florencio C, Couri S, Farinas CS (2012) Correlation between agar plate screening and solid-state fermentation for the prediction of cellulase production by *Trichoderma* strains. Enzyme Res:7. https://doi.org/10.1155/2012/793708

Chapter 20

Test of Efficacy of Cellulases for Biomass Degradation

Henning Jørgensen

Abstract

Testing of cellulases on real biomass samples is required to do a true assessment of their efficacy for biomass degradation. Cellulase enzymes belong to a number of different glycosyl hydrolase families, all with different activity, specificity and modes of action. The concerted and synergistic action of these different cellulases determines the efficacy for plant cell wall deconstruction and cellulose hydrolysis. However, the plant cell wall of lignocellulosic materials is a very complex matrix and the efficacy of a cellulase preparation to degrade lignocellulosic materials is influenced by many factors. In this chapter, two protocols for testing efficacy of cellulases on pretreated biomass samples are described. The first protocol describes a small-scale setup employing low solids concentration that easily enables the testing of a larger number of samples. The second protocol describes a method for testing the efficacy of cellulases at conditions more closely resembling industrial conditions, i.e., high solids concentrations. Both protocols can be used to test the cellulases under a variety of substrate types, substrate concentrations, enzyme loadings and process conditions. The protocols can also be used to evaluate different feedstocks.

Key words Cellulases, Enzyme assays, Lignocellulose, Cellulose, High solids hydrolysis

1 Introduction

Methods for testing the efficacy of cellulase preparations is an important and crucial part of developing better cellulase preparations for cost-effective exploitation of lignocellulosic biomass in biorefineries for production of fuels and chemicals. Cellulases encompass a rather large and diverse number of enzymes belonging to several different glycosyl hydrolase families. For efficient and complete hydrolysis of lignocellulosic substrates, a cocktail of appropriate cellulases is needed as the different cellulases work in synergism to hydrolyze the cellulose [1]. In addition, the efficacy of cellulases on a complex biomass samples will also be indirectly influenced by the presence of other enzymes, e.g., hemicellulases [2].

A number of soluble surrogate or synthesized substrates do exist and are routinely used to easily assess specific cellulases activities, e.g., carboxymethyl cellulose to measure endoglucanases

Mette Lübeck (ed.), *Cellulases: Methods and Protocols*, Methods in Molecular Biology, vol. 1796,
https://doi.org/10.1007/978-1-4939-7877-9_20, © Springer Science+Business Media, LLC, part of Springer Nature 2018

(GH families 5, 6, 7, and 12) or methylumbelliferyl (MU) and paranitrophenyl (PNP) glycosides to measure cellobiohydrolases (GH families 6 and 7) and β-glucosidases (GH family 1–3). To measure the efficacy of a mixture of cellulases, the filter paper assay is commonly used. However, the results obtained on filter paper cannot directly be used to predict the performance on a real complex biomass material. When comparing different cocktails, the best performing cocktail based on filter paper activity may even not be the best on a given sample of pretreated biomass. This is partly also due to the effect of other enzyme activities commonly present in many cellulase preparations that indirectly assist the cellulases. One example is the positive effect of hemicellulases, which will remove hemicellulose moieties covering the cellulose surface and hindering the access of cellulases to the cellulose [3, 4]. Another example is the ability of lytic polysaccharide monooxygenases (LPMOs) to work synergistically with the "classic" cellulases to boost the cellulose hydrolysis [1, 5, 6]. The action of LPMOs does however require an electron donor and availability of molecular oxygen. Whereas ascorbic acid has been used as electron donor in many studies on pure cellulose substrates, the presence of lignin and low molecular lignin derived molecules appears to be involved in delivering electrons in natural biomass samples [7, 8]. The classical filter paper assay does not contain any electron donor, and the LPMO enzymes will have little activity, if any. The effect of eventual LPMOs in the cellulase preparation will therefore not be taken into account in the filter paper assay. This could have huge implication for the evaluation and selection of new cellulase cocktails. To make a useful assessment of various cellulase preparations, they have to be compared on relevant substrates and eventually also at process relevant conditions.

The advances in high throughput automated screening for new enzymes combined with the ease of performing protein engineering have increased the interest for high throughput (HTP) systems for testing the efficacy of new cellulases and cellulase cocktails. Several options for measuring cellulase efficacy on complex biomass substrates in (semi) HTP format have been presented [9, 10] and some systems even combine it with a HTP pretreatment systems [11–13]. Generally, these systems requires expensive and maintenance demanding robotic systems, which may be beyond the reach of many research laboratories [14]. Handling and distributing the pretreated material is still an issue compromising the reproducibility of such systems. Solids dispensing systems are available, but drying of the pretreated material should be avoided as it can result in irreversible pore collapse and have a negative effect on hydrolysis [15]. Automated liquid dispensing of biomass slurries is also difficult, as the biomass particles will quickly sediment and make accurate dispensing of a given amount of substrate difficult.

The test conditions can also be relevant to consider in relation to evaluate the potential of various cellulase preparations for given industrial processes. Commonly, hydrolysis tests are performed with low solids concentrations, in the order of 1–10% dry matter (DM), due the difficulty of mixing the material at higher solids concentrations. It is also common to work with washed materials to avoid interference and inhibition of the cellulases by soluble sugars and low molecular inhibitors formed during the pretreatment [16]. Industrial conditions will clearly not resemble this, as for example solid concentrations are more likely in the 15–25% DM range [17]. Part of evaluating the efficacy of new cellulase preparations should therefore involve tests under conditions more closely resembling industrial conditions. This will allow for evaluating robustness and sensitivity toward inhibitors and high product concentrations that are the result of using high initial solid concentrations and unwashed materials. The activity of LPMOs may also be better assessed on unwashed material as low molecular weight lignins are involved the beneficial electron shuttling taking place in real biomass samples [8, 18].

In this chapter, two protocols both with never dried material, are presented. For both protocols, it is possible to work with washed or unwashed material depending on the purpose of the test. One protocol is designed for use with low solid concentration (1–2% DM) and will be suitable for testing a large number of samples. The other protocol is designed to be used for hydrolysis at high solid concentrations, typically above 20% DM. It uses the principle of gravimetric mixing that has proven to work well for high solid enzymatic hydrolysis [19]. In this setup, the assay is scaled down to work in bottles placed inside a tumbler mixing system. The tumbler mixer is a homemade system, but can rather easily be replicated. Performing enzymatic hydrolysis of biomass at high solid concentration do require some special precautions with respect to how to sample and calculate yields [20], which is also explained in the protocol. In the text, the two protocols are referred to as "low solids system" and "high solids system."

2 Materials

2.1 Plant Material

Many different pretreatment options exist and it is beyond the scope of this chapter to give a detailed account of how to prepare the pretreated material and the analysis of the composition of this material for content of cellulose, hemicellulose, and lignin. The substrate type (type of feedstock/plant material and pretreatment method) to use should be related to the specific application— ideally use same material as the cellulase preparation is intended for, or select one closely resembling it. It is important that pretreated biomass is not dried prior to storage and use, but rather stored frozen in tightly sealed containers or bags (*see* **Note 1**).

2.2 Enzymes	Enzymes can be either monocomponent recombinant enzymes or mixtures thereof, crude culture preparations or commercial cellulases. The later may also contain other enzyme activities, e.g., hemicellulases. Enzymes are added on an mg of protein per g of dry solids or cellulose basis (*see* **Notes 2** and **3**). Several different assays to determine protein concentration exist. However, crude preparations and commercial preparations may contain salts, low molecular weight (MW) peptides and media components, cofactors, inhibitors, stabilizers such as sugars and additives that may interfere with the protein assay. We use the ninhydrin assay, which appears less affected by these type of compounds [21, 22].

2.3 Equipment and Reagents

2.3.1 Ninhydrin Protein Assay

1. 13.5 M NaOH diluted in MilliQ water.
2. Glacial acetic acid.
3. 50% ethanol (v/v) mixed with MilliQ water.
4. 2% solution ninhydrin reagent (ready solution for example from Sigma).
5. Standard stock solution of 1 mg/mL bovine serum albumin (BSA) either buy ready-made or prepared in MilliQ water.
6. 1.5 or 2.0 mL safe-lock Eppendorf tubes or 1.5 mL tubes with screw-cap.
7. Vortex mixer.
8. Autoclave with a program 121 °C, 20 min for liquids.
9. Heating block 100 °C.
10. 96-well microtiter plates.
11. Spectrophotometer measuring microtiter plates at 570 nm.

2.3.2 Preparing the Biomass Material

1. Large glass beaker.
2. Buchner funnel.
3. Filter paper, e.g., Fisherbrand P8 filter paper.
4. Plastic bag—large enough to cover the entire Buchner funnel.
5. Moisture analyzer, e.g., Sartorius MA35 (*see* **Note 4**), and drying pans (or) convection oven at 105 °C and aluminum pans or glass beakers for drying.

2.3.3 Hydrolysis—Low Solids System

1. 50 mM sodium acetate buffer pH 5.0 (*see* **Note 5**).
2. Eventual antibiotics, e.g., tetracycline (*see* **Note 6**).
3. Analytical balance.
4. Kitchen blender, e.g., BL-1200, Wilfa, Norway.
5. Magnetic stirrer.
6. Glass beaker.
7. Magnetic stirrer bar.

   8. Moisture analyzer, e.g., Sartorius MA35.

   9. 2 mL microcentrifuge tubes.

  10. Thermomixer or similar heating block with agitation of microcentrifuge tubes.

  11. Water bath or heating block at 100 °C.

  12. Centrifuge for microcentrifuge tubes.

  13. Syringe filters 0.45 μm.

  14. Syringes.

2.3.4 Hydrolysis—High Solids System

1. Analytical balance, 4 decimal.

2. 1 M sodium acetate buffer pH 5.0 (*see* **Note 5**).

3. 50 mM sodium acetate buffer pH 5.0 (*see* **Note 5**).

4. Eventual antibiotics, e.g., tetracycline (*see* **Note 6**).

5. 20 mL polyethylene liquid scintillation vials with screw caps.

6. Insulation tape 19 mm wide.

7. Tumble mixer (*see* **Note 7**).

8. Water bath at 100 °C.

9. 15 mL tubes.

10. Syringe filters 0.45 μm.

11. Syringes.

*2.3.5 Sugar Measurement—HPLC (See **Note 8**)*

1. HPLC system consisting of isocratic pump, refractive index detector, Phonomenex Rezex ROA column (or similar, e.g., Bio-Rad Aminex HPX-87H).

2. Eluent solution 5 mM H_2SO_4 in MQ-water.

3. Sugar stock solution: 10 g glucose, 10 g xylose, 10 g arabinose, and 5 g cellobiose dissolved in 1 L MQ-water.

---

3 Methods

3.1 Protein Determination with Ninhydrin Assay

The **steps 6–10** have to be done in fume hood.

1. Dilute BSA stock solution to obtain minimum five points calibration curve in the range 50–500 mg/mL. A BLANK sample with only water is included.

2. Samples/enzymes are appropriately diluted in MilliQ water. Eventually do different dilutions to obtain readings that are within the range of the calibration curve (*see* **Note 9**).

3. Add into individual safe-lock Eppendorf tupes 40 μL of sample, standard and BLANK, respectively. All samples and standards are measured in triplicate.

4. Add 60 µL 13.5 M NaOH and vortex.

5. Autoclave tubes 20 min, 121 °C.

6. After cooling to room temperature, neutralize base with 100 µL glacial acetic acid and vortex.

7. Add 200 µL 2% ninhydrin reagent and vortex.

8. Heat for 20 min at 100 °C, in heating block in fume hood. Be aware of popping caps and close them immediately, or place a heavy item on the top.

9. Cool to room temperature and add 1000 µL 50% ethanol.

10. Vortex well and transfer 200 µL sample to a 96-well microtiter plate.

11. Measure absorbance at 570 nm.

12. Subtract BLANK values from all standard and sample values.

13. Construct calibration curve based on the measured values of the standards in an x/y-diagram and calculate linear trend line.

14. Calculate protein content using the trend line; remember to multiply with dilution factor.

3.2 Preparing Biomass for Hydrolysis

To avoid inhibitors and residual soluble sugars in the pretreated biomass from influencing the enzyme performance, the pretreated biomass is washed prior to usage. However, for some application studies it can be relevant to use whole slurry/unwashed material to investigate how the enzymes perform in a more industrial relevant substrate. This could especially be relevant if screening for inhibitor tolerant enzymes or enzymes less sensitive to product inhibition. However, special care to ensure correct pH is needed when using unwashed material as it can have very extreme pH values depending on the pretreatment method. The use of washed or unwashed material is therefore depending on the exact application.

1. If the solids content of the pretreated material is low (less than 15–20% DM), vacuum filter in a Buchner funnel with a coarse filter paper. Otherwise, proceed to 2.

2. The pretreated material or filter cake from **step 1** is washed by mixing well the biomass with five weight volumes of distilled or demineralized water in a beaker.

3. Vacuum filter in a Buchner funnel with a coarse filter paper.

4. Wash the filter cake with additional five weight volumes of water directly in the Buchner funnel.

5. Place a large plastic bag over the Buchner funnel and secure it around the neck of the funnel. The vacuum will suck out all air, and the bag will help pressing out more water from the filter cake.

6. Store the washed material in tightly sealed bags to avoid drying (*see* **Note 1**).

7. Determine the DM content in the material either directly on a moisture analyzer, or by oven drying overnight. For oven drying, weight out in tarred aluminum pans or glass beakers the material and dry overnight at 105 °C in convection oven, and weigh again. The DM percentage is determined by subtracting the tared weight and dividing the dry solid weight by the wet solid weight. The measurement is done in minimum duplicate and before each test.

3.3 Hydrolysis Efficacy at Low Solids Concentration (1% Dry Matter)

For tests with the given protocol it is advisable to work at 1–2% dry solids concentration. One tube is used for each sampling point, and it is advisable to do at least duplicates of each sampling point, i.e., if sampling at 0, 4, 8, 24, 48, and 72 h then 12 tubes need to be prepared. This example is based on 1% (w/w) DM solids concentration (*see* **Note 3**), 10 mg protein per g DM enzyme loading (*see* **Note 10**) and 1 mL reaction volume in 2 mL microcentrifuge tubes.

1. Calculate the amount of biomass needed to prepare a 2% (w/w) DM biomass slurry. For example, if the material has 30% DM, then 6.67 g is needed per 100 g total mass (100 mL).

2. Prepare the biomass slurry by mixing the calculated amount of biomass, in this case 6.67 g, with 93.33 g of 50 mM sodium acetate buffer pH 5.0 (*see* **Note 5**).

3. Homogenize in a kitchen blender for 3 min at medium setting, thereby making a homogenous slurry that can be pipetted into the tubes (*see* **Note 11**).

4. The slurry is always prepared fresh.

5. Transfer the slurry to a beaker, add magnetic stirrer bar and place it on a magnetic stirrer (*see* **Note 12**).

6. Pipette 0.5 mL of the 2% (w/w) biomass slurry into 2 mL microcentrifuge tubes. Cut off a few mm of the pipet tip to ensure it does not block.

7. Check the true dry solids concentration. If a moisture analyzer is available, pipet 0.5 mL slurry onto the drying pan. Based on this measurement, calculate the amount of liquid needed to reach the target dry solids concentration of 1%. Otherwise, it is advisable to pipet 0.5 mL of slurry into preweighted tubes and dry overnight at 105 °C in convection oven. In this case, use the measured true solids concentration to correct for eventual deviations in dry solids content during the data treatment.

8. Calculate the amount of cellulase needed based on the measured protein concentration and the desired cellulase loading. In this case, 0.1 mg protein is needed in each tube.

9. Dilute the cellulase preparation prior to distributing it into the tubes (*see* **Note 9**). If the protein concentration is 150 mg protein per g cellulase preparation, then 0.667 mg of cellulase preparations is needed per tube. Diluting the cellulase preparation 100-fold (e.g., 1 g cellulase preparation diluted to a total of 100 g with buffer) means that 67 µL has to be added (now assuming the density of the diluted solution is 1 g/mL). Dilute with same buffer as used in the hydrolysis, here 50 mM sodium acetate buffer pH 5.0.

10. Calculate the amount of 50 mM sodium acetate buffer pH 5.0 needed to reach a final volume of 1 mL and a solids concentration of 1% DM. In this case 433 µL (1000 µL – 500 µL – 67 µL).

11. Add buffer to all tubes.

12. Add the diluted cellulase preparation to all tubes.

13. The tubes are agitated with vortex mixing at 1250 rpm using a thermomixer at 50 °C (*see* **Note 13**).

14. At each time point remove tubes and place them in a water bath or heating block at 100 °C for 10 min to inactivate the enzyme activity. A time "0 h" samples is taken right after addition of cellulases.

15. After cooling to room temperature, centrifuge the tubes at 10,000 × *g* for 5 min

16. Filter the supernatant using a 0.45 µm syringe filter into new tubes.

17. The tubes can now be stored in freezer until further analysis (*see* Subheading 3.5)

18. Calculate the glucose yield as % of theoretical glucose yield:

$$\text{Percent glucose yield} = \frac{[\text{glucose}] - [\text{glucose}_{t0}]}{1.111 \cdot F_{\text{cellulose}} \cdot [\text{biomass}]} \cdot 100\%$$

Where [glucose] is the glucose concentration (g/L) at a certain time point; [glucose_{t0}] is the glucose concentration (g/L) in the time "0 h" sample; 1.111 is the correction factor taking into account the hydration when hydrolyzing cellulose; $F_{\text{cellulose}}$ is the cellulose content (g/g) in the dry biomass as determined by a compositional analysis; [biomass] is the dry solids concentration (g/L).

19. The efficacy of cellulase preparations can be compared based on percent glucose yield obtain at a certain time point for a given enzyme loading. Another option is to do the test with different enzyme loadings for each cellulase preparation, e.g., five different enzyme loadings, and then plot the percent glucose yields as function of enzyme loading. The enzyme loading required to

obtain a certain percentage of glucose yield can be used to compare the efficacy of different cellulase preparations. With the last methodology, it may not be necessary to do different time points during the hydrolysis, but rather select one fixed time for the experiment, e.g., 72 h.

3.4 Hydrolysis Efficacy at High Solids Concentration (20% Dry Matter)

The protocol for high solids concentrations can be used up to 30–35% DM solids content (*see* **Note 14**). One tube is used for each sampling point, and it is advisable to do at least duplicates of each sampling point, i.e., if sampling at 0, 4, 8, 24, 48, and 72 h then 12 tubes need to be prepared. This example is based on 20% (w/w) DM solids content (*see* **Note 3**), 10 mg protein per g DM cellulase loading (*see* **Note 10**) and 10 g final reaction mass in 20 mL bottles.

3.4.1 Hydrolysis

1. Based on the DM of the biomass, calculate the amount of biomass needed to give 2.0 g dry solids in each bottle, e.g., if the material contains 30% dry matter, then 6.67 g is needed.

2. Add the calculated amount of biomass into the bottles.

3. Add 500 μL of 1 M sodium acetate buffer pH 5.0 (*see* **Note 5**) to each bottle.

4. Calculate the amount of cellulase needed based on the measured protein concentration and the desired cellulase loading. In this case, 20 mg protein is needed in each bottle.

5. Dilute the cellulase preparation prior to distributing into the bottles (*see* **Note 9**). If the protein concentration is 150 mg protein per g cellulase preparation, then 133 mg of cellulase preparation is needed per bottle. Diluting it tenfold in 50 mM sodium acetate buffer pH 5.0 means that 1.33 mL has to be added (now assuming the density of the diluted solution is 1 g/mL), which can be accurately distributed.

6. Calculate the amount of water needed to make up the total mass to 10 g. In this case 1.5 mL.

7. Add water to all bottles—briefly shake bottles by hand.

8. Add the diluted cellulase solution as the last step.

9. Close tightly the bottles and secure the caps by warping around insolation tape (*see* **Note 15**).

10. Place the bottles in the tumbler/mixer preheated to 50 °C (*see* **Note 13**) and rotating 20 rpm (*see* **Note 7**).

11. At each time point, remove bottles and place them in a water bath at 100 °C for 10 min to inactivate the enzyme activity.

12. The bottles can now be stored in freezer until further analysis.

The high solids concentration results in slurries or hydrolysates with densities well over 1 g/mL, and a large fraction of the sample is insoluble solids. Only sampling the liquid will result in biased results unless doing appropriate corrections. This procedure can be used for samples from experiments with high solids concentrations to avoid overestimation of yield [20]. The unit g glucose per kg hydrolysate is used because the initial dilution is done on the basis of weighing out the sample. After the initial dilution, it is assumed that the density of the diluted sample is close to 1 g/mL. The concentration obtained from the HPLC (normally given in g/L) is therefore assumed to be equal to the same concentration in g/kg. The initial solids concentration is also given as g DM per kg hydrolysate.

1. After thawing of the bottles mix well by shaking or on vortex mixer.

2. Transfer around 1 g of hydrolysate to a 15 mL tube—record the exact weight with 4 decimals.

3. Dilute to a total of 10 g by adding MilliQ water or the eluent used for HPLC analysis (e.g., 5 mM H_2SO_4)—record the exact final weight with 4 decimals.

4. The dilution factor can be calculated as final weight divided by weight of hydrolysate (*see* **Note 16**).

5. Mix tubes well.

6. Centrifuge at $5000 \times g$ for 10 min.

7. Filter the supernatant using a 0.45 μm syringe filter into new tubes.

8. The samples are now ready for analysis (*see* Subheading 3.5).

9. Calculate the glucose yield as % of theoretical glucose yield:

$$\text{Percent glucose yield} = \frac{[\text{glucose}] - [\text{glucose}_{t0}]}{1.111 \cdot F_{\text{cellulose}} \cdot [\text{biomass}]} \cdot 100\%$$

Where [glucose] is the glucose concentration (g/kg hydrolysate) at a certain time point; [glucose_{t0}] is the glucose concentration (g/kg hydrolysate) in the time "0 h" sample; 1.111 is the correction factor taking into account the hydration when hydrolyzing cellulose; $F_{\text{cellulose}}$ is the cellulose content (g/g) in the dry biomass as determined by a compositional analysis; [biomass] is the dry solids concentration (g/kg hydrolysate).

10. The efficacy of cellulase preparations can be compared based on percent glucose yield obtain at a certain time point for a given enzyme loading. Another option is to do the test with different enzyme loadings for each cellulase preparation, e.g., five

different enzyme loadings, and then plot the percent glucose yield as function of enzyme loading. The enzyme loading required to obtain a certain percentage of glucose yield can be used to compare the efficacy of the cellulase preparations. With the last methodology, it may not be necessary to do different time point during the hydrolysis, but rather select one fixed time for the experiment, e.g., 72 h.

3.5 Sugar Measurement—HPLC

1. Prepare standards for calibration curve—with this setup, do six different dilutions of the sugar stock solution, e.g., undiluted, 2, 5, 10, 20, and 40-fold diluted. This gives a glucose calibration curve ranging from 0.25 g/L to 10 g/L. Dilute with HPLC eluent.

2. Samples from 3.3 or 3.4 may need further dilution—dilution is done with HPLC eluent.

3. With the given HPLC setup, sugars are eluted with 5 mM H_2SO_4 as mobile phase at a flow rate of 0.6 mL/min, column temperature of 80 °C. Detection is done by refractive index detector.

4. Run standards and samples on HPLC.

5. Calculate the concentration of glucose (and other sugars) in the samples based on the calibration curves and take into account the dilution factor.

4 Notes

1. Drying of the material may result in irreversibly collapse of pores of the biomass, also termed hornification, thereby reducing the accessibility of enzymes to the substrate. During freezing, water may also sublimate if not packed in tightly seal containers or bags. Whole slurry samples from for example acid-catalyzed pretreatment have typically very low pH and high content of inhibitors and are stable for even long periods at 5 °C. Washed samples are more prone to contamination and microbial growth.

2. In especially older literature, it was common to add cellulases on the basis of filter paper activity (FPU per g substrate). As the filter paper activity does not necessarily reflect the efficacy of a cellulase preparation on a given biomass sample very well, it is getting more used to add cellulases based on protein amount. The drawback of using protein amount as the basis is that cellulase preparations often contain substantial amounts of other proteins without cellulase activity, which will lower the apparent activity of the cellulase preparation.

3. The substrate loading can be calculated based on total dry solids content or cellulose content. Both methods have pros and cons. If dosing based on cellulose content, the total solids content may vary among different substrates and give differences in viscosity, mass transfer, etc. On the other hand, this method ensures the same relative amount of cellulase per amount of cellulose is added in all cases. In our view, it is not critical which basis is used, but cellulose content of the biomass should always be measured and reported.

4. Moisture content in the pretreated material can be determined using a moisture analyzer or alternatively moisture can be determined by drying in an oven over night at 105 °C. It is critical to mix the material well before sampling as liquid might drain of during storage and result in more moist material in the bottom. It is also advisable to do the measurement in duplicate on the moisture analyzer.

5. The choice of buffer depends on the optimum pH of the enzyme system to be tested. Most fungal cellulases have optimum in the range 4.8–5.2, whereas some bacterial cellulases have optimum around 7. For pH around 4.8–5.2, it is common to use a sodium acetate or sodium citrate buffer, and both appears frequently used when searching the literature. From our experience, there is no significant difference using either acetate or citrate. From an applied perspective, acetate will naturally be present in the material as it is released during pretreatment from acetylated hemicelluloses. If doing fermentation tests, high concentrations of acetate can be inhibitory, and in these cases citrate buffers might be a better choice.

6. If care is taken to work aseptically, e.g., by using sterile tips, tubes and sterile filtered buffer solutions, it is our experience that contamination is not an issue when performing the hydrolysis at 50 °C. If problems with contaminations are observed (e.g., observed as lactate production or low glucose yields) it can be useful to add antibiotics when preparing the buffer solution, e.g., tetracycline (final concentration 10 mg/L).

7. The tumble mixer is based on a rotating drum in which the bottles randomly tumble around. The drum has to be in a temperature controlled incubator. We have made a few different versions of this for our lab. One option is to rebuild a hybridization oven by replacing the normal rotisserie with a homemade box or drum. Another option is to use a rotator, refit it with a drum, and place it in an oven. Our experience is that rotating at around 20 rpm is sufficient. An important aspect is that the drum must not be overloaded—bottles have to tumble freely.

8. The measurement of released glucose (and other sugars) can also be determined by various colorimetric assays, e.g., reducing sugar assays such as the dinitrosalicylic acid assay (DNS), or more specific by enzymatic assays such as the glucose oxidase assay, which only measures glucose. These assays have the advantage that they can be used for rapidly measuring large numbers of samples, e.g., for screening studies. On the other hand, these assays lack the possibility of simultaneous measuring more monosaccharides (xylose, mannose, arabinose) commonly being released during hydrolysis of lignocellulosic biomass as cellulase preparations often contain side activities, e.g., hydrolyzing hemicellulose. This can especially be a limitation of assays measuring reducing sugars, which will quantify the total amount of released sugar and thereby not give a true measurement of cellulase efficacy. For most purposes, an HPLC or IC method is preferable. The indicated system is just one option. Several different columns are available that enables separation of the most common sugar in lignocellulosic hydrolysates.

9. Especially commercial cellulase preparations can be very viscous and therefore difficult to pipet accurately. In addition, they commonly have densities around 1.15–1.25 g/mL. We therefore routinely do a first dilution by weighting out, e.g., 1 g of cellulase preparation and diluting it to a total of 20 g in MilliQ water. By recording the exact weight, the dilution factor can be accurately determined. This stock solution is then diluted further by normal pipetting, and it can be assumed that the density of this solution is equal to water. As the initial dilution is based on weight (1 g) and given the high density of the cellulase preparations, we use the unit "mg protein per g cellulase preparation" for the protein concentration. For the cellulase hydrolysis tests, we do a similar initial dilution of the cellulase preparation by weight out a given amount of cellulase preparation and diluting it in the same buffer as used in the assay. This makes the addition of small amounts of enzyme more accurate. Make the dilution just prior to usage, as the dilute enzyme solution may not be stable.

10. We use 10 mg enzyme protein per g DM (corresponds roughly to 20–25 mg enzyme protein per g cellulose) as a base case enzyme loading. Much depending on the biomass material and the cellulase preparation, other enzyme loadings can be used. It can also be useful to run the test with different enzyme loadings in a range between 5 and 50 mg enzyme protein per g DM.

11. It is important that the biomass slurry is homogeneous without large particles that will block the pipet tip. The actual

homogenization time needed is biomass depended and should be investigated prior to doing the assay.

12. To ensure consistent distribution of the biomass into the tubes it is important to keep the slurry well agitated/stirred on magnetic stirrer during this step to avoid the biomass from sedimentation.

13. The actual temperature will depend on the cellulase cocktail. 50 °C is commonly used for fungal cellulase preparations.

14. One limitation of upper DM can be the DM of the pretreated material. When designing the experimental setup take into account the volume of buffer and enzymes to be added, which will lower the DM content of the final mixture.

15. One set of bottles has to be removed from the tumbler at each sampling point. As all bottles are randomly mixed we use tape with different colors to identify the bottles that have to be removed at the different time points, e.g., all green after 24 h.

16. It may be necessary to dilute the sample even more, so the indicated 10 g total is just a suggested value. However, it is recommended to do minimum a tenfold dilution in this way to minimize the issues of high density of the hydrolysate and the high insoluble solids content.

Acknowledgment

The author acknowledges all present and former colleagues and students at The Biomass Science and Technology Group, Department of Geosciences and Natural Resource Management, University of Copenhagen and Center for Bioprocess Engineering, Department of Chemical Engineering, Technical University of Denmark for their contribution to develop, validate, and implement the protocols presented in this chapter. Many thanks to Dr. David Cannella from The Biomass Science and Technology Group, Department of Geosciences and Natural Resource Management, University of Copenhagen for reviewing the manuscript.

References

1. Horn SJ, Vaaje-Kolstad G, Westereng B et al (2012) Novel enzymes for the degradation of cellulose. Biotechnol Biofuels 5:45. https://doi.org/10.1186/1754-6834-5-45

2. Dien BS, Ximenes EA, O'Bryan PJ et al (2008) Enzyme characterization for hydrolysis of AFEX and liquid hot-water pretreated distillers' grains and their conversion to ethanol. Bioresour Technol 99:5216–5225

3. Gao DH, Uppugundla N, Chundawat SPS et al (2011) Hemicellulases and auxiliary enzymes for improved conversion of lignocellulosic biomass to monosaccharides. Biotechnol Biofuels 4:5. https://doi.org/10.1186/1754-6834-4-5

4. Selig MJ, Knoshaug EP, Adney WS et al (2008) Synergistic enhancement of cellobiohydrolase performance on pretreated corn Stover by

addition of xylanase and esterase activities. Bioresour Technol 99:4997–5005

5. Harris PV, Welner D, McFarland KC et al (2010) Stimulation of lignocellulosic biomass hydrolysis by proteins of glycoside hydrolase family 61: structure and function of a large, enigmatic family. Biochemist 49:3305–3316

6. Payne CM, Knott BC, Mayes HB et al (2015) Fungal cellulases. Chem Rev 115:1308–1448

7. Rodriguez-Zuniga UF, Cannella D, Giordano RDC et al (2015) Lignocellulose pretreatment technologies affect the level of enzymatic cellulose oxidation by LPMO. Green Chem 17:2896–2903

8. Westereng B, Cannella D, Agger JW et al (2015) Enzymatic cellulose oxidation is linked to lignin by long-range electron transfer. Sci Rep 5. https://doi.org/10.1038/srep18561

9. Berlin A, Maximenko V, Bura R et al (2006) A rapid microassay to evaluate enzymatic hydrolysis of lignocellulosic substrates. Biotechnol Bioeng 93:880–886

10. Chundawat SP, Balan V, Dale BE (2008) High-throughput microplate technique for enzymatic hydrolysis of lignocellulosic biomass. Biotechnol Bioeng 99:1281–1294

11. Selig MJ, Tucker MP, Sykes RW et al (2010) Lignocellulose recalcitrance screening by integrated high-throughput hydrothermal pretreatment and enzymatic saccharification. Ind Biotechnol 6:104–111

12. Lindedam J, Bruun S, Jørgensen H et al (2014) Evaluation of high throughput screening methods in picking up differences between cultivars of lignocellulosic biomass for ethanol production. Biomass Bioenergy 66:261–267

13. Gao XD, Kumar R, DeMartini JD et al (2013) Application of high throughput pretreatment and co-hydrolysis system to thermochemical pretreatment. Part 1: dilute acid. Biotechnol Bioeng 110:754–762

14. Himmel ME, Decker SR, Johnson DK (2012) Challenges for assessing the performance of biomass degrading biocatalysts. In: Himmel ME (ed) Biomass conversion: methods and protocols. Humana Press, Totowa, NJ, pp 1–8

15. Jeoh T, Ishizawa CI, Davis MF et al (2007) Cellulase digestibility of pretreated biomass is limited by cellulose accessibility. Biotechnol Bioeng 98:112–122

16. Rasmussen H, Tanner D, Sørensen HR et al (2017) New degradation compounds from lignocellulosic biomass pretreatment: routes for formation of potent oligophenolic enzyme inhibitors. Green Chem 16:516–547

17. Larsen J, Haven MØ, Thirup L (2012) Inbicon makes lignocellulosic ethanol a commercial reality. Biomass Bioenergy 46:36–45

18. Cannella C, Jørgensen H (2014) Do new cellulolytic enzyme preparations affect the industrial strategies for high solids lignocellulosic ethanol production? Biotechnol Bioeng 111:59–68

19. Jørgensen H, Vibe-Pedersen J, Larsen J et al (2007) Liquefaction of lignocellulose at high-solids concentrations. Biotechnol Bioeng 96:862–870

20. Kristensen JB, Felby C, Jørgensen H (2009) Determining yields in high solids enzymatic hydrolysis of biomass. Appl Biochem Biotechnol 156:127–132

21. Haven MØ, Jørgensen H (2014) The challenging measurement of protein in complex biomass-derived samples. Appl Biochem Biotechnol 172:87–101

22. Mok YK, Arantes V, Saddler JN (2015) A $NaBH_4$ coupled ninhydrin-based assay for the quantification of protein/enzymes during the enzymatic hydrolysis of pretreated lignocellulosic biomass. Appl Biochem Biotechnol 176:1564–1580

Part V

Bioinformatics for Improving Cellulases

Chapter 21

Homology Modeling for Enzyme Design

Wimal Ubhayasekera

Abstract

Homology modeling is a very powerful tool in the absence of atomic structures for understanding the general fold of the enzyme, conserved residues, catalytic tunnel/pocket as well as substrate and product binding sites. This information is useful for structure-assisted enzyme design approach for the development of robust enzymes especially for industrial applications.

Key words Homology modeling, Cellulase, Cellobiohydrolase, Protein structure and function, Cel7 enzymes

1 Introduction

The rapid development in the field of genome sequencing and protein research has generated an enormous amount of biochemical and genomic data over the recent decades. Due to the large number of genome projects, availability of a multitude of protein sequences is exponentially increasing. The sequencing and biochemical data alone are not sufficient to understand the insights of enzymes. Therefore, knowledge of the three-dimensional (3D) structures plays a vital role in understanding the nature of the protein, which is also crucial to modify the protein to improve the characteristics. Structure-assisted protein design can be used to engineer an enzyme to enhance pH and thermal stability, to lower the product inhibition, to improve the catalytic rate, and to boost other biophysical and biochemical properties. Thus, a 3D model of a protein is indispensable for enzyme designing to generate robust industrial enzymes with improved properties compared to the natural variants.

X-ray crystallography, nuclear magnetic resonance spectroscopy (NMR), cryo-electron microscopy (Cryo-EM), and modeling are the methods used in 3D structural studies of proteins. X-ray crystallography, NMR, and EM are time-consuming, slow but accurate methods compared to modeling. Among these, X-ray

Mette Lübeck (ed.), *Cellulases: Methods and Protocols*, Methods in Molecular Biology, vol. 1796,
https://doi.org/10.1007/978-1-4939-7877-9_21, © Springer Science+Business Media, LLC, part of Springer Nature 2018

crystallography is a well-established method to obtain higher resolution structures without a size limitation compared to NMR, which is useful in structural studies of proteins that cannot be crystallized. The cryo-EM method can be used to elucidate 3D structures of molecular complexes in near native conditions. Except modeling, all the other methods totally rely on the amount of protein that is available. For a complete understanding of an enzyme's activity, conformational changes, and substrate/inhibitor/product/cofactor binding it is essential to determine the structures of the apo form, mutants, different complexes in the presence of substrate, product, inhibitors, and cofactors, etc. The function of the protein is dependent on its structure. The knowledge on structure and function relationships is necessary for the designing of robust enzymes.

We routinely use homology modeling to help design expression constructs, to understand genomic data, and as a general method of consolidating the available sequence and structural data. This is a powerful tool in designing changes/mutations to improve enzymatic activities and stability. Homology modeling of *Phanerochaete chrysosporium* cellobiohydrolases helped us understand the structural and functional differences among its six isoenzymes, including in one case a possible change of the catalytic mechanism [1]. For *Piromyces* sp. strain E2 endoglucanase (Cel9A) catalytic domain and dockerin models were useful in getting information about the inverting catalytic mechanism and substrate binding of the protein, as well as suggesting the start and end points of the structural/functional modules of the protein for over-expression experiments [2]. Homology modeling studies of Cel6A enzymes of *Piromyces* sp. E2 (Cel6A(e2)) and *Piromyces equi* (Cel6A(pe)) provided a good understanding of the structural topology (dockerins and catalytic modules) and the functions suggesting that these enzymes are processive cellobiohydrolases having classical single-displacement (inverting) mechanism [3]. Homology model of *Aspergillus saccharolyticus* beta-glucosidase suggested that it possesses a retaining catalytic mechanism with more open catalytic pocket possibly having faster substrate accessibility as well as removal of product compared to other beta-glucosidases [4].

I have selected *Coniophora puteana* (GenBank [5] sequence identifier: BAH59084.1) cellulase (Cp_Cel7A) to work through the process described in this chapter. Homology modeling of Cp_Cel7A, gave us several ideas to improve the enzyme. However, it is necessary to work on actual experimental structural studies to understand the structure–function relationship of this enzyme with different modifications. Cp_Cel7A has two insertions and one deletion compared to the homology modeling template structure *P. chrysosporium* Cel7D (Pc_Cel7D) [6]. Modification at the tunnel

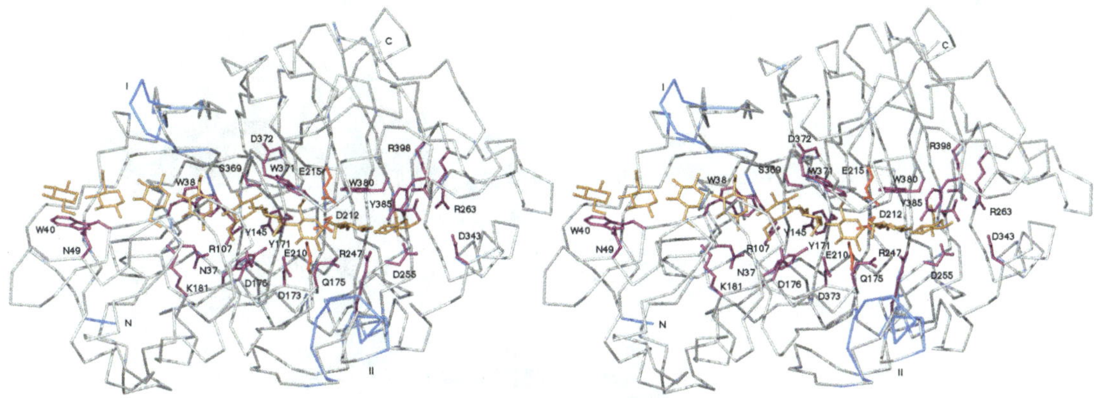

Fig. 1 Stereorepresentation of superimposed C-alpha trace of the Pc_Cel7D structure (gray) (PDB code: 1Z3T) and Cp_Cel7A model (slate blue) showing the insertions and conserved residues important for substrate/product binding (maroon) and catalysis (red). A cellulose chain is modeled to visualize the conserved residues along the catalytic tunnel. Figure 1 was prepared using O [15] and Molray [22]

entrance of Cp_Cel7A from insertion 1 (Residues: Ser100, Gln101, and Lys102) (Fig. 1) and Thr81His can change the properties of substrate binding/recognition. Insertion 2 (Gly243, Asp244, Tyr245, Gly247) may affect the product removal from the product site as well as product inhibition. This loop is shorter in Pc_Cel7D but longer in *Hypocrea jecorina* Cel7A (Hj_Cel7A). In comparison to Pc_Cel7D, Hj_Cel7A has been engineered to stabilize this loop region by introducing a new disulfide bridge and a mutation to reduce hydrogen bonding with the product (PDB code: 1Q2B & 1Q2E) [7]. Further this loop stabilization has enhanced the activity on amorphous and crystalline cellulose. The results have suggested that this loop plays a role in the processive crystalline cellulose degradation [7]. Structure-assisted mutant designing of Hj_Cel7A (PDB code: 1EGN) has revealed the mutation E223S/A224H/L225V/T226A/D262G (corresponding residues in Pc_Cel7D A218/A219/F220/T221/D251; in Cp_Cel7A A221/A222/Y223/T224/D258) causes a shift in the pH optimum and weaker product binding [8]. Through directed evolution Hj_Cel7A thermal stability has been increased by 10.4 °C in melting temperature (PDB code: 5OA5). It has been observed that this Cel7A variant (FCA398) has 18 locally located mutations, which have no effects on the backbone of the enzyme [9]. *Humicola grisea* var. thermoidea cellobiohydrolase Cel7A (PDB code: 4CSI) has exhibited a higher melting temperature (Tm) compared to Hj_Cel7A [10].

Hj_Cel7A catalytic residues (Glu212-Asp214-Glu217) have been mutated in order to confirm and understand the catalytic mechanism. Corresponding Pc_Cel7D and Cp_Cel7A residues

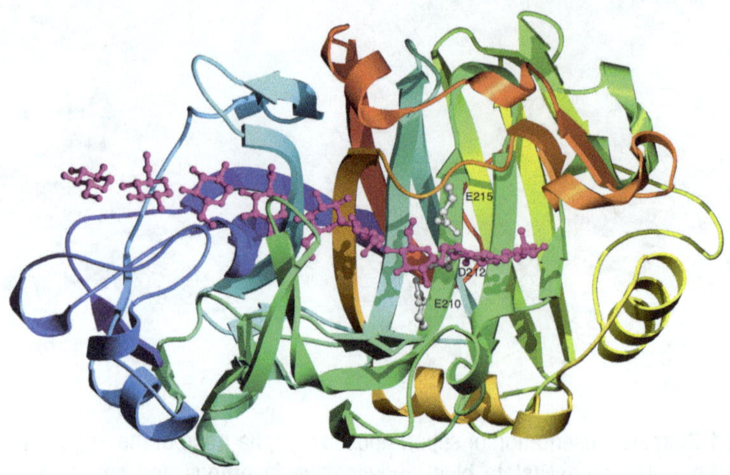

Fig. 2 Ribbon cartoon representation of the Cp_Cel7A homology model showing the overall fold. Conserved residues important for catalysis and modeled cellulose chain into the catalytic tunnel are shown in light gray and magenta respectively. Figure 2 was prepared using O [15], Molscript [21], and Molray [22]

are Glu207-Asp209-Glu212 and Glu210-Asp212-Glu215, respectively (Fig. 2). Based on the structural information it has been proposed that Cel7 cellobiohydrolases possess retaining catalytic mechanism. Complex structures of Hj_Cel7A with cellotetraose, cellopentaose, cellohexaose, product, and inhibitors as well as Pc_Cel7D complex structures with cellobiose, a thio-linked substrate analog, methyl 4-S-beta-cellobiosyl-4-thio-beta-cellobioside (GG-S-GG), and inhibitors revealed the important residues for substrate binding and catalysis confirming the retaining mechanism [6, 11, 12]. Crystal structures of Hj_Cel7A complexed with xylotriose, xylotetraose, and xylopentaose revealed that the enzyme inhibition by xylose has a major binding mode at the entrance of the substrate-binding tunnel of the enzyme. Each xylose residue is shifted ~2.4 Å toward the catalytic center compared with binding of cello-oligosaccharides. The structures further suggested an alternative xylose-binding mode at substrate binding sites −2 and −1 in the vicinity of the catalytic site [13].

Therefore, the model can help us to compare Cp_Cel7A with other Cel7 enzymes to understand the residues involved in substrate/product binding and catalysis. We can deduce a scheme to improve thermal stability, catalytic rate, and pH optimum of Cp_Cel7A by understanding the structure function relationship through comparing different structures in relation to available biochemical and biophysical data. This task is easily achievable by superimposing similar structures and comparing them. However, in the absence of structures a homology model is a better substitute and plays a significant role. A list of necessary observational steps in designing enzymes is mentioned below:

1. Investigate the changes in the catalytic and substrate/product binding amino acid residues in relation to the structure and function of the enzyme

2. Conformational changes upon ligand, product and substrate binding

3. Unstructured parts of the protein structure

4. Locations of the cofactors, ions, and water binding sites

5. Stable and unstable parts of the structure

6. Catalytic mechanism

7. Glycosylation sites

8. Any other property or effect due to structure–function relationship

Homology modeling refers to in silico construction of a 3D model of a protein from its amino acid sequence using a highly identical experimental 3D structure as the template. Homology modeling of proteins can be a very useful method of analyzing them, provided that the structure of a sufficiently similar protein is known. This approach allows to understand the general fold of the polypeptide chain and the catalytic mechanism by comparing the model to available similar structures, to find the residues that are important for substrate/product binding, catalysis and then to use this information for protein engineering. The model will give insights into the location of N- and C- termini of the polypeptide chain to add tags for purification purposes and to identify the boundaries of domains accordingly when designing new constructs for the expression of the enzyme.

The accuracy of a homology model is generally correlated with the pairwise amino acid sequence identity of the protein of interest and the known structure, which is used as the model template. For example, if the sequence identity is more than 50% between two proteins, 90% or more of the residues in their core structure will be expected to be very similar [14]. Similarity between the template structure and the protein of interest is one of the critical factors of getting an accurate homology model. A point of noteworthy is that a homology model is a model (or a 3D representation of a polypeptide chain) but not an experimental structure. Therefore, it is wise to limit the modifications of the model to reduce the number of errors. For example, deletions and insertions can introduce errors to the model, so it is appropriate to locate them in loop regions. Similarly, overspeculation and overinterpretation of a model can be misleading and unfitting.

Homology modeling consists of six main steps.

1. Finding similar structures.

2. Generation of the pairwise and multiple sequence alignments.

3. Preparation of template and input files.

4. Creation of the model.

5. Modeling insertions and deletions.

6. Improve the packing in the interior of the protein.

2   Materials

The following programs have been used under the Mac OS X operating system for the homology modeling and other model manipulations described in this chapter.

Graphics program O [15]—model generation, structural comparison, viewing structures, model building (fixing rotamers, deletions, and insertions) and preparing figures. This program can be obtained from Prof. Alwyn Jones (http://xray.bmc.uu.se/alwyn/TAJ/Home.html). Refer to O manual for commands (http://xray.bmc.uu.se/~alwyn/A-Z_of_O/A-Z_frameset.html).

Uppsala Software Factory (USF) programs (programs and manuals are available at http://xray.bmc.uu.se/usf/) [16] – SOD to generate O-macro and O data block for homology modeling in O, MOLEMAN—coordinate file manipulation, LSQMAN—to superimpose structures/models.

3   Methods

3.1   Finding Similar Structures

1. Find out the boundaries of your sequence (e.g., Cp_Cel7A) with SIGNALP [17] and BLAST (Basic Local Alignment Search Tool) [18]. Then, find the different properties of your protein such as theoretical pI, molecular weight, N-glycosylation sites, and O-glycosylation sites (Fig. 3) using different servers/programs (e.g., listed in Expasy; www.expasy.ch).

2. Run BLAST search of the enzyme against nonredundant protein sequences and collect a set of sequences with varying identity to your enzyme.

3. A multiple sequence alignment with a large set of amino acid sequences similar to your enzyme helps to locate the conserved residues/regions.

```
>BAH59084.1 cellulase (Cp_Cel7A) [Coniophora puteana]
MFPKSILLAFAFAAATSAQQIGTSTAETHPTLTWSQCTSSGCTTESSGSVVLDANWRWLHTVDGYTNCY
TGNEWDTTICTSAEVCAEQCALDGADYEGTYGITTSGDALTLKFVTQSSQKNVGSRVYLMADDTHYQMF
NPLNQEFSFTVDVSQLPCGLNGALYFSQMDADGGLSKYSTNKAGAQYGTGYCDSQCPRDIKFINGVANL
QNWTSTSTNSGTGSLGSCCSEMDVWEANSISAAYTPHPCSVNGQTECTGADCGGDYGRYAGVCDPDGCD
FNSYRMGDTTFYGSGETVDTSQPFTVVTQFLTSDNTTTGTLSEIRRLYVQNGKVIQNSNTDISGLSTYN
SITDDYCTAQKTAFGDTDSFSSHGGLAKMGDSFAAGVVLVLSVWDDYAAQMLWLDSDYPTTADASTPGV
ARGTCATTSGAPADVESSAANAQVIYSNIKFGDIGTTYSA
```

Fig. 3 SignalP analysis suggested that the mature protein starts from QQIG (GlnGlnIleGly). The signal peptide is underlined

4. Run BLAST with your mature enzyme sequence (Fig. 3) to search for similar structures in the PDB (Protein Data Bank) [19]. This will generate a list of PDB codes. For example, the resulted similar structures showed identity of 37–69%. *P. chrysosporium* Cel7D (Pc_Cel7D) structures (PDB codes: 1GPI, 1H46 [1], 1Z3T, 1Z3V, 1Z3W [6] have the highest identity (69%) to *Coniophora puteana* cellulase (Cp_Cel7A) [20] with 95% query coverage. This confirms that Cp_Cel7A has only the catalytic module. Pc_Cel7D structure with PDB code 1Z3T was selected as the template for homology modeling.

5. Obtain the structures/coordinates (1z3t.pdb file) from the PDB and the amino acid sequences of the structures. These structures will be useful in a later stage to compare with your model to the related protein structures.

6. Protein structure with the highest identity and large coverage can be used as the template to build the homology model (*see* **Note 1**).

3.2 Generation of the Sequence Alignments

1. Align your amino acid sequence and the template sequence (*see* **Note 2**) (Fig. 4).

3.3 Preparation of Template and Input Files

1. Superimpose the structures you downloaded on your template structure either using LSQMAN or O. This can be used to locate the boundaries of the different domains of the enzyme (*see* **Note 3**).

2. Read the template PDB file into O (*see* **Note 4**).

3. Carefully check the alignment for deletions and insertions (*see* **Note 5**). For example, according to the sequence alignment in Fig. 4 there are two insertions and one deletion in Cp_Cel7A compared to Pc_Cel7D. Locate the insertion and deletion sites in the Pc_Cel7D structure. First insertion is between residues 99Ser and 100Asn and is in a loop. This insertion can be moved between 98Gly and 99Ser or keep it between 99Ser and 100Asn. For this exercise, I built the insertion between 99Ser and 100Asn. The second insertion is between 239Ala and 240Arg residues. It is also in a loop but it cannot be moved between 240 and 241 as it can obstruct the catalytic tunnel. It cannot be moved between 238 and 239 as it is a helical part of the structure. Therefore, second insertion is between 239Ala and 240Arg residues as predicted in the sequence alignment (Fig. 4). The last two residues of Pc_Cel7D are missing in the Cp_Cel7A. We can simply delete the last two residues from Pc_Cel7D sequence and structure (*see* **Note 6**).

```
Pc_Cel7D      1 EQAGTNTAENHPQLQSQQCTTSGGCKPLSTKVVLDSNWRWVHSTSGYTNCYTGNEWDTSL
Cp_Cel7A      1 QQIGTSTAETHPTLTWSQCTSSGCTTESSGSVVLDANWRWLHTVDGYTNCYTGNEWDTTI


Pc_Cel7D     61 CPDGKTCAANCALDGADYSGTYGITSTGTALTLKFVTGS---NVGSRVYLMADDTHYQLL
Cp_Cel7A     61 CTSAEVCAEQCALDGADYEGTYGITTSGDALTLKFVTQSSQKNVGSRVYLMADDTHYQMF


Pc_Cel7D    118 KLLNQEFTFDVDMSNLPCGLNGALYLSAMDADGGMSKYPGNKAGAKYGTGYCDSQCPKDI
Cp_Cel7A    121 NPLNQEFSFTVDVSQLPCGLNGALYFSQMDADGGLSKYSTNKAGAQYGTGYCDSQCPRDI


Pc_Cel7D    178 KFINGEANVGNWTETGSNTGTGSYGTCCSEMDIWEANNDAAAFTPHPCTTTGQTRCSGDD
Cp_Cel7A    181 KFINGVANLQNWTSTSTNSGTGSLGSCCSEMDVWEANSISAAYTPHPCSVNGQTECTGAD


Pc_Cel7D    238 CA----RNTGLCDGDGCDFNSFRMGDKTFLGKGMTVDTSKPFTVVTQFLTNDNTSTGTLS
Cp_Cel7A    241 CGGDYGRYAGVCDPDGCDFNSYRMGDTTFYGSGETVDTSQPFTVVTQFLTSDNTTTGTLS


Pc_Cel7D    294 EIRRIYIQNGKVIQNSVANIPGVDPVNSITDNFCAQQKTAFGDTNWFAQKGGLKQMGEAL
Cp_Cel7A    301 EIRRLYVQNGKVIQNSNTDISGLSTYNSITDDYCTAQKTAFGDTDSFSSHGGLAKMGDSF


Pc_Cel7D    354 GNGMVLALSIWDDHAANMLWLDSDYPTDKDPSAPGVARGTCATTSGVPSDVESQVPNSQV
Cp_Cel7A    361 AAGVVLVLSVWDDYAAQMLWLDSDYPTTADASTPGVARGTCATTSGAPADVESSAANAQV


Pc_Cel7D    414 VFSNIKFGDIGSTFSGTS
Cp_Cel7A    421 IYSNIKFGDIGTTYSA--
```

Fig. 4 Sequences of Pc_Cel7D catalytic module and Cp_Cel7A have been aligned using. Clustal W web server at the EMBnet (https:/embnet.vital-it.ch/software/ClustalW.html) and BOXSHADE (https:/embnet.vital-it.ch/software/BOX_form.html) was used to generate the figure. Conserved residues are shown in red

4. To edit the alignment, you need to get the alignment in PIR (Protein Information Resource) format and copy it into your SOD input files. Then, edit the sequence according to the sensible positions in the template. It is important to maintain the format of the sequence alignment (60 characters in a row). For example, the sequence alignment will not be edited as there are no changes in the locations of deletions and insertions.

5. Edit the two SOD input files (description of the input files are given below).

6. Run the SOD program (run sod < whatever.inp > whatever. log) in a terminal window. This log file helps you to figure out what has happened as well as the problems with the output file generation. It is always better to use a sensible name (e.g., sod_1Z3T_to_Cp_Cel7A.inp; 1Z3T is the PDB code and Cp_Cel7A is *Coniophora puteana* cellulase).

```
run sod < sod_1Z3T_to_Cp_Cel7A.inp > sod_1Z3T_to_Cp_Cel7A.log
```

Output file: sod_1Z3T_to_Cp_Cel7A.omac

```
run sod <sod_1Z3T_pair_Cp_Cel7A.inp> sod_1Z3T_pair_Cp_Cel7A.log
```

Output file: 1Z3T_pair_Cp_Cel7A.odb

7. Running SOD will generate two files: An O-macro with mutation information of the structure to generate the model (*whatever.omac*) and an O data block with coloring information of the model (*whatever.odb*).

8. If you have deleted parts of the template sequence in your alignment, edit the PDB file removing the deleted amino acid coordinates in a text editor such as gedit, nedit, emacs, and TextWrangler. Then save and read it into O and name it as mentioned in the "molnam" in the SOD input files.

3.4 Creation of the Model

1. You need to start O and read your edited PDB file. I usually read the same file twice in its original PDB code and in the "molnam" you gave in the SOD input file. Some of the useful O commands for homology modeling are given in Table 1.

Table 1
Common O commands for homology modeling (Refer the manual at http:/xray.bmc.uu.se/alwyn/A-Z_of_O/A-Z_frameset.html)

Function of the command	Command
Read a pdb file to O	pdb-r
Write a pdb file from O	pdb-w
List the atoms in the molecule	sam-list-sequence/ s-l-s
Renumber	sam-ren
To save (you need to give a sensible name and use the extension "o" for identification purposes at a later stage (e.g., whatever.o)	save
Center on atom	ce-at residue number (e.g., ce-at 15)
To read an O data block	read name-of-the-o-datablock.odb (e.g., read mymenu.odb)
To read a omacro to the program	@whatever.omac
To regularize	refi_zone
To read stereochemistry parameter file	read stereo_chem.odb
To clear flags	clear_flags
To clear text	clear_id
To quit the program	quit
To save and quit	stop

All the commands are entered in the terminal window that you are running O. It is useful to check the pull-down menus in O before start modeling.

In a terminal window type "ono" to start O program then keep pressing ENTER button until you get the O window.
To read coordinate files into O "pdb-r".
Give the name of the pdb file "1z3t.pdb".
Give an O associated molecule name: "1z3t".
Now this PDB file has other molecules such as water, ligand, and ions. Therefore, for modeling we need a clean PDB file. You can write a PDB file from O. (It is easier than editing it in an editor). You can type "s-l-s" command to check the atoms in the 1z3t molecule. The Pc_Cel7D molecule starts with the residue A1 (PCA1) and the last amino acid residue is A431 (Ser431). As we have deleted last two residues in the alignment for Cp_Cel7A model, we need only up to Gly429. We can write the required PDB file as follows.
Type "pdb-w", then give a file name (e.g., 1z3t-for-Cp_Cel7A-model.pdb) and specify the molecule and residue range: "What molecule: 1z3t" and "Residue range [all molecule]: a1 a429."
Then press ENTER button until the file is written. Now you can read this file as the model file into O. We specified the "mol-nam" in SOD input file as "coni."

```
pdb-r
Util> PDB file: 1z3t-for-coni-model.pdb
Util> O associated molecule name: coni
```

Press ENTER button until the molecule is read to O

Check the molecule with "s-l-s" command. Residues in the molecule are named A1, A2, etc. We do not need this chain name. So we need to renumber without the chain name.

2. After reading the PDB file into O, you need to renumber the molecule using "sam-ren" command and number the structure starting from 1.

```
Sam-rename
Sam> What molecule [CONI ]:
Sam> Residue range [all molecule]:
Sam> NEW name of FIRST residue [A1 ]:1
```

Check the molecule with "s-l-s" command.

3. Remember to save the O file. It is better to give "o" extension (e.g., whatever.o) to separate O files from others.

```
save
Enter file name: coniophora-20171030.o
```

When a molecule is read into O it creates the C-alpha chain. You can visualize all atoms from the display pull-down menu. You can use the pull-down menu to activate several commands. (check O manual for more commands). You can display the "User menu" from pull-down "menus." There you can have several commands for centering, saving and to clear commands (clear flags).

4. Now, you are ready to model the protein. Add the O-macro file (*whatever.omac*) into O by "*@whatever.omac*".
   Add the sod output omacro file to generate the model.
   This command will mutate the amino acids in the template structure as shown in the sequence alignment.

```
@sod_1Z3T_to_Cp_ce17A.omac
Save
```

Now rename the molecule

```
sam-rename
sam> What molecule [CONI ]:
sam> Residue range [all molecule]:
sam> NEW name of FIRST residue [1 ]:
```

You can check the atoms of your model

```
s-l-s
```

From the display pull-down menu read the "Ca atoms" and "All atoms."

5. This will change the template sequence with your sequence as you have aligned. It is worth to note that the sequence alignment is the single important factor in homology modeling.

3.5 Modeling Insertions and Deletions

1. Build the insertions and fix the deletions. We have a three-residue insertion after residue 99 (Fig. 5a). The new residues are from 100 to 102. The following procedure describes how to fix an insertion. The graphics program O uses the main- and side-chain databases, so the modeling is only a rough approximation of the likely truth.

Change the active molecule to "CONI" using "mol" command and type "coni."

```
sel-on coni ;
sel-off coni 99 102
leg-loop coni 97 104
```

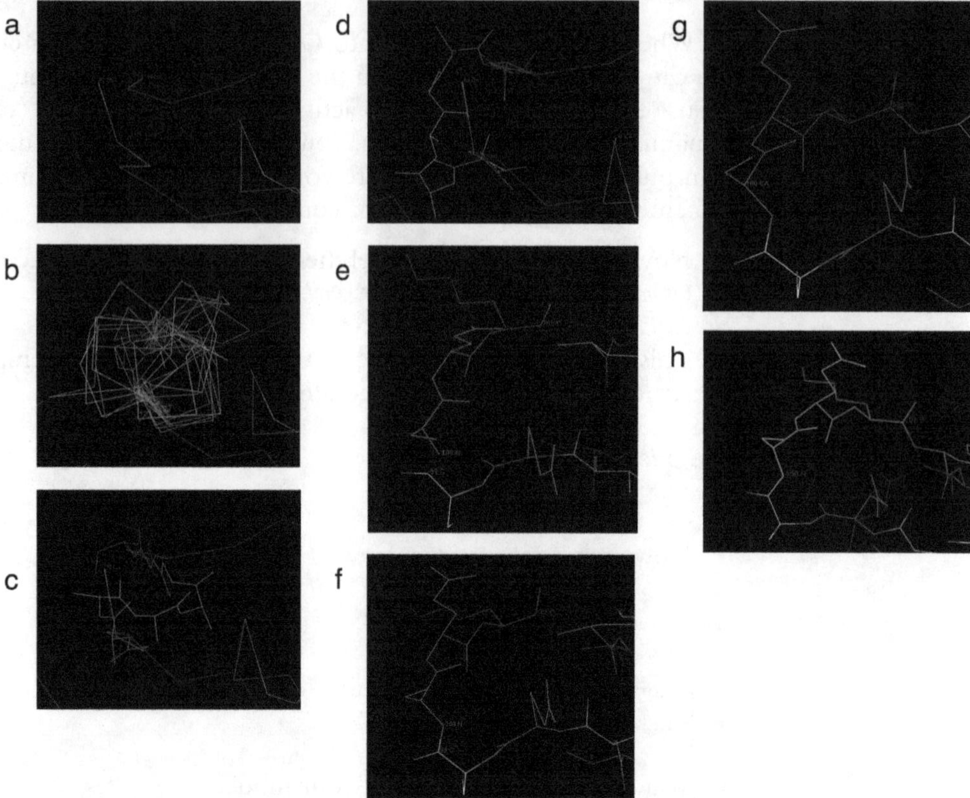

Fig. 5 Steps illustrating modeling of an insertion in the graphics program O. (**a**) Place of the insertion on the C-alpha trace of the model. (**b**) Possibilities for the insertion modeling. (**c**) Insert model that can clash with the other residues. (**d**) An example for a good insert. (**e**) The selected insert (**f, g**). Connecting the insert to the protein (**h**). The modeled and regularized insert is shown in cyan

Here O gives a list of possibilities for the insertion (Fig. 5a) and choose the best fit (Fig. 5c) avoiding possible clashes (Fig. 5b) to the main chain residues.

```
yes
sel-on coni ;
leg-auto-sc coni 99 102
yes
```

Display pull-down menu > Molecule CONI >.
Display pull-down menu > Ca atom.
Display pull-down menu > All atoms.
save
Now you can see the new inserted residues (Fig. 5e) and need to connect this peptide to the main chain. Switch on All atoms from the "Object Menu" > "CONI_All".
Bones pull-down menu > Make bond.

Now click on 99 C (C atom on carboxyl group of residue 99) and then click on the N (N of the amino group) on residue 100 (Fig. 5f).

Bones pull-down menu > Make bond.

Now click on 102 C (C atom on carboxyl group of residue 102) and then click on the N (N of the amino group) on residue 103 (Fig. 5g).

Now click on the "Regularize zone" command in the "User Menu" and define a zone for regularization, e.g., from residues 96 to 105.

Now check for any clashes of the side chains. If there are clashes you can change the rotamer of the inserted residues. To change rotamers click on the "Rotamers" command in the "User Menu" and click on a side chain you want to change. You can select the best rotamer using arrow keys and then click on "yes" in the "User Menu."

Save.

You can find these commands in the pull-down menus. You can color the inserted residues in another color for you to locate the insertions easily (Fig. 5h). Display pull-down menu > Paint mol. > click on the green square on the top left corner to keep that menu visible (click on red square to remove the menu) > click on cyan > click on Paint obj. zone > click on residues 100 and 103. The loop chosen is the best fit to the CA coordinates of residues 98, 99, 102, and 103 (i.e., ignoring each residue next to the insertion). The new and abutting residues have the most common rotamer conformation.

By using C-alpha trace (CONI_CA) in the "Object Menu" locate the residue 242 and fix the insertion as described above. This insertion is located closer to the catalytic tunnel. Therefore, remember not to build the loop into the catalytic tunnel. Here, you can get help from the different cellulase structures you superimposed. For example (4csi.pdb) has an extended loop at this location forming the catalytic tunnel. In fact, we must guide ourselves using these superimposed structures in modeling. You may need to change the active molecule into "CONI" by typing "mol" then "coni" before fixing the insertion.

To fix a deletion you need to move the amino acids closer in the deletions and make the peptide bonds between N (amino terminal) in the next amino acid and carboxyl C in the previous amino acid. To fix the stereochemistry regularize residues in both sides of the deletion using "Regularize zone" command from the pull-down menu > User Menu > Master Menu > Rebuild OR Rebuild pull-down menu > Regularize > Define zone. Now click on the residue range either sides of the deletion.

3.6 Improve the Packing in the Interior of the Enzyme

1. Change the rotamers of the model as those are defined in the structure to minimize errors and to improve packing in the interior of the protein.

2. You can write out your model from O using "pdb-w" command. It is possible to prepare different representation sketches in O (Refer to the O manual).

3. Read this PDB file (your-model.pdb) into a molecular viewer such as swiss pdb viewer, pymol, and molsoft and prepare representations of your model.

3.7 SOD Input Files

There are two SOD input files to generate O-macro (*whatever. omac*) and O data block (*whatever.odb*).

Example File-1 (*sod_1Z3T_to_Cp_Cel7A.inp*)

File-1 generates a macro that can be read into graphics program O to build the model by replacing the existing amino acids of the template structure with those of the interested protein. The content of the File-1 is shown below. The lines 2, 4, and 6 need to be edited with a sensible name for the output O-macro file (e.g., *sod_1Z3T_to_Cp_Cel7A.omac*), molecular name containing four characters to use in O (e.g., *coni*) and the location of the library file of the SOD program respectively. Do not leave spaces in the file names. In the installation process of the USF programs, it is necessary to create a folder with library files of the programs. The PIR format aligned and edited/adjusted sequences are copied below the "SEQUE" line. The first sequence is the protein you are going to model and the second sequence is the template.

```
task homo
outfil sod_1Z3T_to_Cp_Cel7A.omac
format expl (60a1)
molnam coni
ofirst 1
libfil /Applications/usf/lib/sod.lib
keep all
replace all
delete yes
insert yes
SEQUE
QQIGTSTAETHPTLTWSQCTSSGCTTESSGSVVLDANWRWLHTVDGYTNCYTGNEWDTTI
CTSAEVCAEQCALDGADYEGTYGITTSGDALTLKFVTQSSQKNVGSRVYLMADDTHYQMF
NPLNQEFSFTVDVSQLPCGLNGALYFSQMDADGGLSKYSTNKAGAQYGTGYCDSQCPRDI
KFINGVANLQNWTSTSTNSGTGSLGSCCSEMDVWEANSISAAYTPHPCSVNGQTECTGAD
CGGDYGRYAGVCDPDGCDFNSYRMGDTTFYGSGETVDTSQPFTVVTQFLTSDNTTTGTLS
EIRRLYVQNGKVIQNSNTDISGLSTYNSITDDYCTAQKTAFGDTDSFSSHGGLAKMGDSF
AAGVVLVLSVWDDYAAQMLWLDSDYPTTADASTPGVARGTCATTSGAPADVESSAANAQV
IYSNIKFGDIGTTYSA
```

```
EQAGTNTAENHPQLQSQQCTTSGGCKPLSTKVVLDSNWRWVHSTSGYTNCYTGNEWDTSL
CPDGKTCAANCALDGADYSGTYGITSTGTALTLKFVTGS---NVGSRVYLMADDTHYQLL
KLLNQEFTFDVDMSNLPCGLNGALYLSAMDADGGMSKYPGNKAGAKYGTGYCDSQCPKDI
KFINGEANVGNWTETGSNTGTGSYGTCCSEMDIWEANNDAAAFTPHPCTTTGQTRCSGDD
CA----RNTGLCDGDGCDFNSFRMGDKTFLGKGMTVDTSKPFTVVTQFLTNDNTSTGTLS
EIRRIYIQNGKVIQNSVANIPGVDPVNSITDNFCAQQKTAFGDTNWFAQKGGLKQMGEAL
GNGMVLALSIWDDHAANMLWLDSDYPTDKDPSAPGVARGTCATTSGVPSDVESQVPNSQV
VFSNIKFGDIGSTFSG
```

Example File-2 (*sod_1Z3T_pair_Cp_Cel7A.inp*)

This file can generate an O data block with coloring information of the homology model. If you change the color of the model and need to change it back to the original colors indicating conserved residues (yellow) and mutations (magenta) you can read this data block into O. Therefore, it is advisable to prepare this file together with the input file generating O-macro. The content of File-2 is shown below. The lines 2, 4, 5, 7, and 8 need to be edited with a sensible name for the output O data block file (e.g., *1Z3T_pair_Cp_Cel7A.odb*), remarks mentioning your task, molecular name containing four characters used in File-1 (e.g., *coni*) and changes you have made (e.g., renumbering the template and the molecular name that was given to the template to generate the model) respectively. The PIR format aligned and edited/adjusted sequences are copied below the "SEQUE" line. The first and the second sequences belong to the model and the template respectively as shown in File-1. There should be a space (one or more empty lines) between the sequences.

```
task pair
outfil 1Z3T_pair_Cp_Cel7A.odb
format expl (60a1)
remark sod to compare 1Z3T with Cp_Cel7A 'pair'
molnam coni
reference 1
name 1 1Z3T_renumbered
name 2 coni
SEQUE
QQIGTSTAETHPTLTWSQCTSSGCTTESSGSVVLDANWRWLHTVDGYTNCYTGNEWDTTI
CTSAEVCAEQCALDGADYEGTYGITTSGDALTLKFVTQSSQKNVGSRVYLMADDTHYQMF
NPLNQEFSFTVDVSQLPCGLNGALYFSQMDADGGLSKYSTNKAGAQYGTGYCDSQCPRDI
KFINGVANLQNWTSTSTNSGTGSLGSCCSEMDVWEANSISAAYTPHPCSVNGQTECTGAD
CGGDYGRYAGVCDPDGCDFNSYRMGDTTFYGSGETVDTSQPFTVVTQFLTSDNTTTGTLS
EIRRLYVQNGKVIQNSNTDISGLSTYNSITDDYCTAQKTAFGDTDSFSSHGGLAKMGDSF
AAGVVLVLSVWDDYAAQMLWLDSDYPTTADASTPGVARGTCATTSGAPADVESSAANAQV
IYSNIKFGDIGTTYSA


EQAGTNTAENHPQLQSQQCTTSGGCKPLSTKVVLDSNWRWVHSTSGYTNCYTGNEWDTSL
CPDGKTCAANCALDGADYSGTYGITSTGTALTLKFVTG---SNVGSRVYLMADDTHYQLL
```

```
KLLNQEFTFDVDMSNLPCGLNGALYLSAMDADGGMSKYPGNKAGAKYGTGYCDSQCPKDI
KFINGEANVGNWTETGSNTGTGSYGTCCSEMDIWEANNDAAAFTPHPCTTTGQTRCSGDD
CA----RNTGLCDGDGCDFNSFRMGDKTFLGKGMTVDTSKPFTVVTQFLTNDNTSTGTLS
EIRRIYIQNGKVIQNSVANIPGVDPVNSITDNFCAQQKTAFGDTNWFAQKGGGLKQMGEAL
GNGMVLALSIWDDHAANMLWLDSDYPTDKDPSAPGVARGTCATTSGVPSDVESQVPNSQV
VFSNIKFGDIGSTFSG
```

3.8 MOLEMAN to Obtain the Amino Acid Sequence of the Structure

In a terminal window type "run moleman" to run it.

```
>run moleman
Option ? (READ_pdb_file) [PRESS ENTER]
Input PDB file ? (in.pdb)
```
[GIVE YOUR PDB FILE HERE e.g. 1z3w.pdb]

You can give any command when you have "Option?" in MOLEMAN.

To get the PIR sequence file (of the pdb file) [TYPE THE COMMAND AND PRESS ENTER] 'PIR_sequence_file'

```
Amino acid residue names ?
```
[PRESS ENTER]
```
One letter codes ?
```
[PRESS ENTER]
```
PIR output file name ?
```
Write '1z3w.pir' [PRESS ENTER]
```
PIR sequence name ? (SEQUENCE)
```
[PRESS ENTER]
```
PIR sequence title ? (Any title)
```
[PRESS ENTER]
```
QUIT
```

Now MOLEMAN has written the amino acid sequence of the PDB in PIR format "1z3w.pir". MOLEMAN can be used for other manipulations of the PDB file such as renumbering of atoms, amino acids residues and to obtain Ramachandran plot. You can find the program manual from the USF website. You can find different commands by typing "?" when you run any USF program. If you renumber atoms or residues or modify the PDB file and want to rewrite it, you can write the PDB file coordinates as below (*see* **Note** 7).

```
Option ? (RENUm) write
Output PDB file ? (out.pdb) 1z3w-mm.pdb
REMARK at start of file ? (MoleMan PDB file)
Copy all REMARK, HEADER etc. cards from input ? (Y)
Which chain to write (** = any and all) ? (**)
Residue range to write (0 0 = all molecule) ? ( 0 0)
You may output All atoms, only Main-chain atoms,
a Poly-alanine (Gly intact), a poly-Serine,
(Gly and Ala intact) or a poly-Glycine
Which option do you want (All/M/P/S/G) ? (A)
Write HYDROGEN atoms (Y/N) ? (N)
Force consecutive atom numbering (Y/N) ? (Y)
X-PLOR needs OT1 and OT2, but O hates them
If your file contains OT1/2 you may either
keep them, or replace them by O/OXT
```

```
Write X-PLOR OT1/2 ? (Y/N) ? (N)
Cell : ( 87.382 46.578 98.505 90.000 102.580 90.000)
CCP4 requires CRYST, SCALE and ORIGX cards
X-PLOR does not like them at all
Therefore: reply Y for CCP4 and N for X-PLOR :
Write CRYST, SCALE, ORIGX cards (Y/N) ? (Y)
Nr of atoms written : ( 3417)
Option ? (WRITe) quit
```

3.9 Superimposing Structures with LSQMAN

You can run LSQMAN to superimpose two molecules on top of each other.

We are going to superimpose 4d5p.pdb [13] on top of 1z3w.pdb.

In a terminal window type "run lsqman".

```
> read m1 1z3w.pdb
```

This means read 1z3w.pdb as m1 (LSQMAN consider 1z3w.pdb as the molecule 1).

```
> read m2 4d5p.pdb
```

>list (This command shows you the molecules you have read into LSQMAN.)

>brute (Type "brute" to start the superimposition).

Example: LSQMAN > brute

```
Mol 1 ? (*) m1
Chain 1 ? (A)
Mol 2 ? (M1) m2
Chain 2 ? (A)
Fragment length ? ( 50)
Fragment step ? ( 25)
Min nr or fraction of residues to match ? ( 100)
```

If the sequence used is short, then the fragment length and fragment step can be adjusted to shorter fragments. This command generates an alignment matrix.

Then type "imp" to improve the alignment as described below.

```
LSQMAN > imp
Mol 1 ? (M1)
Range 1 ? (A1-999)
Mol 2 ? (M1) m2
Range 2 ? (A1) *
```

Then, you need to apply this matrix and move one molecule (in this case 4d5p.pdb) on the other (on 1z3w.pdb).

```
LSQMAN > apply
Bring Mol 2 on top of Mol 1 ...
Mol 1 (operator to use) ? (M1)
Mol 2 (to be moved) ? (M1) m2
```

When you type the command "apply," the program will ask you the molecules and it says that it will bring Mol 2 on Mol 1. Therefore, you need to specify the fixed molecule and moving molecule respectively.

```
LSQMAN > write
Which mol ? (M1) m2
File name ? (4d5p.pdb) 4d5p-lsq.pdb
```

Then the superimposed molecule needs to be written. I always give the same name with –lsq (e.g., 4d5p-lsq.pdb), which denotes that it has been superimposed on another molecule. It is important to structure your files to minimize confusion as well as to trace back to it if necessary.

4   Notes

1. The identity between the interested protein and the amino acid sequence of the template needs to be more than 40% in order to build a reliable homology model.

2. Amino acid sequence given in the PDB can be slightly different than your template structure. Sometimes structures are not complete due to untraceable electron density. Therefore, it is better to use the exact amino acid sequence of the structure. You can use MOLEMAN to generate the sequence of the structure. It is very important to obtain the correct alignment of the sequences, as it is the determinant of the correctness of the model.

3. You may need to remove parts from your sequence as well as the template structure (only possible from N- and C- termini) to obtain a better sequence alignment. It is necessary to observe the protein structure with the sequence alignment to understand where to adjust the sequence alignment. If there are missing residues in the template structure they need to be modeled before using it as the template. This can be done as described for fixing insertions in the model.

4. If there is more than one molecule in the template PDB file (i.e., more than one molecule in the asymmetric unit of the

crystal lattice) we need to write down the best molecule. Usually the molecules are identical but sometimes they can differ. Therefore, you can choose more complete molecule as your template (Here, you can use BLAST search results to guide you). We only need the coordinates of the protein but not the water and other molecules in the structure such as ligands, ions, DNA, RNA, and other interacting protein molecules.

5. Your experience with viewing, understanding and modifying protein structures is valuable at this step. You need to adjust the insertions and deletions into loop regions of the structure where there is space to do so, especially in the vaguely aligned regions. Sometimes the computer generated sequence alignment can be wrong. Therefore, it is always necessary to check the location of the changes in the structure and adjust them in the aligned SOD input sequences.

6. Superimposition of several related structures sometimes show different conformations in the loop you are interested in. Therefore, observation of superimposed structures can generate a fairly good idea about the placement of the insertion.

7. I use default values and the same name of the input file with –mm (e.g., 1z3w-mm.pdb) to have a consistency in the file naming and to keep it simple.

References

1. Muñoz IG, Ubhayasekera W, Henriksson H et al (2001) Family 7 cellobiohydrolases from *Phanerochaete chrysosporium*: crystal structure of the catalytic module of Cel7D (CBH58) at 1.32 angstrom resolution and homology models of the isozymes. J Mol Biol 314 (5):1097–1111

2. Steenbakkers PJM, Ubhayasekera W, Goossen JAM et al (2002) An intron-containing glycoside hydrolase family 9 cellulase gene encodes the dominant 90 kDa component of the cellulosome of the anaerobic fungus *Piromyces* sp strain E2. Biochem J 365:193–204

3. Harhangi HR, Freelove AC, Ubhayasekera W et al (2003) Cel6A, a major exoglucanase from the cellulosome of the anaerobic fungi *Piromyces* sp E2 and *Piromyces* equi. Biochim Biophys Acta 1628(1):30–39

4. Sorensen A, Ahring BK, Lübeck M et al (2012) Identifying and characterizing the most significant beta-glucosidase of the novel species *Aspergillus saccharolyticus*. Can J Microbiol 58 (9):1035–1046

5. Benson DA, Karsch-Mizrachi I, Lipman DJ et al (2004) GenBank: update. Nucleic Acids Res 32:D23–D26

6. Ubhayasekera W, Muñoz IG, Vassella A et al (2005) Structures of *Phanerochaete chrysosporium* Cel7D in complex with product and inhibitors. FEBS J 272(8):1952–1964

7. von Ossowski I, Ståhlberg J, Koivula A et al (2003) Engineering the exo-loop of *Trichoderma reesei* cellobiohydrolase, Cel7A. A comparison with *Phanerochaete chrysosporium* Cel7D. J Mol Biol 333(4):817–829

8. Becker D, Braet C, Brunner H et al (2001) Engineering of a glycosidase Family 7 cellobiohydrolase to more alkaline pH optimum: the pH behaviour of *Trichoderma reesei* Cel7A and its E223S/A224H/L225V/T226A/D262G mutant. Biochem J 356(Pt 1):19–30

9. Goedegebuur F, Dankmeyer L, Gualfetti P et al (2017) Improving the thermal stability of cellobiohydrolase Cel7A from *Hypocrea jecorina* by directed evolution. J Biol Chem 292 (42):17418–17430

10. Momeni MH, Goedegebuur F, Hansson H et al (2014) Expression, crystal structure and cellulase activity of the thermostable cellobiohydrolase Cel7A from the fungus *Humicola grisea* var. *thermoidea*. Acta Crystallogr D Biol Crystallogr 70(Pt 9):2356–2366

11. Divne C, Ståhlberg J, Teeri TT, Jones TA (1998) High-resolution crystal structures reveal how a cellulose chain is bound in the 50 angstrom long tunnel of cellobiohydrolase I from *Trichoderma reesei*. J Mol Biol 275 (2):309–325

12. Ubhayasekera W (2005) Structural studies of cellulose and chitin active enzymes, Dissertation, Swedish University of Agricultural Sciences, Uppsala, Sweden. https://pub.epsi lon.slu.se/772/

13. Momeni MH, Ubhayasekera W, Sandgreen M et al (2015) Structural insights into the inhibition of cellobiohydrolase Cel7A by xylo-oligosaccharides. FEBS J 282(11):2167–2177

14. Chothia C, Lesk AM (1986) The relation between the divergence of sequence and structure in proteins. EMBO J 5(4):823–826

15. Jones TA, Zou JY, Cowan-Jacob SW, Kjeldgaard M (1991) Improved methods for building protein models in electron density maps and the location of errors in these models. Acta Crystallogr A 47:110–119

16. Kleywegt GJ, Zou JY, Kejldgaard M, Jones TA (2001) Around O. In: Rossmann MG, Arnold E (eds) International tables for crystallography, Vol. F. Crystallography of Biological Macromolecules. Kluwer Academic, Dordrecht, pp 353–356

17. Bendtsen JD, Nielsen H, von Heijne G, Brunak S (2004) Improved prediction of signal peptides: SignalP 3.0. J Mol Biol 340 (4):783–795

18. Altschul SF, Madden TL, Schäffer AA et al (1997) Gapped BLAST and PSI-BLAST: a new generation of protein database search programs. Nucleic Acids Res 25(17):3389–3402

19. Berman HM, Westbrook J, Reng Z et al (2000) The protein data bank. Nucleic Acids Res 28 (1):235–242

20. Henrissat B (1991) A classification of glycosyl hydrolases based on amino-acid-sequence similarities. Biochem J 280:309–316

21. Kraulis PJ (1991) Molscript - a program to produce both detailed and schematic plots of protein structures. J Appl Crystallogr 24:946–950

22. Harris M, Jones TA (2001) Molray-a web interface between O and the POV-ray ray tracer. Acta Crystallogr D Biol Crystallogr 57 (Pt 8):1201–1203

A

B

C

Mette Lübeck (ed.), *Cellulases: Methods and Protocols*, Methods in Molecular Biology, vol. 1796,
https://doi.org/10.1007/978-1-4939-7877-9, © Springer Science+Business Media, LLC, part of Springer Nature 2018

CPSIA information can be obtained
at www.ICGtesting.com
Printed in the USA
LVHW05*0930190718
584225LV00003B/23/P